Kinetics and Mechanism

Second Edition

Kinetics and Mechanism

A Study of Homogeneous
Chemical Reactions

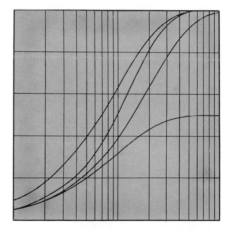

Arthur A. Frost
and
Ralph G. Pearson

Professors of Chemistry
Northwestern University

New York · London, John Wiley & Sons, Inc.

SECOND PRINTING, APRIL, 1962

COPYRIGHT, 1953 © 1961 BY JOHN WILEY & SONS, INC.

LIBRARY OF CONGRESS CATALOG CARD NUMBER: 61–6773
PRINTED IN THE UNITED STATES OF AMERICA

PREFACE

4/19/63 – wiley – $8.76

This revision was made necessary by the substantial advances in chemical kinetics made during the last eight or nine years. The most important gains have been made in the areas of elementary reactions in the gaseous phase and the study of very rapid chemical reactions. The new edition attempts to do justice to these topics at an introductory level.

Since A. A. F. has succumbed to the siren call of quantum mechanics, the revision is almost entirely the work of R. G. P., who must be blamed for its shortcomings.

Thanks are due Dr. E. W. Schlag for reading much of the revised manuscript and for many helpful suggestions.

RALPH G. PEARSON

Evanston, Illinois
January, 1961

v

PREFACE TO THE FIRST EDITION

When we started to write this book we were particularly struck by the fact that the existing textbooks on kinetics treated reaction mechanisms in a rather perfunctory style. There are, of course, excellent books on mechanisms particularly of organic reactions in which some mention of the use and value of kinetics is made. However, there seemed to be no work which showed enough of the intimate relationship between kinetics and mechanism to enable the student to understand exactly how much detail of reaction mechanism can be found from reaction kinetics and to understand what the limitations of the kinetic method of studying mechanism are.

A study of the recent literature will show that the great majority of the work on reaction velocities now being done is primarily concerned with trying to find out exactly in what manner the reactions are proceeding. Thus, while the theories of kinetics, mathematical and experimental details, and the calculation of energetics are all of great importance and, we hope, have not been neglected in this book, a neglect of mechanism would be to ignore the most important application of kinetics. Consequently we have included a great deal of rather detailed stereochemical discussion of the reaction steps.

We have not tried to include a catalogue of all the chief kinds of reactions that may be encountered, but we have tried to select a number of varied and fairly typical examples. Even in this we were regretfully forced to forego discussing a number of topics which might properly be included in a course in reaction kinetics, such as heterogeneous reactions and photochemistry, for example.

We hope that the absence of several such topics will be compensated for by the added material on mechanism and that we are presenting a work which will be useful as a textbook for courses in kinetics on the graduate level, and as a reference book for those interested in the study of the mechanisms of chemical reactions.

We should like to take this opportunity to thank the following persons who contributed in one way or another to the writing and completion of this book: Professors R. L. Burwell, Jr., L. Carroll King, Ronald P. Bell, Louis P. Hammett, Richard E. Powell, Frank H. Seubold, Frank J. Stubbs, Lars Melander; Misses Elaine Strand, Marianne Fält, Mrs. Lenore Pearson, and Mrs. Faye Frost.

ARTHUR A. FROST

RALPH G. PEARSON

Evanston, Illinois
November, 1952

CONTENTS

1.

INTRODUCTION

Kinetics is a part of the science of motion. In physics the science of motion is termed dynamics and is subdivided into kinematics, which treats of the motion of bodies, and kinetics, which deals with the effect of forces on motion. In chemistry, no such distinction is made. Kinetics deals with the rate of chemical reaction, with all factors which influence the rate of reaction, and with the explanation of the rate in terms of the reaction mechanism. Chemical kinetics might very well be called chemical dynamics.

Chemical kinetics with its dynamic viewpoint may be contrasted with thermodynamics with its static viewpoint. Thermodynamics is interested only in the initial and final states of a system; the mechanism whereby the system is converted from one state to another and the time required are of no importance. Time is not one of the thermodynamic variables. The most important subject in thermodynamics is the state of equilibrium, and, consequently, thermodynamics is the more powerful tool for investigating the conditions at equilibrium. Kinetics is concerned fundamentally with the details of the process whereby a system gets from one state to another and with the time required for the transition. Equilibrium can also be treated in principle on the basis of kinetics as that situation in which the rates of the forward and reverse reactions are equal. The converse is not true; a reaction rate cannot be understood on the basis of thermodynamics alone. Therefore, chemical kinetics may be considered a more fundamental science than thermodynamics. Unfortunately, the complexities are such that the theory of chemical kinetics is difficult to apply with accuracy. As a result, we find that thermodynamics will tell with precision the extent of reaction, but only kinetics will tell (perhaps crudely) the rate of the reaction.

Underlying both chemical kinetics and thermodynamics are the more detailed theories of statistical mechanics and the kinetic-molecular theory, which provide alternative viewpoints for understanding microscopic phenomena in terms of atomic and molecular structure and dynamics. Because of the greater rigor of thermodynamic methods, there has been considerable effort in the last thirty years to approach kinetics from the thermodynamic viewpoint, particularly combined with the methods of statistical mechanics. The important feature of this effort is to treat reaction rates as involving an equilibrium between average molecules and high-energy molecules which are aligned and activated ready for reaction, or between molecules in the initial state and in the so-called "transition state" or "activated complex." Even in such a treatment a fundamental problem remains: calculating the rate of decomposition of the activated complex. Only quantum mechanics seems to offer a complete answer.†
In both thermodynamics and kinetics, recourse must eventually be had to quantum mechanics for a calculation of the various energy levels involved.

The science of chemical kinetics may be of interest in itself, as, for example, in determining how changes in environment change the rate of a given reaction. Or, from a practical point of view, one may be interested in the rate of reaction, as in chemical engineering applications. However, of greatest interest to most chemists is the fact that kinetics provides the *most general method of determining the mechanism of reaction.* From a classical point of view (developed chiefly by physical chemists), mechanism of reaction is understood to mean all the individual collisional or other elementary processes involving molecules (atoms, radicals, and ions included) that take place simultaneously or consecutively in producing the observed overall reaction. Compare the mechanism of the two following similar reactions as carried out in the gas phase at elevated temperatures:

$$H_2 + I_2 = 2HI$$
$$H_2 + Br_2 = 2HBr$$

The reaction between hydrogen and iodine is known to take place at bimolecular collisions involving a single molecule of each kind. The reaction of hydrogen and bromine, however, is much more complicated, involving first the dissociation of bromine molecules into atoms followed by reactions between atoms and molecules, as follows:

$$Br_2 \rightleftharpoons 2Br$$
$$Br + H_2 \rightarrow HBr + H \qquad \text{slow}$$
$$H + Br_2 \rightarrow HBr + Br \qquad \text{fast}$$
$$H + HBr \rightarrow H_2 + Br \qquad \text{fast}$$

† See for example references 1, 2, and 3 at end of chapter.

From the above examples, it is obvious that the mechanism cannot be predicted from the overall reaction alone. Consider the formation of water from its elements:

$$2H_2 + O_2 = 2H_2O$$

This reaction is often naively assumed to involve reaction at collisions of two hydrogen molecules and one oxygen molecule. This is definitely not true, the mechanism being sufficiently complicated so that it is not yet completely understood.

The above facts concerning the steps involved in the formation of hydrogen bromide and hydrogen iodide, respectively, are examples of the conclusions that can be drawn from kinetic studies. It must not be inferred, however, that kinetics will invariably give a definite answer concerning the individual steps of a chemical reaction. In general, the experimental results of studying the rate of a reaction as a function of concentrations, temperature, and other operating variables can be interpreted in several ways; that is, there are several conceivable mechanisms consistent with the data. Further experimentation may sometimes eliminate certain of these. However, if one mechanism remains which is in agreement with all the known facts, there is no assurance that it is unique or that new experiments will not add evidence discrediting it.

The difficulty is due to the fact that all such postulated mechanisms are essentially theories. The results of kinetic measurements (and other experiments) furnish facts. The mechanism is a mental model devised to explain the facts. Like any theory, a reaction mechanism currently in vogue may be eliminated later by newer facts or by newer concepts of the structure of matter. In spite of this difficulty, many reactions studied kinetically can be explained by a particular set of simple processes which are so reasonable and so in accord with all chemical experience that we accept them as essentially true. The justification for this becomes apparent when it is observed that a mechanism can successfully predict reaction products or the optimum conditions for running a chemical reaction.

There is, in addition to the classical definition already mentioned, a newer concept of reaction mechanism, developed chiefly by organic chemists. This newer concept includes not only a knowledge of all the individual steps in the overall reaction, but also a *detailed stereochemical picture* of each step as it occurs. This implies a knowledge not only of the composition of the activated complex in terms of the various atoms or molecules of reactants, but also of the geometry of the activated complex in terms of interatomic distances and angles. For example, in the conversion of hypochlorite ion to chlorate it can be shown that the reaction consists of two steps

$$ClO^- + ClO^- \xrightarrow{slow} ClO_2^- + Cl^- \tag{1}$$

$$ClO_2^- + ClO^- \xrightarrow{fast} ClO_3^- + Cl^- \tag{2}$$

with the formation of chlorite ion the slower, or rate-determining, step, and the formation of chlorate ion a rapid process.[4] This can be deduced from kinetic evidence since the rate of formation of chlorate ion is proportional to the square of the hypochlorite concentration. Furthermore, the chlorite ion can be prepared separately and its rate of reaction with hypochlorite shown to be fast. These facts would have sufficed to fix a probable mechanism according to the older definition. The newer approach adds to the mechanism a picture such as (3) for the transfer of an oxygen atom between ions.

$$\begin{bmatrix} Cl\text{-}\!-\!O\text{—}Cl \\ | \\ O \end{bmatrix}^{-2} \rightarrow \begin{bmatrix} Cl\text{—}O \\ | \\ O \end{bmatrix}^{-} + Cl^- \tag{3}$$

This stereochemical representation is guessed at from chemical intuition and experience. It enables the reaction to be classified as a member of a large class of similar reactions, a nucleophilic displacement of one base (Cl⁻) by another (ClO⁻). This classification encourages us to focus attention on what is going on and to understand such phenomena as the effect of pH on the rate of reaction, the effect of changing the halogen from chlorine to, say, bromine. We can better understand any one reaction, in other words, by drawing on a large body of information on similar reactions.[5] The stereochemical picture also suggests the possibility that the mechanism may be more complex in that the oxygen atom which leaves one hypochlorite ion may not be the same as that which appears on the other. Thus the solvent might be involved in the oxygen transfer[6] in some such way as shown in (4). In this particular example the products are the

$$\begin{bmatrix} O\text{—}Cl \\ \\ O\text{—}H \\ | \\ H\text{-}\!-\!O\text{—}Cl \end{bmatrix}^{-2} \rightarrow \begin{bmatrix} O\text{—}Cl \\ | \\ O \end{bmatrix}^{-} + Cl^- + H_2O \tag{4}$$

same whether the solvent is involved or not, and whether (3) or (4) is more accurate may be of theoretical importance only. When we reflect, however, that the hydrolysis of an organic halide may lead to an alcohol with an inversion of original configuration, retention of configuration, or a mixture of the two, then the necessity for a detailed and pictorial representation of a mechanism becomes more apparent.

Whereas kinetics is very useful in determining the individual steps of a reaction, it is rather limited in giving stereochemical details. Therefore, it is usually necessary to refer to other methods of obtaining information about mechanisms, which are incomplete in themselves and are best used in combination with kinetic evidence. Some of these other methods will be mentioned briefly now with further and more complete illustrations to be given later in the text.

The Products of Reaction

The most important circumstantial evidence as to reaction mechanism is the identity of the products formed. This seems so obvious that it is difficult to believe that kinetic studies have frequently been reported in which the exact nature of the products was unknown. Such, however, is the case. As an example of the type of reasoning involved after the products have been identified, consider the alkaline hydrolysis of a simple ester such as ethyl acetate. The formation of alcohol and an acetate ion indicates reaction of a hydroxide ion with an ester molecule breaking one C—O bond

$$CH_3-\overset{\overset{\displaystyle O}{\|}}{C}-OC_2H_5 + OH^- \rightarrow CH_3-\overset{\overset{\displaystyle O}{\|}}{C}-O^- + C_2H_5OH \qquad (5)$$

That the reaction goes as indicated could be proved by a kinetic study which shows that the rate of formation of acetate ion is proportional both to the concentration of ester and to the concentration of hydroxide ion.

Stereochemical Evidence

It is to be expected that considerable information as to the intimate details of a mechanism can be gained from the stereochemical evidence, that is, by examining the stereochemistry of the substances reacting and the substances formed. In the reaction above, if an optically active alcohol, $RR'CHOH$, is used instead of ethyl alcohol in forming the ester, the fact that the original alcohol can be recovered without racemization or inversion after hydrolysis[7] suggests strongly that in both the formation of ester and in its hydrolysis the C—O bond which is involved is the one in which the oxygen atom of the alcohol remains unchanged. It may be said at this point that, so far as reaction mechanism is intended to mean intimate, detailed mechanism, stereochemistry is as important an investigational tool as kinetics.

Use of Isotopes

The above conclusion that the bond is broken between the oxygen and the carbonyl carbon is confirmed by the use of hydroxide ion enriched with O^{18}. The heavy oxygen is then found in the anion of the acid rather than in the alcohol.[8] The increased production of isotopes of the more common elements has led to an increasing application of this technique in studying reactions where mechanisms are not easily determined by other means. Thus the question of solvent participation in the hypochlorite-ion reaction could be approached by the use of heavy oxygen water.[6] To be successful in this case it would be necessary to show that oxygen exchange between hypochlorite ion and the solvent was not so rapid as the rate of formation of chlorite ion.

Detection of Short-Lived Intermediates

A method that has been applied chiefly to reactions involving free atoms or radicals which are extremely reactive, and hence short-lived, is the actual demonstration of their existance. This can be done by adding substances (e.g., NO or I_2) to trap the radicals as stable compounds or by observing some specific physical or chemical property (e.g., mass spectrogram, absorption spectrum, or removal of metallic mirrors). A recent and very powerful tool for the study of free radicals is electron paramagnetic resonance (EPR), which measures directly the concentration of radicals and also supplies specific information about their structure. Closely related are photochemical techniques to produce the suspected atoms or radicals and simultaneous study of the reaction rate.

Refinements of Kinetic Methods

Actually no different in method from ordinary kinetic studies, but supplying additional information as to mechanisms, are such devices as studying the effect of substituents on the rate of a given reaction, or the effect of changing the solvent, ionic strength, etc. For example, in the alkaline hydrolysis of esters mentioned earlier, if a series of substituted ethyl benzoates is studied kinetically the influence of substituents on the rate can be used to elucidate the mechanism. It is found that electron-attracting substituents such as the nitro group increase the rate of hydrolysis, whereas electron-repelling substituents such as methoxyl decrease the rate.[9] The interpretation is that increased positive charge on the carbonyl carbon facilitates the addition of the hydroxide ion whose close approach then pushes off the ethoxide ion.

$$\text{C}_6\text{H}_{11}\text{-C(=O)-OC}_2\text{H}_5 + \text{OH}^- \rightleftarrows \text{C}_6\text{H}_{11}\text{-C(O}^-\text{)(OH)-OC}_2\text{H}_5 \rightarrow$$

$$\text{C}_6\text{H}_{11}\text{-C(=O)-OH} + \text{OC}_2\text{H}_5^- \rightarrow \text{C}_6\text{H}_{11}\text{-C(=O)-O}^- + \text{C}_2\text{H}_5\text{OH} \qquad (6)$$

In a similar way, the effect of changing the dielectric constant of the solvent or the ionic strength can be used in conjunction with modern theories to give strong evidence that the reaction is between a negative ion and a neutral, but polar, molecule. Although the mechanism of basic hydrolysis of an ester seems clear enough from the discussion given, the subject will be taken up again in a later chapter. It will then be seen that certain ambiguities still exist, and that in particular the results mentioned above *cannot* be generalized to cover all cases of ester hydrolysis under basic conditions. This is a general result when a type reaction of a class of compounds is being studied unless the reaction type and the class are quite narrowly defined.

REFERENCES

1. S. Golden, *J. Chem. Phys.*, *17*, 620 (1949).
2. S. Golden and A. M. Peiser, *ibid.*, *17*, 630 (1949).
3. H. Eyring, J. Walter and G. Kimball, *Quantum Chemistry*, John Wiley and Sons, New York, 1944, Chapter 16.
4. F. Foerster and P. Dolch, *Z. Elektrochem.*, *23*, 137 (1917).
5. H. Taube, *Record of Chemical Progress*, *17*, 25 (1956); J. O. Edwards, *J. Chem. Ed.*, *31*, 270 (1954).
6. J. Halperin and H. Taube, *J. Am. Chem. Soc.*, *72*, 3319 (1950); T. C. Hoering, *ibid.*, *80*, 3876 (1958).
7. B. Holmberg, *Ber.*, *45*, 2997 (1912).
8. M. Polanyi and A. L. Szabo, *Trans. Faraday Soc.*, *30*, 508 (1934).
9. K. Kindler, *Ann.*, *450*, 1 (1926); *452*, 90 (1927); *464*, 278 (1928).

2.

EMPIRICAL TREATMENT OF REACTION RATES

Kinds of Systems

A closed system, sometimes called static, is one in which no matter is gained or lost, as in a typical reaction in a liquid phase in a flask or in a closed bomb, or in a gas-phase reaction taking place in a reaction vessel of constant volume. An open system, or flow system, involves gain or loss of matter and is exemplified by reactions of a flowing gas in a heated tube or at a solid catalyst, by a flame, and by living organisms where nutrients and metabolic products are exchanged with the surroundings. Closed systems are more convenient in general than open systems, both for precise rate measurements and for theoretical interpretation. It is to be understood in what follows that closed systems are always being considered unless an open system is explicitly indicated. The theory of the open system may be obtained by generalization of the theory of the closed system. There has been an increasing interest in open systems in recent years.†

A closed or an open system may be either homogeneous or heterogeneous, but the open system is usually heterogeneous and may have concentration gradients in the reaction zone which make such reactions quite different from typical homogeneous reactions. Heterogeneous closed systems are of particular interest in connection with surface catalysis, but, because of their more fundamental value to chemical kinetic theory, homogeneous closed systems will be the principal subject of this treatise.

Isothermal systems are of most significance because temperature can then be considered an independent variable. However, non-isothermal or

† See Chapter 11.

approximately adiabatic systems are sometimes of interest as in connection with rapid exothermic reactions such as flames and explosions where the heat of reaction cannot be conducted away rapidly enough.

Definition of Reaction Rate

The reaction rate is usually defined as the rate of change of concentration of a substance involved in the reaction with a minus or plus sign attached, depending on whether the substance is a reactant or a product. In a general reaction with stoichiometric equation

$$aA + bB = gG + hH$$

the rate may be expressed by

$$-d[A]/dt, \quad -d[B]/dt, \quad +d[G]/dt, \quad \text{or} \quad +d[H]/dt$$

where t is time and brackets mean concentration. The sign is attached so that the rate will be positive numerically. A derivative is used because the rate almost invariably changes as time goes on. The use of concentration in the definition rather than the amount of a substance makes the rate an intensive property, that is, independent of size of system.

In stating numerical values of reaction rate, not only must the time and concentration units be given, but also the particular substance whose concentration is involved must be stated because the reaction may use or produce different numbers of moles of the various reactants or products. For example, in the reaction

$$2H_2 + O_2 = 2H_2O$$

since two moles of hydrogen react for each mole of oxygen,

$$-d[H_2]/dt = -2d[O_2]/dt$$

For the general reaction above it is easily seen that

$$\frac{1}{a}\left(-\frac{d[A]}{dt}\right) = \frac{1}{b}\left(-\frac{d[B]}{dt}\right) = \frac{1}{g}\left(\frac{d[G]}{dt}\right) = \frac{1}{h}\left(\frac{d[H]}{dt}\right)$$

It is possible to avoid such integral conversion factors in relating different derivative rate expressions by defining rate in terms of equivalent concentration rather than molar concentration. If x is the equivalents per liter that have reacted in time t, then dx/dt is a convenient expression for the rate. The definition of the equivalent must be explicitly stated. Such a variable as x may be called the *reaction variable.*†

Variables other than concentration may be used in defining rate,

---

† "Umsatzvariabel" of A. Skrabal, *Homogenkinetik*, Steinkopff, Dresden, 1938.

for examples, pressure as in a gas reaction or optical rotation as in a racemization, inversion, or mutarotation reaction. Such variables are always related to one or more concentrations, usually linearly, and so may be treated as equivalent to concentrations for this purpose. A detailed discussion of such relations is given in the next chapter.

Effect of Concentration on Reaction Rate—Empirical Rate Expressions

The reaction rate at a fixed temperature is a function of the concentrations of some or all of the various components of the system, but usually of only the reactants. If the concentration of a product affects the rate, this effect is called either autoinhibition or autocatalysis. If a substance, neither a reactant nor a product, affects the rate it is called an inhibitor, retarder, sensitizer, or a catalyst, depending on the nature of the effect. The functional relation between rate and concentration is called a rate expression. In general it is not possible to predict the rate expression for a given reaction by just knowing the stoichiometric equation. Although the reactions of hydrogen with iodine and bromine follow similar stoichiometric equations, the rate expressions are of quite different form:

$$\text{For}\quad H_2 + I_2 = 2HI \qquad \frac{d[HI]}{dt} = k[H_2][I_2]$$

$$\text{For}\quad H_2 + Br_2 = 2HBr \qquad \frac{d[HBr]}{dt} = \frac{k[H_2][Br_2]^{1/2}}{1 + k'[HBr]/[Br_2]}$$

This striking difference is accounted for by the different mechanisms of the two reactions mentioned in Chapter 1, the rate expressions being the principal evidence for the postulated mechanisms.

Order of Reaction and Molecularity

Rate expressions that are of the form of a product of powers of concentrations such as

$$-dc_1/dt = kc_1{}^{n_1}c_2{}^{n_2}c_3{}^{n_3}$$

are easier to handle mathematically than expressions of a more complex type, such as that given above for the hydrogen-bromine reaction. For this restricted type of rate expression, and for this type only, there is defined the concept of *order of reaction*, *n*, where

$$n = n_1 + n_2 + n_3 + \cdots$$

the sum of all the exponents of the concentrations. Also each individual exponent is called the *order with respect to that component*. For example, the hydrogen-iodine reaction is a second-order reaction, and the order with respect to each reactant separately is one. For the hydrogen-bromine reaction the concept of order does not apply since its rate expression is not of the restricted form required for this concept.

If conditions for a given reaction are such that one or more of the concentration factors are constant or nearly constant during a "run," these factors may be included in the constant k. In this case the reaction is said to be of *pseudo-nth order* or *kinetically of the nth order* where n is the sum of the exponents of those concentration factors which change during the run. This is the situation for catalytic reactions with the catalyst concentration remaining constant during the run, or if there is a buffering action that keeps a certain concentration such as that of the hydrogen ion nearly constant, or if one reactant is in large excess over another so that during the run there is only a small percentage change in the concentration of the former. Take, for example, the inversion of sucrose† catalyzed by strong acids. The rate is given by

$$-d[S]/dt = k[S][H_2O][H^+]$$

where S stands for sucrose. The reaction is third-order. However, since H^+ is a catalyst and its concentration remains constant during a run and also since $[H_2O]$ is essentially constant when water is the solvent, the reaction is pseudo-first-order. If some inert solvent is used and water is present only as a reactant, the reaction would be pseudo-second-order.

The exponents are usually simple positive integers, but occasionally they may be fractional or even negative, depending on the complexity of the reaction. When the order is one, two, or three, there is some confusion in the older literature with the terms unimolecular, bimolecular, and termolecular. These terms, examples of the *molecularity* of a reaction, are intended to indicate the number of molecules involved in a simple collisional reaction process. Molecularity is a theoretical concept, whereas order is empirical. They are generally different numerically. However, as will be shown later, a bimolecular reaction is usually second-order and a termolecular reaction third-order, but the reverse of these statements is less often true.

The k is called the rate constant, or the specific reaction rate. It is seen from the equation that it has dimensions of

$$[\text{conc.}]^{1-n}[\text{time}]^{-1}$$

Commonly used concentration units are moles/liter, moles/cc, molecules/cc,

† This reaction is of historical interest as it is considered to be the first one studied kinetically. See reference 1 at end of chapter.

pressure in mm Hg or in atmospheres, etc. The time unit may be
seconds, minutes, etc. For a first-order reaction a rate constant is expressed
typically in sec^{-1}, the value being independent of the concentration unit.
For a second-order reaction typical units would be liters $moles^{-1}$ sec^{-1},
conveniently written as M^{-1} sec^{-1}. The second is the preferred unit of
time for any order.

Integrated Forms of Simple Rate Expressions

A rate of reaction is not usually obtained directly from experiment, but
rather a concentration is observed as a function of the time for several

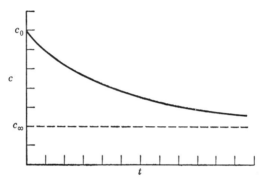

Fig. I. Concentration versus time for a typical reaction.

successive runs under varying conditions. A typical result for a single run
would be as shown in Fig. 1, where the concentration c of a reactant starts
at zero time at some initial value c_0 and decreases more and more slowly,
approaching zero or some equilibrium value c_∞ asymptotically. The rate
at any time would be the negative of the slope of the curve expressed in the
appropriate units. It could be determined approximately by graphical or
numerical methods. However, since it is the concentration that is usually
observed, the integrated form of the rate expression is more convenient to
work with.

A number of typical empirical rate expressions will now be integrated
and discussed, the simplest examples where only one concentration
factor is involved being discussed first.

nth Order Reaction of a Single Component

For this case the form of the rate expression is

$$-dc/dt = kc^n \tag{1}$$

It may be readily integrated after first multiplying by dt and dividing by

c^n to get the variables separated. The limits of integration are taken as $c = c_0$ at $t = 0$ and c at $t = t$, respectively.

$$-dc/c^n = k \, dt$$

$$-\int_{c_0}^{c} dc/c^n = k \int_{0}^{t} dt \qquad (2)$$

For $n = 1$ *first-order* the result is

$$\ln (c_0/c) = kt$$

or

$$\ln c = \ln c_0 - kt \qquad (3)$$

or

$$c = c_0 e^{-kt}$$

To test data for a first-order reaction the logarithmic form is preferable since a plot of $\ln c$ or $\log c$ (base 10) versus t should be linear with a slope

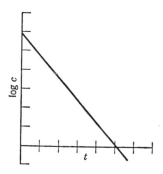

Fig. 2. Linear plot for a first-order reaction.

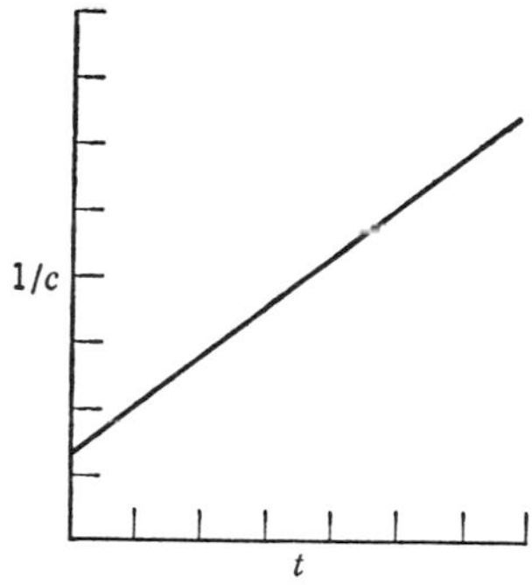

Fig. 3. Linear plot for a second-order reaction.

of $-k$ or $-k/2.303$, respectively. See Fig. 2. The rate constant k may be evaluated from this slope. Alternatively, the equation can be checked numerically by solving for k

$$k = 1/t \ln c_0/c \qquad (4)$$

and substituting successive values of c and t as obtained experimentally. If the k's so calculated show no significant trend during the course of the reaction run, the conclusion is that the reaction is first-order.

For $n = 2$ *second-order* the integration yields

$$1/c - 1/c_0 = kt \qquad (5)$$

According to this a plot of $1/c$ versus t should be linear, and with a positive slope equal to k. See Fig. 3. Or, again, k can be calculated, if the initial concentration is known, for each successive c, t pair and its constancy verified.

For $n = 3$ *third-order* the result of integrating (2) is

$$\tfrac{1}{2}(1/c^2 - 1/c_0^2) = kt \tag{6}$$

In general, for $n \neq 1$ integration gives

$$[1/(n - 1)](1/c^{n-1} - 1/c_0^{n-1}) = kt \tag{7}$$

This is valid for fractional n's as well as integers.

It is sometimes more convenient to use, as the dependent variable x, the decrease in concentration of reactant in time t. Then $c = a - x$, where a is commonly used to indicate the initial concentration, previously c_0, and equations 1 and 3 become

$$dx/dt = k(a - x)^n \tag{8}$$

$$\ln a/(a - x) = kt \qquad \text{First-Order} \tag{9}$$

Also (5) becomes

$$1/(a - x) - 1/a = x/[a(a - x)] = kt \qquad \text{Second-Order} \tag{10}$$

This method is particularly useful in more complicated situations to be discussed shortly where concentrations of several reactants, and perhaps products, appear in the rate expression, in which case they all can be expressed in terms of one variable x, the reaction variable.

Plot Using Dimensionless Parameters

A more general way of presenting these relations is in terms of the dimensionless variables α and τ defined as follows:

$$\alpha = c/c_0 \qquad \text{Relative concentration} \tag{11}$$

$$\tau = kc_0^{n-1}t \qquad \text{Time parameter}$$

By substitution in (3) and (7) there results for *first-order*

$$\ln \alpha = -\tau \tag{12}$$

and *nth*-order ($n \neq 1$)

$$\alpha^{1-n} - 1 = (n - 1)\tau \tag{13}$$

These equations no longer contain c_0 or k explicitly. For any given order there is a unique relation between α and τ. Powell[2] has shown the usefulness of a plot of α versus $\log \tau$ for various orders as diagrammed in Fig. 4. Now suppose that for a given reaction of this type, experimental values of α are plotted versus $\log t$ (τ is not experimental because of the at-first-unknown values of k and n relating τ to t). This experimental curve should match the form of the theoretical curve for a reaction of the same order except that because $\log \tau = \log t + \log kc_0^{n-1}$ the experimental curve will be shifted along the $\log t$ axis by an amount $-\log kc_0^{n-1}$. Therefore the Powell plot gives a unique form for α versus $\log t$ for each order.

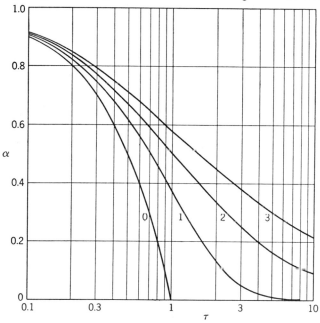

Fig. 4. Fraction remaining, α, as a function of time parameter τ, for zero, first-, second-, and third-order reactions (plot according to R. E. Powell, private communication).

Second-Order Reaction. First-Order with Respect to Each Reactant A and B with Stoichiometric Equation $A + B = \cdots$

$$-dA/dt = kAB \qquad (14)$$

A and B are here used for the concentrations of the corresponding substances. From the stoichiometric equation $dA = dB$, and integration gives

$$A_0 - A = B_0 - B$$

and

$$B = B_0 - A_0 + A \qquad (15)$$

where A_0 and B_0 are the initial concentrations. Multiplying (14) by dt/AB and substituting for B by (15) gives an equation in the two variables A and t, and with these variables separated:

$$-dA/[A(B_0 - A_0 + A)] = k \, dt \qquad (16)$$

Now use the method of partial fractions to write the left side as a sum of two simpler terms. Let

$$\frac{1}{A(B_0 - A_0 + A)} \equiv \frac{p}{A} + \frac{q}{B_0 - A_0 + A} \tag{17}$$

where p and q are constants and the identity symbol means equality of the two sides of the equation for all values of A. The p and q are evaluated by using a common denominator and equating coefficients of like powers of A in the numerators. Equating the numerators gives

$$1 \equiv p(B_0 - A_0 + A) + qA \tag{18}$$

Equating constant terms gives

$$1 = p(B_0 - A_0) \tag{19}$$

and equating coefficients of A

$$0 = p + q \tag{20}$$

Solving (19) and (20) simultaneously for p and q yields

$$p = 1/(B_0 - A_0) \qquad q = -1/(B_0 - A_0) \tag{21}$$

Equation 16 then becomes

$$-\frac{dA}{(B_0 - A_0)A} + \frac{dA}{(B_0 - A_0)(B_0 - A_0 + A)} = k\,dt \tag{22}$$

Integration of each term is simple, and the result is

$$\frac{1}{B_0 - A_0} \ln \frac{A_0}{A} + \frac{1}{B_0 - A_0} \ln \frac{(B_0 - A_0 + A)}{B_0} = kt$$

which reduces to

$$\frac{1}{B_0 - A_0} \ln \frac{A_0(B_0 - A_0 + A)}{B_0 A} = kt$$

or

$$\frac{1}{B_0 - A_0} \ln \frac{A_0 B}{B_0 A} = kt \tag{23}$$

Experimental data may be plotted linearly either by plotting the left side of the equation against t, or with less calculation by plotting just log (B/A) against t. Or data may be tested without plotting by solving the equation for k, substituting successive observed values of A, B, and t, noticing the constancy of k.

In terms of the variable x representing the decrease in concentration of a reactant in a given time, (14) becomes

$$dx/dt = k(a - x)(b - x) \tag{24}$$

where a and b are now used to represent the initial concentrations, and the integrated form, which is equivalent to (23), is

$$\frac{1}{b-a} \ln \frac{a(b-x)}{b(a-x)} = kt$$

or

$$\frac{1}{a-b} \ln \frac{b(a-x)}{a(b-x)} = kt \tag{25}$$

A nomograph to aid in the application of equations 23 or 25 has been constructed by Nord.[3]

Table I

n-PROPYL BROMIDE AND SODIUM THIOSULFATE

(Crowell and Hammett[4])

t, sec	Iodine Titer[a]	k, liters/mole-sec
0	37.63	. . . .
1,110	35.20	0.001658
2,010	33.63	0.001644
3,192	31.90	0.001649
5,052	29.86	0.001636
7,380	28.04	0.001618
11,232	26.01	0.001618
		Av. 0.001637
78,840	22.24	. . . .

[a] In cubic centimeters of $0.02572N$ iodine solution per 10.02-cc sample.

An example of the application of this equation, just one chosen from the many examples found in the literature, is given by Crowell and Hammett.[4] The rate of reaction of n-propyl bromide and other alkyl bromides with thiosulfate ion according to the equation

$$RBr + S_2O_3^{-2} = RSSO_3^- + Br^-$$

was measured by titrating the thiosulfate ion with iodine in the presence of ice water to stop the reaction. Table 1 gives data and the calculated second-order constants k for a run involving $0.0200M$ sodium acetate solution at $37.50°$. The acetate was present to buffer the solution against a possible decrease in pH due to hydrolysis of the alkyl halide and so prevent the decomposition of the thiosulfate. The concentration of the n-propyl bromide was calculated from the titer after a long reaction time, this titer representing the excess of thiosulfate over n-propyl bromide. The constant k was calculated at each time using equation 25. Figure 5 shows a plot of log $[(b - x)/(a - x)]$ versus t, which should be a straight line if this is a second-order reaction.

The left sides of equations 23 and 25 become indeterminant if the initial concentrations are made equal. But this situation of equal initial concentrations is equivalent mathematically to a second-order reaction of one reactant with equation 5 or equation 10 as the solution. This equation may

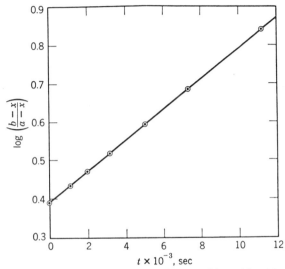

Fig. 5. Second-order plot for the reaction of n-propyl bromide with sodium thiosulfate (data from Crowell and Hammett).

also be obtained from equation 23 or equation 25 by an expansion of the logarithmic term followed by a limiting process.

If the initial concentrations are not quite equal, exact equality being difficult to attain experimentally, another formula may be necessary since the argument of the logarithm in (23) or (25) will be close to unity. Here again an expansion of the logarithm is useful. Following Widequist,[6] let $a = d + s$ and $b = d - s$, where d is the mean initial concentration and $2s$ is the excess of a over b. Substitution in (25) results in

$$\ln\left(1 + \frac{s}{d - x}\right) - \ln\left(1 - \frac{s}{d - x}\right) + \ln\left(1 - \frac{s}{d}\right) - \ln\left(1 + \frac{s}{d}\right) = 2skt$$

Each logarithm is of the form $\ln(1 + y)$, where $y < 1$ at least in the early stages of the reaction. Expansion and collection of terms results in

$$\frac{1}{d - x} - \frac{1}{d} + \frac{s^2}{3}\left[\frac{1}{(d - x)^3} - \frac{1}{d^3}\right] + \cdots = kt \qquad (26)$$

When $s = 0$ this reduces to

$$1/(d - x) - 1/d = kt \qquad (27)$$

which is the equivalent of (10). Since terms in s to the first degree are missing in (26) owing to the use of the mean concentration d, it is apparent that (27) may also be useful when the concentrations are only slightly unequal, that is, when $s \ll d$. In such a case a simple plot of $1/(d - x)$ against the time will be linear with a slope equal to k.

Second-Order Autocatalytic Reaction

Let the stoichiometric equation be $A = B + \cdots$, and the rate expression

$$-dA/dt = kAB$$

In this case $A_0 - A = B - B_0$ so that $B = A_0 + B_0 - A$. Integration, again by partial fractions, yields

$$\frac{1}{A_0 + B_0} \ln \frac{A_0 B}{B_0 A} = kt \qquad (28)$$

Examples of such kinetics are found in the acid-catalyzed hydrolyses of various esters and similar compounds, and in various biochemical

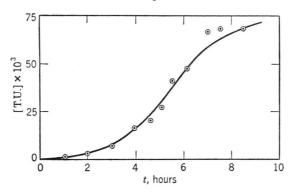

Fig. 6. Autocatalytic conversion of trypsinogen into trypsin (Kunitz and Northrop).

processes. An example from biochemistry is the conversion of trypsinogen into trypsin with the trypsin catalyzing the reaction. Figure 6 shows a plot of data obtained by Kunitz and Northrop.[7] The ordinate is a measure of the trypsin concentration in terms of "hemoglobin tryptic units." The solid curve is calculated from (28) with $A_0 = 0.072$ (T.U.), $B_0 = 0.0003$ (T.U.), and $k = 14.6$ (T.U.)$^{-1}$ hours^{-1}. If it were not for the small trace (B_0) of trypsin present in the original solution the reaction presumably would not take place. The solution of (28) for B as a function of t yields

$$B = \frac{A_0 + B_0}{1 + (A_0/B_0)e^{-k(A_0+B_0)t}} \tag{29}$$

This gives the S-shaped logistic curve typical of autocatalytic reactions and characteristic of many growth processes, including the growth of populations.[8]

Third-Order Reaction with Three Reactants

Stoichiometric equation: $A + B + C = \cdots$.
Rate expression

$$-dA/dt = kABC \tag{30}$$

but

$$A_0 - A = B_0 - B = C_0 - C$$

and

$$B = B_0 - A_0 + A \qquad C = C_0 - A_0 + A$$

Then

$$-dA/dt = kA(B_0 - A_0 + A)(C_0 - A_0 + A) \tag{31}$$

Integration by partial fraction results in

$$\frac{1}{(B_0 - A_0)(C_0 - A_0)} \ln \frac{A_0}{A} + \frac{1}{(A_0 - B_0)(C_0 - B_0)} \ln \frac{B_0}{B}$$

$$+ \frac{1}{(A_0 - C_0)(B_0 - C_0)} \ln \frac{C_0}{C} = kt$$

or

$$\frac{1}{(A_0 - B_0)(B_0 - C_0)(C_0 - A_0)} \ln \left(\frac{A}{A_0}\right)^{(B_0-C_0)} \left(\frac{B}{B_0}\right)^{(C_0-A_0)} \left(\frac{C}{C_0}\right)^{(A_0-B_0)} = kt \tag{32}$$

Third-Order Reaction with Two Reactants

Suppose that $2A + B = \cdots$ and

$$-dA/dt = kA^2B$$

Here $A_0 - A = 2(B_0 - B)$ and $B = B_0 - A_0/2 + A/2$ and

$$-dA/dt = kA^2(B_0 - A_0/2 + A/2)$$

The partial fractions in this case are of the form

$$\frac{1}{A^2(B_0 - A_0/2 + A/2)} = \frac{p}{A^2} + \frac{q}{A} + \frac{r}{B_0 - A_0/2 + A/2}$$

with p, q, and r constants. Integration yields

$$\frac{2}{(2B_0 - A_0)}\left(\frac{1}{A} - \frac{1}{A_0}\right) + \frac{2}{(2B_0 - A_0)^2} \ln \frac{B_0 A}{A_0 B} = kt \tag{33}$$

This equation applies to such reactions as those between nitric oxide and oxygen, chlorine, or bromine.

A different situation exists for a third-order reaction with two reactants if the stoichiometric equation is $A + B = \cdots$ and

$$- dA/dt = kA^2 B$$

In this case $A_0 - A = B_0 - B$ and

$$-dA/dt = kA^2(B_0 - A_0 + A)$$

The result of integration is

$$\frac{1}{B_0 - A_0}\left(\frac{1}{A} - \frac{1}{A_0}\right) + \frac{1}{(B_0 - A_0)^2} \ln \frac{B_0 A}{A_0 B} = kt \tag{34}$$

An example of this is given by Swain.[9] Triphenyl methyl (trityl) chloride reacts with methanol in dry benzene solution as follows:

$$(C_6H_5)_3CCl + CH_3OH = (C_6H_5)_3COCH_3 + HCl$$

Pyridine was added to remove the HCl and so prevent the reverse reaction. The reaction was followed by taking advantage of the fact that pyridine hydrochloride is only slightly soluble in benzene and precipitates out as the reaction proceeds. After a given time, each sample was filtered and unreacted trityl chloride hydrolyzed with water, and the resulting hydrochloric acid titrated with standard sodium hydroxide. A correction was made for the solubility of pyridine hydrochloride. The stoichiometry might lead one to expect a second-order reaction, but Swain suspected that it was really third-order, that is, first-order with respect to the trityl chloride and second-order with respect to the methanol. Both second-order constants k_2 and third-order constants k_3 were calculated for the same runs. The data and calculated results are in Table 2. It is evident that k_3 is a reasonable constant where k_2 is not. The values of k_3 were calculated from the reaction variable form of (34).

$$k = \frac{1}{t(b - a)}\left[\frac{x}{a(a - x)} + \frac{2.303}{(b - a)} \log \frac{b(a - x)}{a(b - x)}\right] \tag{35}$$

where a is the initial moles/liter of methanol, b is the initial moles/liter of trityl chloride, and x is the moles/liter of methanol or trityl chloride that reacts in time t.

Table 2

REACTION OF 0.106M TRITYL CHLORIDE WITH 0.054M METHANOL
IN DRY BENZENE SOLUTION IN THE PRESENCE OF PYRIDINE AT 25°
(Swain[9])
(Run 46, 0.064M pyridine; run 50, 0.108M pyridine; run 47, 0.215M
pyridine. $b/a = 1.963$.)

Run	t, min	x, moles/liter	x_{cor}	k_2	k_3
50	20	0.0010	. . . . . .	. . . . . .	. . . . .
47	22	0.0003	. . . . . .	. . . . . .	. . . . .
46	22	0.0010	. . . . . .	. . . . . .	. . . . .
47	168	0.0067	0.0091	0.0107	0.224
46	174	0.0086	0.0110	0.0127	0.278
47	418	0.0157	0.0181	0.0101	0.234
50	426	0.0165	0.0189	0.0105	0.248
46	444	0.0183	0.0207	0.0115	0.278
50	1,150	0.0294	0.0318	0.0089	0.272
47	1,440	0.0310	0.0334	0.0077	0.252
46	1,510	0.0321	0.0345	0.0080	0.264
50	1,660	0.0330	0.0354	0.0077	0.263
47	2,890	0.0394	0.0418	0.0066	0.296
46	2,900	0.0390	0.0414	0.0064	0.281
50	3,120	0.0392	0.0416	0.0060	0.269
47	193,000	0.0490	0.0514	. . . . . .	. . . . .
					Av. 0.263

Reactions involving equilibria, reverse reactions, side reactions, and consecutive reactions are often met in practice. These situations are discussed in Chapter 8, Complex Reactions.

Effect of Temperature on Reaction Rate

Observed rates or rate constants as a function of temperature T may be of various forms as indicated in Fig. 7. I is most typical and will be discussed in detail. II represents an explosion where the sudden rise in rate occurs at the ignition temperature. III is observed, for example, in catalytic hydrogenations and in enzyme reactions. IV is observed in the oxidation of carbon. V is observed in the nitric oxide-oxygen reaction.

Case I may be called an Arrhenius temperature dependence. II to V are sometimes referred to as anti-Arrhenius. Case I corresponds to the common statement that a reaction rate increases by a factor of 2 or 3 for each 10-degree rise in temperature. It is usually found that a plot of log

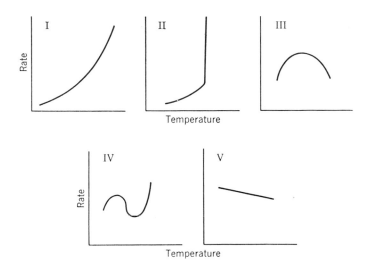

Fig. 7. Various forms for the dependence of rate on temperature.

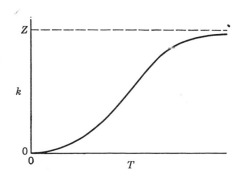

Fig. 8. Arrhenius temperature dependence of a rate constant.

k versus $1/T$ is nearly linear with negative slope. This result is equivalent to the Arrhenius equation

$$d \ln k/dT = E_a/RT^2 \qquad (36)$$

If E_a, the Arrhenius activation energy, is a constant with respect to temperature, integration results in

$$\ln k = -E_a/RT + \text{const.} \qquad \text{or} \qquad k = Ze^{-E_a/RT} \qquad (37)$$

A plot of k versus T according to (37) would have the appearance of Fig. 8 shown here rather than just like I above, so that k would reach the constant value Z asymptotically. For most reaction the accessibles temperature range corresponds to the lower, rising part of the curve. But there are reactions involving free atoms or radicals with very small or zero activation energy such that the upper portion of the curve is approached. However, under such conditions the factor Z is also a function of the

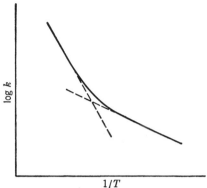

Fig. 9. Transition from homogeneous to heterogeneous reaction.

temperature and a more accurate equation must be used. Such an equation is

$$k = AT^n e^{-E/RT} \tag{38}$$

or

$$\ln k = -E/RT + n \ln T + \ln A \qquad A \text{ a constant} \tag{39}$$

This has been used to treat empirical results by Kassel.[10] The equation also has theoretical justification, the exponent n having a particular value, depending on the kind of theory used and the nature of the reaction considered. (See Chapters 4 and 5.)

In treating experimental data by (37) and (38) the activation energies will differ. It is of importance to know the relation between E_a and E. Differentiate (39) with respect to T, assuming E constant, and set this equal to (36)

$$E/RT^2 + n/T = E_a/RT^2 \qquad \text{and} \qquad E = E_a - nRT \tag{40}$$

The difference of the two activation energies will typically be of hundreds or a few thousands of calories per mole.

If equation 39 is correct with n not zero, a plot of $\log k$ versus $1/T$ will show a slight curvature. There are cases, however, where nonlinearity is much more pronounced. The curve may sometimes be resolved into two parts, each of which approaches linearity. This can result if there are two competing reactions with different activation energies and is often observed where the same reaction may occur both homogeneously and hetero-geneously. The homogeneous reaction usually has a higher activation energy and so it is favored at high temperatures, whereas the heterogeneous reaction predominates at lower temperatures. Such a case is illustrated in Fig. 9.

PROBLEMS

1. The rate constant for the formation of hydrogen iodide from hydrogen and iodine at 781° K is 3.58 when time is in minutes and the concentration is in moles per 22.4 liters. Calculate the rate constant for disappearance of hydrogen in units of liters per mole-second and cubic centimeters per molecule-second.

2. Consider a reaction with stoichiometric equation

$$A + 2B \rightarrow C + D$$

and which is first-order in each of the two reactants. Set up the rate expression in terms of a reaction variable x and integrate. How could this integrated result have been obtained directly from equation 25?

3. The oxidation of certain metals is found to obey the parabolic equation

$$y^2 = k_1 t + k_2$$

where y is the thickness of oxide film at time t. What order could be ascribed to this reaction? How could this be interpreted? [See W. J. Moore, *J. Chem. Phys.*, *18*, 231 (1950).]

4. The decomposition of nitrogen dioxide is a second-order reaction with rate constants k as follows:

$T°$ K	592	603.2	627	651.5	656
k, cc/mole-sec	522	755	1,700	4,020	5,030

Calculate the Arrhenius activation energy, E_a, also E of the equation

$$k = AT^{1/2}e^{-E/RT}$$

Compare with equation 40. [M. Bodenstein, *Z. physik. Chem.*, *100*, 106 (1922).]

5. Show that the Powell plot could be generalized so that a unique curve results for each order of reaction when log c is plotted against log t. By shifting in two dimensions, without rotation, all curves for the same order of reaction should be superimposable. This would presumably make it possible to determine the order of reaction without knowing the initial concentration. Test the method on the data of Chapter 3, Table 1. Is the method practicable?

REFERENCES

1. L. Wilhelmy, *Ann. Physik. Chemie* (Poggendorf), *81*, 413, 499 (1850).
2. R. E. Powell, private communication.
3. M. Nord, *Chem. Ind.*, *64*, 280 (1949).
4. T. I. Crowell and L. P. Hammett, *J. Am. Chem. Soc.*, *70*, 3444 (1948).
5. S. Arrhenius, *Z. physik. Chem.*, *1*, 110 (1887).
6. S. Widequist, *Arkiv Kemi*, *26A*, 2 (1948).
7. M. Kunitz and J. H. Northrop, *J. Gen. Physiol.*, *19*, 991 (1936).
8. H. T. Davis, *Theory of Econometrics*, Principia Press, Bloomington, Ind. (1941).
9. C. G. Swain, *J. Am. Chem. Soc.*, *70*, 1119 (1948).
10. L. S. Kassel, *Kinetics of Homogeneous Gas Reactions*, A.C.S. Monograph 57 (1932), p. 150.

3.

EXPERIMENTAL METHODS
AND TREATMENT OF DATA

The determination of reaction rates by conventional methods reduces to a study of concentrations as a function of time. In these cases a problem in quantitative analysis is always present, and one or more of the countless analytical procedures that have been devised may be needed in a particular kinetics problem. It may be mentioned also that kinetic studies can be made without any measurement of concentration and without any reference to laboratory time. These are by methods designed chiefly for the study of fast reactions, which will be discussed in Chapter 11.

In general, analytical procedures may be divided into two broad categories, chemical and physical. Chemical analysis implies a direct determination of one of the reactants or products by volumetric or gravimetric procedures, the former being preferred because of their rapidity. An important restriction on any chemical method is that it must be rapid compared with the reaction being studied. Or if the method is relatively slow, the reaction must be stopped or frozen by some sudden change such as lowering the temperature, removal of a catalyst, addition of an inhibitor, or removal of a reactant. Chemical methods of analysis have the advantage of giving an absolute value of the concentration.

On the other hand, physical methods of analysis are usually much more convenient than chemical methods. A physical method is one which measures some physical property of the reaction mixture, which changes as the reaction proceeds. There must be a substantial difference in the contribution of the reactants and products to the particular physical property chosen. Common among physical methods are pressure measurements in gaseous reactions; dilatometry, or measurement of volume change; optical methods such as polarimetry, refractometry, colorimetry, and spectrophotometry; electrical methods such as conductivity, potentiometry, polarography, and mass spectrometry. Theoretically any property which changes sufficiently could be used to follow the course of a reaction. Thermal conductivities, solidification temperatures, viscosities (for polymerization reactions), coagulating power towards colloids, and heats of reaction are among the more unusual properties which have been utilized.

In general, a physical method of analysis has the advantage of being rapid so that more experimental points are available in a given time. Measurements can frequently be made in the reaction vessel so that sampling with its attendant errors is eliminated. The system is usually not destroyed by the method nor even perceptibly disturbed. Often it becomes possible to make automatic and continuous recordings of the changes in property. Physical methods, however, have the limitation of not giving absolute values of concentration directly. Furthermore, errors due to side reactions may be enormously magnified. For example, in spectrophotometric studies, small amounts of highly colored impurities or by-products will obscure the desired quantities. For a complete study of any given reaction, more than one method of analysis should be used. It is especially desirable that the stoichiometry of the reaction be verified to be sure that the reaction being studied is one in which the products are known with certainty.

Correlation of Physical Properties with Concentrations

One requirement of any physical measurement as a criterion of extent of reaction is that the property being measured differs appreciably from reactants to products. Another requirement is that the property varies in some simple manner with the concentrations of reactants and products. The most common and useful relationship is that the physical property be a linear function of the concentration. Such a relationship exists, for example, between concentration and electrical conductance, optical density, rotation of polarized light, and pressure of gases. In dilute solutions, many physical properties such as the specific volume, refractive index, vapor pressure, and fluidity become linear functions of the concentration. In practice, of course, many of these linear relationships will break down if applied over too wide a range of concentration, the reasons being not only deviations from ideal behavior but also nonlinearities in the mathematical forms of the ideal laws relating the properties to concentration.

A general equation can be derived for relating a measured physical quantity with concentration if a linear relationship exists. Suppose we have the reaction which goes to completion

$$nA + mB + pC = rZ \tag{1}$$

where Z includes all products. Let λ be the value of the physical property at any time t.

$$\lambda = \lambda_M + \lambda_A + \lambda_B + \lambda_C + \lambda_Z \tag{2}$$

where the first term is the contribution of the medium and the others vary with concentration as, for example,

$$\lambda_A = k_A[A]$$

k_A being a proportionality constant. Letting the initial concentrations of reactants be a, b, and c, respectively, and the reaction variable x be the equivalents reacting in time t, then

$$\lambda = \lambda_M + k_A(a - nx) + k_B(b - mx) + k_C(c - px) + k_Z rx \tag{3}$$

and

$$\lambda_0 = \lambda_M + k_A a + k_B b + k_C c \tag{4}$$

$$\lambda_\infty = \lambda_M + k_B(b - ma/n) + k_C(c - pa/n) + k_Z ra/n \tag{5}$$

where λ_0 and λ_∞ are the initial and final values of λ, and in equation 5 it is assumed that A is the reactant present in limiting amount. Subtracting (4) from (5) gives us

$$\lambda_\infty - \lambda_0 = k_Z \frac{ra}{n} - k_A a - k_B \frac{ma}{n} - k_C \frac{pa}{n} \tag{6}$$

and (4) from (3)

$$\lambda - \lambda_0 = k_Z rx - k_A nx - k_B mx - k_C px \tag{7}$$

so that we may write

$$\lambda - \lambda_0 = x\,\Delta k \qquad \lambda_\infty - \lambda_0 = (a/n)\,\Delta k$$

and

$$\lambda_\infty - \lambda = (a/n - x)\,\Delta k$$

where

$$\Delta k = k_Z r - k_A n - k_B m - k_C p$$

From these we may get the kinetically useful relationships

$$\frac{nx}{a} = \frac{\lambda - \lambda_0}{\lambda_\infty - \lambda_0} \tag{8}$$

$$\frac{a}{a - nx} = \frac{\lambda_\infty - \lambda_0}{\lambda_\infty - \lambda} \tag{9}$$

It is also possible to express $(b - mx)$ and $(c - px)$ in terms of the measured variable. The result is of the form

$$\frac{b}{b - mx} = \frac{(b/a)(\lambda_\infty - \lambda_0)}{(b/a)(\lambda_\infty - \lambda_0) - (m/n)(\lambda - \lambda_0)} \tag{10}$$

Considerable simplification can be obtained by using equivalent concentrations of reactants, so that $b/a = m/n$, etc.

Reactions which do not go to completion can be handled also if the equilibrium constant is known independently. We shall illustrate the use of equations 8 and 9 with several examples taken from the literature for reactions both in solution and in the gas phase.

Reactions in the Gas Phase

Lack of space will not permit an extended discussion of apparatus and experimental methods. Fortunately a number of excellent references are available to the reader in various treatises.[1-8]

The most useful methods for following gaseous reactions are manometric. This may be a direct measurement of pressure in systems where there is a change in the number of moles such as the decomposition of phosgene

$$COCl_2 = CO + Cl_2 \tag{11}$$

or where a reaction product is removed continuously by absorption or condensation. For example, in the reaction

$$H_2 + Cl_2 = 2HCl \tag{12}$$

the acid may be removed by absorption in water. Or the pressure may be read intermittently after removing a product, so that in reaction 12 chlorine and hydrogen chloride may be condensed by liquid air and the residual pressure of the hydrogen measured.

Table 1 gives some results obtained in the decomposition of di-*t*-butyl peroxide by Raley, Rust, and Vaughan.[9] This substance decomposes in a first-order reaction to give essentially acetone and ethane

$$(CH_3)_3COOC(CH_3)_3 = 2(CH_3)_2CO + C_2H_6 \tag{13}$$

so that the pressure should increase to three times its original value as the reaction proceeds. Actually, it is found that there is only a 2.88-fold increase in pressure since a small amount of the peroxide gives other products.

The total pressure listed in Table 1 also includes a small partial pressure due to nitrogen which was used to force the peroxide into the reaction vessel. Nevertheless, we set

$$x/a = (P - P_0)/(P_\infty - P_0) \qquad (a - x)/a = (P_\infty - P)/(P_\infty - P_0)$$

For a first-order reaction $t = (1/k) \ln [a/a - x)]$, which can be expressed also as $t = (1/k) \ln (P_\infty - P_0) - (1/k) \ln (P_\infty - P)$. Figure 1 shows the

Table I

DECOMPOSITION OF DI-t-BUTYL PEROXIDE AT 154.6°
(Raley, Rust, and Vaughan[9])

Time, min	Total Pressure, mm	$P_\infty - P$	$k \times 10^4$ [a] sec^{-1}
0	173.5	318.3	
2	187.3	304.5	
3	193.4	298.3	3.58
5	205.3	286.5	3.39
6	211.3	280.5	3.42
8	222.9	268.9	3.50
9	228.6	263.2	3.45
11	239.8	251.9	3.61
12	244.4	247.4	3.44
14	254.5	237.3	3.32
15	259.2	232.5	3.43
17	268.7	223.1	3.43
18	273.9	217.9	3.60
20	282.0	209.7	3.45
21	286.8	204.9	3.42
∞	491.8		

Av. 3.46 ± 0.07

[a] Rate constants evaluated for successive 3-min intervals.

logarithm of $(P_\infty - P)$ plotted against the time, the linearity confirming the first-order kinetics. Table 1 also shows the first-order rate constants calculated for successive 3-minute intervals from the equation

$$k = \frac{1}{t - t'} \ln \frac{(P_\infty - P')}{(P_\infty - P)}$$

The average of these agrees with the specific rate constant obtained by multiplying the slope of Fig. 1 by 2.303. One of the characteristics of a first-order reaction is that the initial concentration need not be known; the value of P_0 does not necessarily enter into the calculations. Actually in this experiment P_0 was not measured but obtained from an extrapolation back to zero time. The question arises as to whether it would be permissible to use a theoretical value of P_∞ (equal to $3P_0$ based on equation 13) instead of an experimental one. Actually, in this case it makes a difference of only a few per cent in the rate constant over the range of pressures studied kinetically. The error increases, however, as the reaction proceeds so that Fig. 1, for example, would not remain linear. It is generally

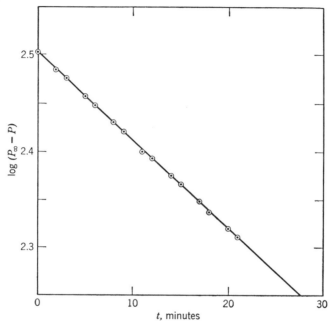

Fig. I. First-order plot for the decomposition of di-*t*-butyl peroxide (Raley, Rust, and Vaughan).

preferable to use an experimental value of the equilibrium reading instead of a theoretical one unless it is believed that some anomaly or error exists in the experimental value. For example, an equilibrium might exist which prevents the reaction from going to completion. This could be taken into account by a suitable correction of the λ_∞ values. If a slow secondary reaction of a product occurred, this could give a false value of λ_∞. In such a case a theoretical value of λ_∞ would be preferred.

As an example of a more complex gaseous reaction which requires a correction to each measured value of the pressure, the results of Kistiakowsky and Lacher[10] on the condensation of acrolein and 1,3-butadiene to form tetrahydrobenzaldehyde may be cited. This is an example of the Diels-Alder reaction and proceeds essentially to completion in the temperature range 155–330°. The reaction is second-order and follows

$$\tag{14}$$

the rate expression

$$dx/dt = k[\text{acrolein}][\text{butadiene}] \tag{15}$$

It is complicated by a simultaneous second-order polymerization reaction of the butadiene:

$$2C_4H_6 = C_8H_{12} \tag{16}$$

This reaction can be studied separately (by omitting acrolein) and is found to be about one tenth as fast as the main reaction. A correction was made

Table 2

CONDENSATION OF ACROLEIN AND BUTADIENE AT 291.2° C

(Kistiakowsky and Lacher[10])

Time, sec	P_{total}, mm	$-\Delta P$, mm	$-\Delta P_{dim}$, mm	$P_{acrolein}$, mm	$P_{butadiene}$, mm	$k \times 10^7$, mm^{-1} sec^{-1}
0	658.2			418.2	240.0	
63	652.1	6.1	0.2	412.3	233.7	9.6
181	641.4	10.7	0.3	401.9	222.7	9.5
384	624.1	17.3	0.5	385.1	204.9	9.9
542	612.2	11.9	0.3	373.5	192.7	9.7
745	598.1	14.1	0.3	359.7	178.3	10.0
925	587.1	11.0	0.3	349.0	167.0	9.7
1,145	574.9	12.2	0.3	337.1	154.5	9.8
1,374	564.1	10.8	0.3	326.6	143.4	9.3
1,627	552.8	11.3	0.2	315.5	131.9	9.9
1,988	539.4	13.4	0.3	302.4	118.2	9.4
						Av. 9.7

for the side reaction even though the error is small. Table 2 gives the values of the total pressure of the system at various times, starting with a known initial pressure of acrolein and of butadiene. From the total drop in pressure for each successive time interval the pressure change for acrolein and for butadiene was calculated in the following way: the change in pressure due to the dimerization of butadiene was calculated approximately from $\Delta P_{dim} = k' (P_{butadiene})^2 \Delta t$ where $P_{butadiene}$ is the partial pressure of butadiene at the start of the time interval and k' is the specific rate constant for dimerization (previously evaluated). Then ΔP_{dim} was subtracted from ΔP_{total} to give the pressure drop due to the Diels-Alder

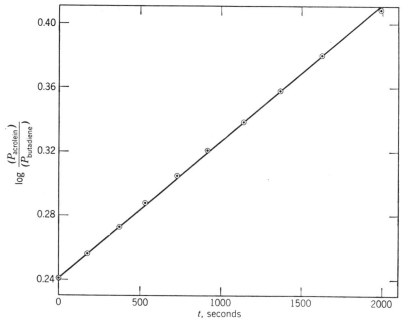

Fig. 2. Second-order plot for the condensation of acrolein with butadiene (Kistiakowsky and Lacher).

reaction. From the stoichiometry of reactions 14 and 16 it then follows that

$$\Delta P_{acrolein} = \Delta P_{total} - \Delta P_{dim} \tag{17}$$

$$\Delta P_{butadiene} = \Delta P_{total} + \Delta P_{dim} \tag{18}$$

so that the new pressure of each at the end of the time interval can be found. This enables the next two columns of Table 2 to be filled out. The second-order rate constant given in the last column was evaluated by using the unintegrated form of (15)

$$\frac{-\Delta P_{acrolein}}{\Delta t} = k(P_{acrolein})(P_{butadiene}) \tag{19}$$

where the pressures are the average partial pressures of each component during the time interval Δt. The constancy of k (average 9.7×10^{-7} mm^{-1}

sec^{-1}) is evidence that the reaction is second-order. Another way of treating the data is to use the integrated form of (15) which, as has been shown in Chapter 2, becomes

$$kt = \frac{1}{(P^0_{\text{acrolein}} - P^0_{\text{butadiene}})} \ln \frac{(P_{\text{acrolein}})}{(P_{\text{butadiene}})} + \text{constant} \qquad (20)$$

Figure 2 shows the result of plotting $\log [(P_{\text{acrolein}})/(P_{\text{butadiene}})]$ against the time. The slope multiplied by 2.30 and divided by $(418.2 - 280.0)$ gives k as 10.9×10^{-7} mm^{-1} sec^{-1}. Part of the difference between 10.9 and 9.7 is due to the dimerization reaction which is included in the rate calculated by equation 20, and the remainder is due to the difference in using integrated and unintegrated forms of the rate equation.

Reactions at Constant Pressure

Though most reactions in closed systems are carried out at constant volume, sometimes it is convenient to work with constant pressure. If there is no change in the total volume as reaction proceeds, there is no difference in the method of mathematical analysis. However, if the volume changes in a gaseous reaction, the effect of this change on the reaction rate must be taken into account.

Consider the first-order gaseous reaction

$$A \rightarrow \nu B \qquad (21)$$

where ν moles are formed from one mole. This increases the total volume and dilutes the reactant A. The change in the volume can be expressed as

$$V = V_0[1 + (\nu - 1)x/a] \qquad (22)$$

if the initial system consists only of gaseous A. The concentration of A is reduced by the ratio of V_0 to V. Putting this result back in the rate equation for the first-order reaction in question gives

$$k_1 t = \int \frac{[1 + (\nu - 1)x/a]\, dx}{(a - x)} \qquad (23)$$

which can be integrated to give

$$k_1 t = \nu \ln \frac{a}{(a - x)} - (\nu - 1)x/a \qquad (24)$$

This result differs appreciably from the usual equation, unless, of course, ν is equal to 1, in which case the two equations become identical. Similar corrections can be applied to reactions of other orders at constant pressure. Hougen and Watson[11] give a list of integrated expressions for a number of simple reaction systems.

Reactions in Solution

Many key references to reactions in solution which have been studied kinetically are to be found in treatises by Moelwyn-Hughes, Skrabal, Amis, and Laidler.[12–15] Reilly and Rae[16] discuss experimental methods and a number of texts describe the instruments that might be applied to kinetic measurements.[17]

Table 3

PHENACYL BROMIDE AND PYRIDINE IN METHYL ALCOHOL

(McGuire[18])

(Temperature 35.0°; molarity 0.0385.)

Time, min	Resistance, ohms	$-R/(R_\infty - R)$
7	45,000	1.019
28	11,620	1.074
53	9,200	1.096
68	7,490	1.120
84	6,310	1.145
99	5,537	1.170
110	5,100	1.186
127	4,560	1.213
153	3,958	1.253
203	3,220	1.330
368	2,182	1.580
∞	801	

k from slope 0.0445 liter mole^{-1} min^{-1}

In addition to chemical methods of analysis and measurements based on the rate of evolution of gaseous products, the three most widely used analytical methods for reactions in solution have been colorimetric or spectrophotometric, conductometric, and dilatometric. An example of each of these will be discussed in some detail.

Whenever a reaction occurs with a change in the number or kind of ions present so that the electrical conductivity changes, measurement of the resistance offers a convenient and accurate means of following the course of the reaction. A combination of a direct-reading a-c bridge and a cathode-ray oscilloscope for a null point indicator is a great improvement over the usual student slide wire bridge and earphones. Temperature control to 0.03°, which is readily attained with commercial thermostats, is sufficiently precise for most kinetic studies and for the accurate measurement of resistance. One restriction on this kind of measurement is that the solvent employed must have a high dielectric constant and be a good solvating medium. Otherwise, the resistance and the concentration do not have a simple relationship.

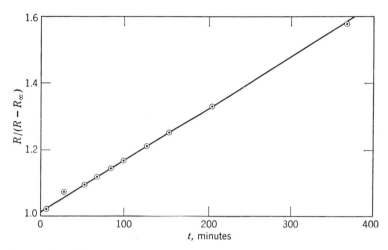

Fig. 3. Reaction of phenacyl bromide and pyridine as second-order with equal initial concentrations (McGuire).

Since resistance is inversely proportional to conductance, which in turn is proportional to concentration, the fundamental equations for kinetic studies using resistance are shown in (25) and (26). Table 3 shows some

$$x/a = \frac{1/R - 1/R_0}{1/R_\infty - 1/R_0} = \frac{(R_0 - R)R_\infty}{R(R_0 - R_\infty)} \tag{25}$$

$$a/(a - x) = \frac{1/R_\infty - 1/R_0}{1/R_\infty - 1/R} = \frac{(R_\infty - R_0)R}{R_0(R_\infty - R)} \tag{26}$$

data[18] for the reaction of pyridine with phenacyl bromide to form a quaternary ammonium salt, phenacylpyridinium bromide, in methyl alcohol solution. Since neutral molecules are forming ions, the resistance

$$\overset{O}{\overset{\|}{C_6H_5-C-CH_2Br}} + C_5H_5N = \overset{O}{\overset{\|}{C_6H_5-C-CH_2-N\overset{+}{C}_5H_5}} + Br^- \tag{27}$$

decreases sharply as the reaction goes on. This is a second-order reaction where the reactants are present in equal concentrations. According to Chapter 2, $1/(a - x)$ should be a linear function of the time. Figure 3 shows that $R/(R - R_\infty)$ is linear with time since the usual second-order equation expressed as

$$t = (1/ak)[a/(a - x)] - 1/ak \tag{28}$$

becomes

$$t = (1/ak)[(R_\infty - R_0)/R_0][R/(R_\infty - R)] - 1/ak \tag{29}$$

in terms of resistance. The specific rate constant may be found by dividing the slope of the line in Fig. 3 by the initial concentration, a, and by the intercept which is $R_0/(R_0 - R_\infty)$. The intercept is very nearly equal to unity since R_0, the resistance at $t = 0$, is very large. The exact value of R_0 has no significance in this particular case. It is obvious that from the intercept and the equilibrium resistance, equations 25 and 26 could be used to calculate x or $(a - x)$ at any time. This is usually unnecessary.

As an example of a spectrophotometric kinetic study, the results of Andrews[19] on the alcoholysis of cinnamal chloride are of interest. The reaction is shown in (30). The cinnamal chloride, having a double bond

$$C_6H_5CH{=}CH{-}CHCl_2 + C_2H_5OH = C_6H_5{-}CH{-}CH{=}CHCl + HCl$$
$$\underset{\displaystyle OC_2H_5}{|} \tag{30}$$

conjugated with the benzene ring, absorbs strongly at 2600 A, whereas the product (1-chloro-3-ethoxy-3-phenyl-1-propene) being unconjugated does not absorb until 2100 A. Accordingly, a measurement of the optical density at 2600 A as a function of time permits the rate of the reaction to be measured. These readings are conveniently made with a commercial photoelectric spectrophotometer which can be thermostatted, or, for relatively slow reactions, samples can be withdrawn from the thermostatted reaction mixture for measurement. Such spectrophotometry is preferable to colorimetric methods because of greater ease of reading and because spectrophotometry applies to compounds which do not absorb in the visible, as in the present example.

Table 4 shows the data of Andrews on the reaction in question. Although the solution contains sodium ethoxide, the reaction velocity is not dependent on its concentration. The specific rate constants are calculated from the first-order equation

$$k = \frac{2.303}{t} \log\left(\frac{D_0}{D}\right) \tag{31}$$

or more properly from

$$k = \frac{2.303}{t} \log \frac{D_\infty - D_0}{D_\infty - D} \tag{32}$$

Since D_∞ is essentially zero, equation 31 is sufficiently precise.

The dilatometric method can be illustrated by the data of Ciapetta and

Kilpatrick[20] on the hydration of isobutene. This reaction is catalyzed specifically by hydrogen ions

$$\underset{\begin{array}{c}|\\CH_3\end{array}}{CH_3-C{=}CH_2} + H_2O \xrightarrow{\;H_3O^+\;} \underset{\begin{array}{c}|\\OH\end{array}}{\underset{\begin{array}{c}|\\CH_3\end{array}}{CH_3-C-CH_3}} \qquad (33)$$

and yields *t*-butyl alcohol. It is a pseudo-first-order reaction since the concentrations of acid and of water remain constant. The reaction is

Table 4

CINNAMAL CHLORIDE IN ABSOLUTE ETHANOL AT 22.6°

(Andrews[19])

([RCl_2] = 2.11 × $10^{-5}M$; [NaOEt] = 0.547M.)

Time, min	Optical Density, D^a at 2600 A	$k \times 10^3$, min^{-1}
0	0.406	
10	0.382	6.0
31	0.338	5.8
74	0.255	6.3
127	0.184	6.0
178	0.143	5.8
1,200	0.01	

^{a} $D = \epsilon lC$ where ϵ is the molar extinction coefficient, l is the thickness of the absorbing medium, and C is the molar concentration.

accompanied by a contraction in volume which can be followed by reading the height of the liquid level in a capillary connected to a rather large volume of reaction mixture. This liquid level changed 8–9 cm during a run, corresponding to a volume change of only 0.02 ml. This small volume change necessitated temperature control to a thousandth of a degree. Tong and Olson[21] have a discussion of errors in the use of dilatometers.

Table 5 shows the dilatometer readings (in arbitrary units) for a number of times arranged in pairs with a fixed interval between them of 2 hours. The reason for doing this will be brought out in a later paragraph. Figure 4 shows the result of plotting log ($V - V_\infty$) against the time. The slope multiplied by 2.303 gives the observed rate constant, which has the units of a first-order constant, but which is more complex since it varies with the concentration of acid. If the observed rate constant is divided by the hydrogen ion concentration, a second-order rate constant is obtained which remains constant for different acid concentrations, providing that the ionic strength is held fixed.

Though the easiest physical measurements to use in kinetic studies are those in which a linear relationship exists with concentration, it is obvious

Table 5

HYDRATION OF ISOBUTENE IN PERCHLORIC ACID SOLUTION

(Ciapetta and Kilpatrick[20])

$(T = 25°$; $[HClO_4] = 0.3974M$; $[isobutene] = 0.00483M$;
$k_{obs} = 1.322 \times 10^{-2}$ min^{-1}; $V_\infty = 12.16$.)

Time,	Dilatometer Reading, V	
min	At t	At $t + 120$ min
0	18.84	13.50
5	18.34	13.42
10	17.91	13.35
15	17.53	13.27
20	17.19	13.19
25	16.86	13.12
30	16.56	13.05
35	16.27	13.00
40	16.00	12.94

that other methods could be adapted by suitable calibration or by theoretical equations if equations of sufficient validity are available. Electromotive force measurements are suitable in spite of the logarithmic relationship with concentration, though this factor tends to reduce the accuracy. A method described by Swain and Ross[22] involves potentio-metric measurements of concentration cells to get concentrations directly. One electrode is in the reaction mixture and the other is an adjustable standard.

Determination of the Order of Reaction: The Rate Expression

Assuming that data have been obtained giving concentrations at various times, the problem next arises of determining the order of a reaction with respect to all participants. This is equivalent to obtaining the form of the rate expression governing the reaction. In general, it is the rate expression that is desired and not the reaction order, since the rate expression exists even for reactions of no simple order. The rate expression provides the important clue to the mechanism of the reaction.

There is no general method, unfortunately, for finding the order of a reaction. Usually a trial-and-error procedure is used based upon intelligent guesses. These guesses come from the stoichiometry of the reaction or from assumptions concerning its mechanism. The assumed expression is integrated when possible to give a relation between concentration and

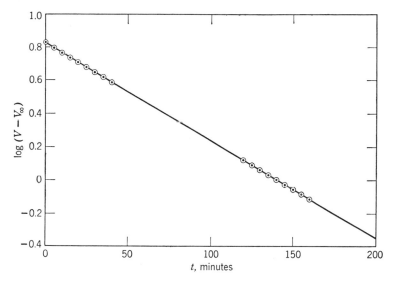

Fig. 4. Hydration of isobutene plotted as pseudo-first-order reaction (Ciapetta and Kilpatrick).

time. This relation is tested with the experimental data by numerical or graphical methods. The procedure is repeated until the assumed rate expression closely reproduces the data.

Fractional-Life Period Methods

In addition, there are several direct methods for getting the order with respect to each reactant, providing that certain restrictions can be met. In the event that the rate expression is of the form

$$dx/dt = k(a - x)^n \tag{34}$$

the half-life method (or other fractional life) can be used. The half-life period is defined as the time required for one-half of a given reactant to be used up. Notice that this is not one-half of the time required to complete the reaction, which is presumably infinite except for a zero-order reaction.

Substituting $(a - x) = a/2$ in the integrated forms of (34) gives for $n = 1$

$$t_{1/2} = (\ln 2)/k \tag{35}$$

for $n \neq 1$

$$t_{1/2} = \frac{(2^{n-1} - 1)}{k(n - 1)a^{n-1}} \tag{36}$$

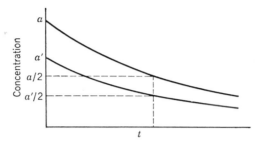

Fig. 5. Illustration of the independence of concentration on the half-life for a first-order reaction.

The dependence of the half-life period on the initial concentration is of particular importance. It is evident that for all values of n

$$t_{1/2} = f(n, k)/a^{n-1} \tag{37}$$

where f is some function of n and k and therefore constant for a given reaction at constant temperature. This relation includes the usually

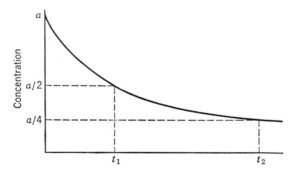

Fig. 6. Two successive half-life periods in a second-order reaction. $t_2 = 3t_1$.

special first-order case, which has the unusual feature of a $t_{1/2}$ independent of a. Putting (37) in logarithmic form yields

$$\log t_{1/2} = \log f - (n - 1) \log a \tag{38}$$

A log-log plot of $t_{1/2}$ versus a should be linear with a slope of $1 - n$.

By applying (38) to any two pairs of $t_{1/2}$, a data, say $t_{1/2}$ and a, and $t_{1/2}'$ and a', and subtracting one equation from the other, the $\log f$ term drops out. Solving for n gives the Noyes equation (39). With this equation the

$$n = 1 + \frac{\log t_{1/2}' - \log t_{1/2}}{\log a - \log a'} \tag{39}$$

order of reaction can be calculated directly from data on two runs at sufficiently different initial concentration, as in Fig. 5.

Equations 35 to 39 can be generalized to apply to t_y, the time for the fraction reacted to equal y, when the concentration has dropped from a to $a(1 - y)$. Equation 35 becomes

$$\log t_y = \log f - (n - 1) \log a \tag{40}$$

and, in general, a log-log plot of t_y versus a should be linear. Equation 39 becomes equation 41.

$$n = 1 + \frac{\log t_y' - \log t_y}{\log a - \log a'} \tag{41}$$

So far the discussion of fractional-life periods has depended on the experimental observation of two or more runs with different initial concentrations. But it is evident that two or more successive time intervals in a *single run* may be used in a similar manner, the concentration resulting at the end of one time interval being considered the initial value for a new time interval. Figure 6 illustrates such an experiment. It is obvious that, if t_1 corresponds to the concentration $a(1 - y)$, then t_2 corresponds to the concentration $a(1 - y)^2$. Also $t_y' = t_2 - t_1$ so that equation 41 becomes equation 42. For the case illustrated in

$$n = 1 + \frac{\log [(t_2/t_1) - 1]}{\log [1/(1 - y)]} \tag{42}$$

Fig. 6, $t_2/t_1 = 3$ and $y = \frac{1}{2}$ so that

$$n = 1 + \frac{\log 2}{\log 2} = 2$$

Because rate data may be more accurate at the beginning of a reaction, a value of y less than one-half should be used. On the other hand, y should not be too small since, owing to the log ratio in (38) tending towards an indeterminate form, slight experimental errors are greatly amplified in calculating n. A suitable value of y might be 0.2, in which case t_1 and t_2 would be the times for the concentration of reactant to have fallen to 0.80 and 0.64 of its original value.

It should be remembered that this treatment is valid only if the rate expression is given by (34). If this is not true, equations such as (41) and

(42) will still give a value of n, but one which will change with the concentration. Whenever the order by this method turns out to be fractional it may be expected that further investigation will show that a more complex rate expression is needed.

The Powell plot of Chapter 2 may be considered a generalization of these fractional-life methods.

Suppose that the rate expression is now given by

$$\frac{dx}{dt} = k[A]^{n_a}[B]^{n_b}[C]^{n_c} \tag{43}$$

where [A], [B], and [C] are the concentrations of substances A, B, and C, respectively. There are certain conditions which, if satisfied, make for simplification in the determination of the order.

Suppose that there is effectively constant concentration of all components except one, which for convenience is taken as component A. This is the case of a pseudo-n_ath-order reaction as discussed in Chapter 2. The effectively constant concentration may be obtained by having a large excess of all except the one component. Then during a reaction run [B] and [C] and their powers may be taken as approximately constant and lumped in with the rate constant k to form a new effective rate constant k' where

$$k' = k[B]^{n_b}[C]^{n_c}$$

so that

$$dx/dt = k'[A]^{n_a} \tag{44}$$

This is of the same form as (34) and all the equations 34 to 42 apply to this case, but with the reinterpretation of the symbols, in particular, k becomes k' the effective rate constant, and n becomes n_a the pseudo-order of the reaction for the given conditions, which also is the order with respect to A.

In principle, it is possible in this way to get the order with respect to each reactant (and product, if it affects the rate) by making each one successively the least concentrated component.

A second method of simplifying the solution of equation 43 is to choose the initial concentration of each reactant so that all concentrations throughout the run are in constant proportion to each other. This will be true if equivalent amounts of the reactants are used. For example, if in the reaction

$$2NO + O_2 = 2NO_2$$

the molar concentration or partial pressure of NO is twice as great as that of O_2, then neither reactant is in excess. For a general reaction under these conditions, the reactant concentrations are in constant proportion during the run.

$$[B] = p[A] \qquad [C] = q[A] \qquad \text{etc.}$$

where p and q are constants determined from the stoichiometric equation of the reaction. Then (39) becomes

$$dx/dt = kp^{n_b}q^{n_c}[A]^{n_a+n_b+n_c} \tag{45}$$

or

$$dx/dt = k''(a - x)^n$$

where the new constant k'' is $kp^{n_b}q^{n_c}$ and $n = n_a + n_b + n_c$ is the total order of the reaction. Equation 45 is again of the same form as (34) and the previous discussion again applies. A different constant, k'', now results, and the total order of the reaction with respect to all reactants rather than to just one is obtained. There is an important restriction in using this method. No product of the reaction must influence the rate, for, if it did, its concentration could not be proportional to the concentration of a reactant during a run.

Initial Rate as a Function of Initial Concentrations

The most obvious use of equation 43 is to make a direct comparison between rate and concentration. This can be done approximately by taking as a measure of the derivative dx/dt the corresponding ratio of finite increments, $\Delta x/\Delta t$. For reasonable accuracy the fraction reacting should be no more than, say, 0.1. Then, by making a run at each of two different initial concentrations of any one component, say, B, the other concentrations remaining constant, the data will enable the determination of the order of reaction with respect to that component. Let the two rates and corresponding initial concentrations be $(dx/dt)_1$, $(dx/dt)_2$ and $[B]_1$, $[B]_2$. Then

$$(dx/dt)_1 = (k[A]^{n_a}[C]^{n_c})[B]_1{}^{n_b}$$
$$(dx/dt)_2 = (k[A]^{n_a}[C]^{n_c})[B]_2{}^{n_b}$$

Dividing, taking the logarithm, and solving for n_b yields (46). This

$$n_b = \frac{\log(dx/dt)_1 - \log(dx/dt)_2}{\log[B]_1 - \log[B]_2} \tag{46}$$

equation is due to van't Hoff.

Varying successively the initial concentration of each component will give the order with respect to each component, following which the rate constant k may be evaluated approximately from any one run. Equation 46 is also applicable to the simpler case of one variable concentration, but the methods already discussed are more accurate and convenient.

An example of the use of equation 46, as well as of equation 41, is found in the work of Klute and Walters[23] on the thermal gas phase decomposition of tetrahydrofuran. This reaction is complex, yielding a variety of products. The pressure-time data do not correspond to a simple order of reaction. Accordingly, a number of runs were made, and the maximum rates, $(\Delta P/\Delta t)_{max}$, and the half-lives were determined for different starting pressures. A plot of log $(\Delta P/\Delta t)_{max}$ against log P_0 gave a straight line of slope equal to 1.5, and log $t_{1/2}$ plotted against log P_0 gave a straight line with a slope of -0.55. This indicated that the reaction was of the 1.5 order.

Rate Expression Determined Directly from Rate

When integration is not feasible, or when a preliminary search for the rate expression is being made, it may be helpful to determine the value of the rate (that is, the value of dx/dt) from concentration-time data and plot it against concentration. Since the unintegrated rate expression (apart from the derivative) is usually simpler mathematically than the integrated form, it may be possible to see a relation between the rate and concentration. Such a relationship, for example, a power series in the concentration, may be checked by suitable graphing. Sprauer and Simons[24] and Pearlson and Simons[25] have used this method in analyzing the kinetics of alkylation of toluene with t-butyl chloride catalyzed by hydrogen fluoride.

$$t\text{-}C_4H_9Cl + C_7H_8 \xrightarrow{\text{HF}} t\text{-}C_4H_9C_7H_7 + HCl \tag{47}$$

The reaction was run in toluene solution, and the rate of evolution of HCl gas was found by measuring the pressure. The pressure was plotted against time and dP/dt found by measuring the tangent for various values of the pressure. The tangent may be found by simply laying a straightedge along the curve at various points and finding its slope in units of P and t, or an optical device may be used. Pearlson and Simons used a mirror which, when oriented normal to the curve, will show a reflection that is a smooth continuation of the curve. The tangent is then drawn perpendicular to the normal. Frampton[26] uses a simple prism to find the normal to the curve. Less accurately, small finite differences, $\Delta P/\Delta t$, could be set equal to dP/dt for the pressure equal to the midpoint of the pressure interval.

Sprauer and Simons then plotted the rate dP/dt against the concentration of the remaining t-butyl chloride which was expressed as $(P_\infty - P)$, the final pressure minus the pressure at any time being proportional to $(a - x)$. The resulting curves were found to be hyperbolic, which suggested the rate expression

$$dP/dt = m(P_\infty - P)/[n - (P_\infty - P)]$$

where m and n are constants. This equation was integrated and the constants m and n found by fitting the data to the integrated form.

Experiments in which the initial pressure of HCl was changed by adding HCl to the reaction mixture and in which the catalyst concentration was changed showed that m and n were functions of the catalyst concentration and the initial pressure of HCl, respectively. A mechanism was postulated by Sprauer and Simons which led to the rate expression

$$d[\text{HCl}]/dt = \frac{k_1[\text{H}^+]k_2[\text{toluene}][t\text{-butyl chloride}]}{k_3[\text{HCl}] + k_2[\text{toluene}]} \tag{48}$$

which was in satisfactory agreement with the empirical rate equation if m and n were given suitable values.

Yost and Hayward[27] give another example of using the rates directly measured for various concentrations to support an assumed mechanism. The rate expression in this case is one which is difficult to verify in the integrated form.

Plot of t/p versus t. A method recently devised by Wilkinson[28] would appear to have an advantage over other methods just described for determining the order of a simple reaction.

Equation 7 of Chapter 2 may be put into the form

$$(1 - p)^{1-n} = 1 + (n - 1)Kt \tag{49}$$

where p is $1 - \alpha$, the fraction reacted, and $K = kC_0^{n-1}$. If $(1 - p)^{1-n}$ is expanded by the binomial theorem and terms higher than the second degree in p are discarded, (49) can be put in the form

$$\frac{t}{p} = \frac{1}{K} + \frac{nt}{2} \tag{50}$$

If t/p is plotted against t, this is the equation of a straight line with slope of $n/2$ and intercept $1/K$. Because (46) is only approximate, actual plots might be expected to be curves with a limiting slope of $n/2$. Over the range up to $p = 0.4$ the lines for actual data are remarkably straight with apparent slopes rather close to $n/2$. At least it is easy to distinguish the order of the reaction to the nearest half integer.

After the order is obtained, the intercept will yield a K which can then be used to obtain an approximate rate constant.

Evaluation of Rate Constants

After the order of a reaction or the form of the rate expression has been determined, the next step for the experimental data is the calculation of the best values of the various rate constants which may appear in the rate equation. Complex reactions involving more than one rate constant require special techniques, each case usually being sufficiently different so that no general method will apply. In Chapter 8 a number of examples of complex reactions will be discussed in detail.

For present purposes we shall consider only reactions involving a single rate constant, that is, simple first-, second-, or third-order reactions where the integrated form of the rate expression is readily available. An excellent article by Roseveare[29] contains a mathematical analysis of the problem of determining rate constants. In general, three methods have been widely used: the calculation of a rate constant for each experimental point, using the integrated equation; the calculation of a rate constant for adjacent pairs of points; and the graphical method, using the slope to find the rate constant. To illustrate with a first-order reaction, the rate constant can be found from equations

$$k = (1/t) \ln [a/(a - x)] \tag{51}$$

$$k = [1/(t - t') \ln [(a - x)'/(a - x)] \tag{52}$$

or from the plot of $\ln (a - x)$ versus t by getting the best value of the slope

$$k = -\text{slope} = -[\Delta \ln (a - x)]/\Delta t \tag{53}$$

The first method suffers from the disadvantages of placing a very high weight on the value of the initial concentration and being very sensitive to errors when x is either small or nearly equal to a. The second method, if used for all pairs of successive point and the resulting k's are averaged, tends to approach the procedure of finding k from (52), using only the first and last experimental points, which are generally the least accurate. Roseveare gives methods of weighting the data which remove most of the objections but considerably increase the labor involved. The same arguments hold true for second- and third-order reactions.

The easiest and best method appears to be the graphical one, which plots a suitable function of the concentration against the time and finds the rate constant from the slope (if a straight line is achieved) and the intercept. The initial concentration, or the initial reading if some physical property is measured, may be known, in which case it furnishes an extra point on the curve, or it may be found from the intercept. The best straight line through the experimental points can usually be found by inspection, that is, by moving a transparent straightedge until it appears to fit the data with minimum deviation. In doubtful cases, standard numerical methods such as least squares can be used to find the slope. The graphical methods readily show trends and deviations from straight-line behavior. Isolated points which are obviously in error can be eliminated at once. The chief exception to the graphical method is when the accuracy of the data exceeds that of plotting on a reasonable scale. This rarely happens.

Roseveare's weighting formulas indicate that for a first-order reaction the most accurate points are to be found at about two-thirds of complete reaction. For a second-order reaction with equivalent concentrations, the best points are at about one-half reaction. This is true if the error in time is small, as it usually is except for very rapid reactions. Consequently, in graphing or in direct calculation, less attention should be paid to the initial points and the final ones in determining the slope or rate constant.

The accepted rate constant should never be based upon a single kinetic run, even if a number of points are taken. A minimum of three runs with different initial concentrations should be made, and the rate constant found from a consideration of the separate constants found for each run. Such a procedure is also useful in checking the assumed order of the reaction. A method which can be used for comparing data from several different runs is to write $F(c) = kt$ where $F(c)$ is some function of the concentration, that is, for a first-order reaction $F(c) = \ln [(a/(a - x)]$. It is then possible to plot $F(c)$ for several runs against time. A single straight line should be obtained with an intercept at the origin and a slope equal to the rate constant. This procedure is helpful when only a few determinations can be made in each run, and when the points tend to scatter.

Methods Where the Final Reading is Unknown

Guggenheim[30] has described a method for finding the rate constant for a first-order reaction where x, the concentration reacted, is determined directly, but where a, the initial concentration, or, what is equivalent, the value of x at infinite time, is not known. The method is equally applicable to a first-order reaction followed by a physical measurement where the equilibrium or final reading cannot be made. If times t_1, t_2, t_3, etc., and $t_1 + \Delta$, $t_2 + \Delta$, $t_3 + \Delta$, etc., are selected where Δ is a constant increment, then the following equations are true:

$$(\lambda_1 - \lambda_\infty) = (\lambda_0 - \lambda_\infty)e^{-kt_1} \tag{54}$$

$$(\lambda_1' - \lambda_\infty) = (\lambda_0 - \lambda_\infty)e^{-k(t_1 + \Delta)} \tag{55}$$

where λ_1 and λ_1' are readings of a suitable physical property at t_1 and $t_1 + \Delta$, respectively, and the usual first-order equation is written in exponential form. Similar equations would be true for t_2 and $t_2 + \Delta$. Subtracting (55) from (54) gives

$$(\lambda_1 - \lambda_1') = (\lambda_0 - \lambda_\infty)e^{-kt_1}(1 - e^{-k\Delta}) \tag{56}$$

or

$$kt_1 + \ln (\lambda_1 - \lambda_1') = \ln [(\lambda_0 - \lambda_\infty)(1 - e^{-k\Delta})]$$

$$= \text{a constant} \tag{57}$$

which can be generalized by dropping the subscript 1. Before this method can be used it must be certain that the reaction is simple first-order, since certain other more complex reactions (reversible and concurrent first-order reactions) will give apparent rate constants by this method. The interval Δ should be two or three times as great as the half-life period of the reaction for accuracy.

This method was used by Ciapetta and Kilpatrick for the data given in Table 5. Two sets of reading with a constant interval of 120 minutes are given. When the logarithms of $(V_t - V_{t+120})$ are plotted against the time, a straight line is obtained with a slope equal to $-k/2.303$. The rate constant found in this way is the same as that found by plotting log $(V - V_\infty)$ against the time. The infinity value can be measured experimentally, but it is somewhat more convenient to eliminate it by the Guggenheim device.

Roseveare also gives an analogous method for a second-order reaction with equivalent concentrations. His final equation is

$$k = \frac{[(\lambda_2 - \lambda_1) - (\lambda_3 - \lambda_2)]^2}{2(t_2 - t_1)(\lambda_3 - \lambda_1)(\lambda_2 - \lambda_1)(\lambda_3 - \lambda_2)} \tag{58}$$

where λ_1, λ_2, and λ_3 refer to three readings at times t_1, t_2, and t_3 which are separated by a constant interval. The units of k are in terms of λ. Sturtevant[31] has generalized the method to apply to unequal concentrations and gives a way of reducing the cumbersome calculations.

PROBLEMS

1. The rate of iodination of nitroethane in the presence of pyridine according to the equation

$$C_2H_5NO_2 + C_5H_5N + I_2 \rightarrow C_2H_4INO_2 + C_5H_5NH^+ + I^-$$

was followed by measuring the change in electrical conductivity. The accompanying data were obtained at 25° C in a water-alcohol solvent.

   Original concentrations: nitroethane and pyridine, each $0.1M$; iodine $0.0045M$.

Time, min	Resistance, ohms
0	2,503
5	2,295
10	2,125
15	1,980
20	1,850
25	1,738
30	1,639
∞	1,470

Determine the apparent order of the reaction and the apparent rate constant in suitable units. Assuming that the reaction is first-order in nitroethane and first-order in pyridine, convert to the correct second-order constant.

2. The thermal decomposition of dimethyl ether in the gas phase has been studied by measuring the increase in pressure.

$$(CH_3)_2O \rightarrow CH_4 + H_2 + CO$$

Some measurements made at 504° C and an initial pressure of 312 mm of ether are as follows:

Time, sec	390	777	1,195	3,155	∞
Pressure increase, mm	96	176	250	467	619

Calculate the rate constant after determining the order of the reaction. [C. N. Hinshelwood and P. J. Askey, *Proc. Roy. Soc.*, A115, 215 (1927).]

3. The substance 3,3'-dicarbazyl phenyl methyl chloride exists in an alkaline solution partly as a negative ion and partly as an anhydro base which is neutral. The two forms have different colors, the negative ion being green and absorbing strongly at 7300 A and the anhydro base being red and absorbing strongly at 5000 A. One or both forms combine with water or methanol to form colorless carbinols or ethers. The accompanying data were collected, a photoelectric spectrophotometer and a water-acetone solvent at 25° C being employed. What conclusions can be drawn concerning the rate of disappearance of each of the colored forms? What conclusions can be drawn concerning whether both forms are reacting to form carbinol?

Original concentrations: NaOH, $2.03 \times 10^{-2}M$; 3,3'-dicarbazyl phenyl methyl chloride $\simeq 10^{-5}M$.

5,000 A		7,300 A	
Time, min	$E - E_\infty{}^a$	Time, min	$E - E_\infty$
3.7	0.320	2.7	0.562
5.7	0.246	4.7	0.423
7.7	0.188	6.7	0.333
9.7	0.143	8.7	0.243
11.7	0.109	10.7	0.191
13.7	0.084	12.7	0.148
15.7	0.065	14.7	0.111
17.7	0.049	16.7	0.087
19.5	0.036	18.7	0.065
23.7	0.023	21.7	0.045

^{a} E refers to optical density and E_∞ to the density for complete reaction. The latter value is essentially zero. [G. E. K. Branch and B. M. Tolbert, *J. Am. Chem. Soc.*, 69, 523 (1947).]

4. The cleavage of diacetone alcohol by alkali to form acetone (see Chapter 12) can be followed with a dilatometer since there is a substantial increase in

volume in concentrated solutions as the reaction proceeds. The following data were collected by Åkerlof at $25°$ C. Calculate the rate constant after determining the order of the reaction.

Original concentrations: KOH in water, $2N$; diacetone alcohol, 5% by volume. [G. Åkerlof, *J. Am. Chem. Soc.*, *49*, 2955 (1927).]

Time, sec	Cathetometer Reading
0.0	8.0
24.4	20.0
35.0	24.0
48.0	28.0
64.8	32.0
75.8	34.0
89.4	36.0
106.6	38.0
133.4	40.0
183.6	42.0
∞	43.3

5. The decomposition of nitrogen pentoxide in the gas phase can be followed manometrically but is complicated by the reversible dissociation of the product nitrogen tetroxide

$$2N_2O_5 \rightarrow 2N_2O_4 + O_2$$

$$N_2O_4 \rightleftharpoons 2NO_2$$

The latter reaction rapidly reaches equilibrium, and the equilibrium constant can be independently measured [E. and L. Natanson, *Wied. Ann.*, *24*, 454 (1885)]. Its value at $25°$ C is 97.5 mm. From the accompanying data of F. Daniels and E. H. Johnston [*J. Am. Chem. Soc.*, *43*, 53 (1921)] calculate the rate constant for the decomposition of nitrogen pentoxide. Temperature, $25°$ C.

Time, min	Pressure, mm
0	268.7
20	293.0
40	302.2
60	311.0
80	318.9
100	325.9
120	332.3
140	338.8
160	344.4
∞	473.0

Use may be made of the following table which shows the fraction of nitrogen tetroxide dissociated as a function of the pressure of oxygen in the system, assuming the decomposition of the pentoxide to be the only source of both substances.

Pressure O_2, mm	Fraction Dissociated
5	0.761
25	0.496
50	0.386
100	0.292
150	0.247

6. Ammonium nitrite breaks down to give water and nitrogen and the rate can be followed by measuring the volume of nitrogen evolved. An aqueous solution containing ammonium ions and nitrite ions also contains ammonia and nitrous acid molecules formed by hydrolysis (nitrous acid is weak, $K_a = 6.3 \times 10^{-4}$). Also such ions as H_2ONO^+ and NH_2^- are possible. From the following data, collected by J. H. Dusenberry and R. E. Powell [*J. Am. Chem. Soc.*, *73*, 3266 (1951); see also A. T. Austin et al., *J. Am. Chem. Soc.*, *74*, 555 (1952)], decide the order with respect to each of the possible reactants (some combinations may be kinetically indistinguishable, see Chapter 12, the ammonium cyanate-urea conversion):

$$NH_4^+ + NO_2^- \rightarrow N_2 + 2H_2O$$

Temperature, 30°C

pH constant at 2.85

Total Nitrous Acid, moles/liter	Total Ammonia, moles/liter	Initial Rate $\times 10^8$, moles/liter-sec
0.00904	0.395	128
0.00896	0.197	64
0.00916	0.098	34.9
0.00924	0.049	16.6
	pH constant at 2.93	
0.0940	0.186	643
0.0507	0.196	338
0.0249	0.196	156
0.0100	0.198	65
0.0049	0.198	32.6
0.0024	0.198	17.5

(Total nitrous acid means summation of all forms in solution.)

Total nitrous acid constant at 0.047M and total ammonia constant at 0.197M:

pH	0.43	0.96	1.81	2.95	3.68	4.11	4.33	5.00	6.18
Initial rate $\times 10^8$, moles/liter-sec	517	513	371	305	140	52	31	9.2	0.6

7. A gaseous reaction takes place between A and B. A is in great excess. The half-life, $t_{1/2}$, as a function of initial pressures at 50° C is as follows:

p_A, mm	500	125	250	250
p_B, mm	10	15	10	20
$t_{1/2}$, min	80	213	160	80

(a) Show that the rate expression is

$$\text{rate} = k p_A p_B^2$$

(b) Evaluate the rate constant for concentration units of moles/liter and time in seconds.

8. Test the Wilkinson method for finding the order of a reaction using data from the literature.

REFERENCES

1. A. Farkas and H. W. Melville, *Experimental Methods in Gas Reactions*, Macmillan and Co., London, 1939.
2. E. W. R. Steacie, *Atomic and Free Radical Reactions*, Reinhold Publishing Corp., New York, 1946.
3. C. N. Hinshelwood, *The Kinetics of Chemical Change*, Oxford University Press, Oxford, 1941.
4. W. A. Noyes and P. A. Leighton, *Photochemistry of Gases*, Reinhold Publishing Corp., New York, 1941.
5. A. F. Trotman-Dickenson, *Gas Kinetics*, Academic Press, New York, 1955.
6. N. N. Semenov, *Some Problems of Chemical Kinetics and Reactivity*, Pergamon Press, London, 1958. Translated by J. E. S. Bradley.
7. R. N. Pease, *Equilibrium and Kinetics of Gas Reactions*, Princeton University Press, Princeton, 1942.
8. F. Daniels, J. H. Mathews, J. W. Williams, and Staff, *Experimental Physical Chemistry*, McGraw-Hill Book Co., New York, 4th edition, 1949.
9. J. H. Raley, F. F. Rust, and W. E. Vaughan, *J. Am. Chem. Soc.*, 70, 98 (1948).
10. G. B. Kistiakowsky and J. R. Lacher, *J. Am. Chem. Soc.*, 58, 123 (1936).
11. O. A. Hougen and K. M. Watson, *Chemical Process Principles*, Part III, John Wiley and Sons, New York, 1947, pp. 834ff.
12. E. A. Moelwyn-Hughes, *Kinetics of Reactions in Solution*, Oxford University Press, Oxford, 1946.
13. A. Skrabal, *Homogenkinetik*, Steinkopff, Dresden and Leipzig, 1941.
14. E. S. Amis, *Kinetics of Chemical Change in Solution*, The Macmillan Co., New York, 1949.
15. K. J. Laidler, *Chemical Kinetics*, McGraw-Hill Book Co., New York, 1950.
16. J. Reilly and W. N. Rae, *Physical-Chemical Methods*, Vol. III, Methuen and Co., London, 1943.
17. A. Weissberger, *Physical Methods of Organic Chemistry*, Interscience Publishers, New York, especially Vol. VIII.
18. W. J. McGuire, M.S. thesis, Northwestern University, Evanston, Ill., 1949.
19. L. J. Andrews, *J. Am. Chem. Soc.*, 69, 3062 (1947).

20. F. G. Ciapetta and M. Kilpatrick, *J. Am. Chem. Soc.*, *70*, 639 (1948).
21. L. K. J. Tong and A. R. Olson, *J. Am. Chem. Soc.*, *65*, 1704 (1943).
22. C. G. Swain and S. D. Ross, *J. Am. Chem. Soc.*, *68*, 658 (1946).
23. C. H. Klute and W. D. Walters, *J. Am. Chem. Soc.*, *68*, 506 (1946).
24. J. W. Sprauer and J. H. Simons, *J. Am. Chem. Soc.*, *64*, 648 (1942).
25. W. H. Pearlson and J. H. Simons, *J. Am. Chem. Soc.*, *67*, 352 (1945).
26. V. L. Frampton, *Science*, *107*, 323 (1948).
27. P. Hayward and D. M. Yost, *J. Am. Chem. Soc.*, *71*, 915 (1949).
28. R. W. Wilkinson, private communication.
29. W. E. Roseveare, *J. Am. Chem. Soc.*, *53*, 1651 (1931).
30. E. A. Guggenheim, *Phil. Mag.*, *2*, 538 (1926).
31. J. M. Sturtevant, *J. Am. Chem. Soc.*, *59*, 699 (1937).

4.

ELEMENTARY PROCESSES:
KINETIC THEORY OF GASES

The elementary processes which are the units out of which actual reactions are built include unimolecular, bimolecular, and termolecular processes in which reaction occurs either spontaneously, at a collision between two molecules, or at a collision among three molecules, respectively. In the gas phase, collisions involving four or more molecules are so improbable that they are of no importance in chemical kinetics. There is no known reaction suspected of having a molecularity greater than three. In the liquid state, molecules are always in close proximity so that such multiple collisions where the solvent participates are more likely.

Other elementary processes of importance in chemical kinetics are photochemical changes, involving either absorption or emission of radiation, and energy transfers between molecules at bimolecular collisions, especially those in which there is energy transfer between translation and vibration.

Before considering these processes in detail from the standpoint of kinetic theory, it will be necessary to present some of the general concepts of the kinetic theory of gases.[1,2]

The Distribution Law of a Component of Velocity

Assuming that the molecules move independently of each other except for the short duration of each collision, there will be a definite distribution of velocities that can be calculated, disregarding the collisions except for their furnishing of a mechanism for giving a random distribution. The distribution laws of velocity can be easily obtained as special cases of the Boltzmann distribution of energy. According to the latter, the probability that a molecule is in a level of energy ϵ_i and statistical weight g_i is proportional to $g_i e^{-\epsilon_i/kT}$, where k is Boltzmann's constant and T the absolute temperature. The probability may be represented by dn/n_0, the fractional number of molecules with the desired energy.

Consider first the distribution of a given component of velocity, say $\dot{x}$. The energy is translational with $\epsilon_i = \frac{1}{2}m\dot{x}^2$, where m is the mass of a molecule. For a continuous range of energy as in this case the statistical weight is the volume in phase space in units of h^f for f degrees of freedom. Here

$$g_i = m \, d\dot{x} \, dx/h$$

with h as Planck's constant. Therefore, for the fractional number of molecules with x component of velocity between $\dot{x}$ and $\dot{x} + d\dot{x}$ and x coordinate between x and $x + dx$

$$dn/n_0 = A \frac{m \, d\dot{x} \, dx}{h} e^{-m\dot{x}^2/2kT}$$

where A is the proportionality constant. A may be determined by integrating with respect to $\dot{x}$ and x over their complete ranges $-\infty$ to $+\infty$ and 0 to a, respectively, where a is the length of the corresponding edge of the container, assumed to be rectangular.

$$\int \frac{dn}{n_0} = 1 = (Aam/h) \int_{-\infty}^{+\infty} e^{-m\dot{x}^2/2kT} \, d\dot{x}$$

$$= (Aam/h)(2\pi kT/m)^{1/2}$$

Therefore,

$$A = (h/a)[1/(2\pi mkT)^{1/2}]$$

and

$$dn/n_0 = (m/2\pi kT)^{1/2} e^{-m\dot{x}^2/2kT} d\dot{x} \tag{1}$$

This distribution, which has been integrated to include all positions x, has a maximum at zero $\dot{x}$ and falls off as $\dot{x}$ increases positively or negatively as shown in Fig. 1.

For all three components in specified ranges $\dot{x}$ to $\dot{x} + d\dot{x}$, $\dot{y}$ to $\dot{y} + d\dot{y}$, and $\dot{z}$ to $\dot{z} + d\dot{z}$ the probabilities multiply and therefore

$$dn/n_0 = (m/2\pi kT)^{3/2} e^{-mc^2/2kT} d\dot{x} \, d\dot{y} \, d\dot{z} \tag{2}$$

where $c^2 = \dot{x}^2 + \dot{y}^2 + \dot{z}^2$ is the square of the magnitude of the velocity vector.

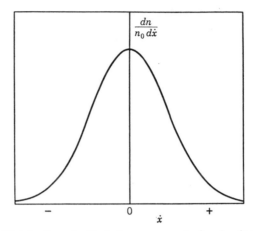

Fig. I. Distribution of a Cartesian component of molecular velocity.

Distribution of Magnitude of Velocity. Maxwell Distribution Law

The distribution of magnitude of velocity without regard to direction is often desired. The Cartesian coordinates of velocity of equation 2 may be replaced by spherical polar coordinate velocity variables c, θ, ϕ, where

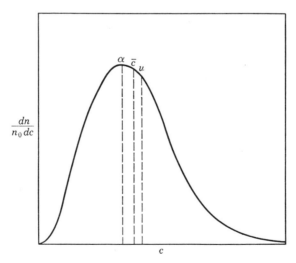

Fig. 2. Distribution of magnitude of molecular velocity, c, for motion in three dimensions. α, most probable velocity; $\bar{c}$, average velocity; u, root-mean-square velocity.

c is the magnitude of velocity and θ and ϕ colatitudinal and longitudinal angles giving the direction of velocity. The "volume" element $d\dot{x}\,d\dot{y}\,d\dot{z}$ becomes $c^2 \sin\theta\,dc\,d\theta\,d\phi$ and integrating over θ from 0 to π and ϕ from 0 to 2π gives

$$\int_0^\pi \int_0^{2\pi} \sin\theta\,d\theta\,d\phi = 4\pi$$

Therefore, the distribution without regard for direction is

$$dn/n_0 = 4\pi(m/2\pi kT)^{3/2}c^2 e^{-mc^2/2kT}dc \tag{3}$$

This is the common form of the Maxwell distribution law and has the graph shown in Fig. 2. The maximum shifts to higher c values as T is increased or as m is decreased.

Most Probable, Average, and Root-Mean-Square Velocities

These particular velocities associated with the distribution law enter into various applications.

The most probable velocity α may be obtained from (3) by differentiating and setting equal to zero to obtain the maximum. The result is

$$\alpha = (2kT/m)^{1/2} \tag{4}$$

The average velocity $\bar{c}$ is obtained by multiplying (3) by c and integrating from 0 to ∞

$$\bar{c} = (8kT/\pi m)^{1/2} = (2/\pi^{1/2})\alpha \simeq 1.128\alpha \tag{5}$$

The root-mean-square velocity u or $(\overline{c^2})^{1/2}$ is obtained by multiplying (3) by c^2, integrating to obtain $\overline{c^2}$, and then taking the square root.

$$u = (3kT/m)^{1/2} = (\tfrac{3}{2})^{1/2}\alpha \simeq 1.224\alpha \tag{6}$$

This root-mean-square velocity is the one of importance in the kinetic theory equation for the pressure of an ideal gas

$$PV = \tfrac{1}{3}nmu^2$$

The pressure or force per unit area is dependent on the rate of change of momentum, which involves for a given molecule the number of collisions with the wall as determined by its velocity. The change in momentum per collision is also proportional to the molecular velocity. Thus a mean-square velocity is needed. For the bimolecular collision number to be discussed below, the average velocity is wanted.

Collision Number per Molecule per Second

Consider the molecules as rigid spheres with diameters σ_A and σ_B for two types of molecules A and B, with n_A and n_B the corresponding numbers of molecules per cubic centimeter. A bimolecular collision will be defined as the situation when there is contact of the surfaces of the two spheres. In order to calculate the number of collisions per second suffered by a molecule of type A, consider such a molecule as moving in an arbitrary direction with a *mean relative velocity* $\bar{r}$ relative to a molecule of type B. If the center of molecule B is at a position within a distance $\sigma_{AB} = (\sigma_A + \sigma_B)/2$ of the line of flight of the center of molecule A during the passage of A, a collision will result (see Fig. 3). The total number of collisions of molecule A with those of type B per second can then be estimated from the volume swept out by a sphere of radius σ_{AB} multiplied by the number of

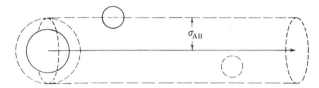

Fig. 3. Cylinder space swept out by a molecule. Molecules whose centers are within cylinder would undergo collision.

type B molecules per cubic centimeter, n_B. The required volume is $\pi\sigma_{AB}^2 \bar{r}$ and the collision number $\pi\sigma_{AB}^2 \bar{r} n_B$.

The appropriate mean relative velocity $\bar{r}$ is not quite the same as the mean velocity $\bar{c}$ of a single kind of molecule but is slightly greater. It can be calculated by using the product of two distribution laws (2), one distribution for each kind of molecule. The velocity coordinates are changed to the velocity of the center of gravity and the relative velocity r where

$$r^2 = (\dot{x}_A - \dot{x}_B)^2 + (\dot{y}_A - \dot{y}_B)^2 + (\dot{z}_A - \dot{z}_B)^2 \tag{7}$$

The expression is then integrated over all directions and velocities, after multiplying by r. The result is the same as $\bar{c}$ except for the presence of μ the reduced mass instead of m, where

$$\mu = \frac{m_A m_B}{m_A + m_B}$$

$$\bar{r} = (8kT/\pi\mu)^{1/2} \tag{8}$$

It is of interest to notice that in case $m_A = m_B = m$

$$\mu = m/2 \quad \text{and} \quad \bar{r} = (2)^{1/2}\bar{c}$$

The formula for the collision number (7) after substitution of (8) is

$$(8\pi kT/\mu)^{1/2}\sigma_{AB}^2 n_B \tag{9}$$

For gases at standard conditions of temperature and pressure there are about 10^{10} collisions per molecule per second.

Immediately derivable from the collision number expression are two other concepts, the mean time between collisions, which is the reciprocal of (9), and the mean free path (mean distance traveled between successive collisions), l, which in the case of like molecules is

$$l = \frac{\bar{c}}{(16\pi kT/m)^{1/2}\sigma^2 n} = \frac{1}{(2)^{1/2}\pi\sigma^2 n} \tag{10}$$

For standard conditions with typical gas molecules $l \simeq 10^{-5}$ cm and is inversely proportional to pressure at constant temperature. In gas reactions at low pressures, where the mean free path may become the same order of magnitude as the size of the container, there are often wall effects which complicate the reaction.

The mean free path enters into the formula for gas viscosity, η

$$\eta = \tfrac{1}{3}\bar{c}\rho l \tag{11}$$

where ρ is the density. Measurement of viscosity, therefore, enables calculation of molecular diameters through equations 10 and 11. Molecular diameters can also be obtained from data on heat conductivity, diffusivity, and van der Waals' b.

Number of Bimolecular Collisions per Cubic Centimeter per Second

Equation 9 gives the number of collisions per molecule of type A. Multiplying by n_A then gives the number of bimolecular collisions Z_{AB}' between *unlike* molecules of type A and B per cubic centimeter per second.

$$Z_{AB}' = (8\pi kT/\mu)^{1/2}\sigma_{AB}^2 n_A n_B \tag{12}$$

In collisions between like molecules in a single gas the subscripts may be dropped and $\mu = m/2$. Also a factor of one-half must be applied, since otherwise the formula would count each collision twice. Then for *like* molecules the collision number Z' is

$$Z' = 2(\pi kT/m)^{1/2}\sigma^2 n^2 \tag{13}$$

For a typical gas at standard conditions this number is approximately 10^{28} collisions per cubic centimeter per second.

Rate of a Bimolecular Reaction

If reaction occurred at every collision between given reacting molecules, the rate would be much greater than is usually observed. This is seen most easily from (9). With a single molecule making 10^{10} collisions per second, the half-life period of the reaction under standard initial conditions would be of the order of 10^{-10} second which is smaller than results usually observed by a factor of 10^{-15} to 10^{-20}. Furthermore, the temperature dependence of the collision number, $T^{1/2}$, does not agree with the Arrhenius equation. However, the collision-number formulas 12 and 13 do predict correctly the second-order nature of the reaction.

Let q be the fraction of collisions that are effective, q being a function of T and dependent on the nature of the reacting molecules. The rate of reaction according to collision theory is then for unlike molecules

$$qZ_{AB}' = q(8\pi kT/\mu)^{1/2}\sigma_{AB}{}^2 n_A n_B \tag{14}$$

for like molecules

$$qZ' = q2(\pi kT/m)^{1/2}\sigma^2 n^2 \tag{15}$$

The rate is here defined as the number of effective collisions per cubic centimeter per second. For unlike molecules this is the same as $-dn_A/dt$ or $-dn_B/dt$, but for like molecules is $-\frac{1}{2} dn/dt$ if two like molecules react for each effective collision.

The Bimolecular Rate Constant

By dividing out the concentration factors in the above formulas for rate, there will result the corresponding rate constants or specific reaction rates. Designating such constants by k and, to avoid confusion, replacing Boltzmann's constant k by R/N, where R is the gas constant and N is Avogadro's number, yields for unlike molecules, or reactions of type $A + B \rightarrow \cdots$

$$k = q\sqrt{8\pi RT/N\mu}\,\sigma_{AB}{}^2 = q\sqrt{8\pi RT(1/M_A + 1/M_B)}\,\sigma_{AB}{}^2 \tag{16}$$

and for like molecules, or reactions of type $2A \rightarrow \cdots$,

$$k = q2\sqrt{\pi RT/M}\,\sigma^2 \tag{17}$$

where M_A, M_B, M are molecular weights.

Factors Determining Effectiveness of Collisions

1. Orientation of molecules—steric factor, p. For any molecular species more complicated than a free atom, it is apparent that reaction is not to be expected at a collision unless the molecules are so oriented relative to each other that the groups reacting or the bonds to be shifted are relatively close. A steric factor, p, is considered to represent the fraction of collisions that have the proper orientation for the colliding molecules and is expected to range from a value near unity down to 0.1 or 0.01, depending on the complexity of the molecules, but unfortunately there appears to be no satisfactory way to predict the value. Hence p is determined empirically and because of this it may be considerably in error, since it will contain all the errors of simplification involved in the collision theory, such as the use of the rigid sphere model and the neglect of the internal degrees of freedom.

2. Restriction on bimolecular association reactions. Consider a reaction of the type $A + B \rightarrow AB$ or $2A \rightarrow A_2$. To form a stable molecule in such a reaction there is always an evolution of energy. If the product molecule has several vibrational degrees of freedom, the energy to be evolved in forming the new bond or bonds may be distributed among the other degrees of freedom and so stabilize the molecule until the extra energy is removed by later collisions with other molecules. In the association of atoms to form diatomic molecules, however, there is only the one vibrational degree of freedom, and the energy remains to cause the molecule to dissociate again in one-half a period of vibration ($\sim 10^{-13}$ second). It might be supposed that the molecule could be stabilized by emission of radiation, but this process is "forbidden" for the formation of symmetrical diatomic molecules and too slow for the formation of others. Association of atoms and of some simple radicals like OH do not take place as bimolecular reactions but require termolecular collisions where the third molecule may be of any kind and serves to remove excess energy in such forms as translation, vibration, or possibly rotation.

3. Quantum restrictions. If the electronic state of the system changes during the reaction (e.g., a change of multiplicity), the reaction will be slow, other things being equal. Reactions so restricted are the decomposition of nitrous oxide, certain reactions forming ions and excited molecules such as would occur in flames and explosions, and possibly some *cis-trans* isomerizations. These are called non-adiabatic reactions.[†] Reactions in which two free radicals combine to form a molecule, or in which a molecule dissociates into two free radicals, are not restricted even though a formal change in multiplicity is involved. However, for the recombination reaction a steric factor of one-fourth is expected because of the three to one ratio of repulsive to attractive electron spin states.

In bimolecular collisions involving energy transfer, the exchange of vibrational and translational energy is often found to be restricted. Measurements on the dispersion of sound indicate that such transfers become less likely the greater the spacing of the vibrational energy levels, and in particular when $h\nu > kT$, where ν is the vibration frequency. Also if the molecule is small and in a low vibrational level, energy transfer is slow. A study of the fluorescent spectra of highly excited molecules shows

† For a more complete discussion see reference 3 at the end of this chapter.

that energy transfer becomes much more efficient upon collision, if the molecule is in a high vibrational level. There seems to be little restriction to transfer between rotational and translational energies.[4]

4. Activation energy. The energy requirement is by far the most important of the factors determining the effectiveness of collision. That excess energy at a collision is expected if the collision is to be fruitful for reaction is shown by a comparison of bond distances between atoms in a

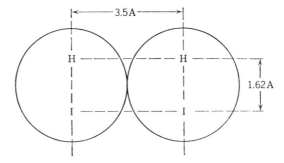

Fig. 4. Relative positions of atoms in hydrogen iodide molecules undergoing a possible normal collision.

normal molecule with the distances between atoms in different molecules undergoing collisions. Consider the reaction

$$2HI \rightarrow H_2 + I_2$$

The relative positions of the atoms in the HI molecules at an ordinary collision might be diagrammed as in Fig. 4. The distance between H and I nuclei in an HI molecule is 1.62 A.U. The molecular diameter of HI as found by gas viscosity measurements is 3.5 A.U. This latter distance would also be the distance between the two H atoms or the two I atoms in the colliding molecules. This distance is large compared with the internuclear distances of the product molecules, 0.76 A.U. for H_2 and 2.66 A.U. for I_2; therefore, the collisions must be sufficiently energetic to cause a compression of the molecules so that the atoms to be bonded approach more closely their normal distance in the product molecules.

In the case of endothermic reactions there is an additional energy requirement, since, for successful reaction, sufficient energy must be supplied in the collision to put the product molecules at least in their lowest energy levels.

A complete theory of collisional reaction rate would have to include a detailed treatment of the forces and motions involving all the atoms of the colliding molecules. This has never been carried out from the kinetic theory standpoint. Instead, it is usually assumed that for reaction the energy must be greater than a certain minimum value E, the activation energy, for reaction to occur. E is found empirically, although in principle it could be calculated theoretically on the basis of quantum mechanics. The energy which is to be greater than E is not the total kinetic energy of the colliding molecules nor even the relative kinetic energy, but, in a bimolecular collision, it is presumably only that part of the relative kinetic energy of approach of the two molecules in a direction along their line of centers at the moment of collision. Kinetic energy associated with motion perpendicular to the line of centers would correspond to a sideswiping or a rotation and would not be expected to be effective in reaction. The next section gives a detailed treatment of this situation and also provides a more exact derivation of equation 12 for bimolecular collision number.

Number of Bimolecular Collisions with Relative Kinetic Energy along the Line of Centers at Time of Contact Greater than E

Consider the following particular type of collision between two different molecules, assumed to be rigid spheres, of mean diameter σ_{AB} and mass m_A and m_B, respectively. Let the relative velocity of molecule A with

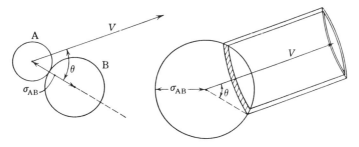

Fig. 5. Collisions between molecules with angle θ to $\theta + d\theta$ between line of centers and relative velocity vector. Molecules shown at left. Cylindrical shell swept out shown at right.

respect to molecule B be between V and $V + dV$, and let the angle between the vector V and the line of centers of the two molecules at the moment of contact be between θ and $\theta + d\theta$ (see Fig. 5). The various orientations of the two molecules in collisions of this type correspond to the center of molecule B being anywhere in a ring-shaped surface on a sphere of radius σ_{AB} drawn about molecule A. This ring-shaped surface has circumference $2\pi\sigma_{AB} \sin \theta$, width $\sigma_{AB} \, d\theta$, and therefore an area of $2\pi\sigma_{AB}^2 \sin \theta \, d\theta$. As molecule A moves through space relative to B, the ring-shaped surface sweeps out a volume of $V2\pi\sigma_{AB}^2 \sin\theta\cos\theta\,d\theta$ per second, where the $\cos\theta$ enters the expression in order to get the component of the surface area normal to V. The number of collisions of this type per cubic centimeter per second may now be obtained by multiplying this volume swept out by the number of pairs of molecules A and B having relative velocities from V to $V + dV$.

The desired number of pairs may be obtained with the aid of the distribution law (2) applied to molecules A and B, separately. Multiplying the two distributions together yields

$$dn_A\,dn_B = n_An_B(m_Am_B/4\pi^2k^2T^2)^{3/2}e^{-(m_Ac_A^2+m_Bc_B^2)/2kT}\,d\dot{x}_A\cdots d\dot{z}_B \quad (18)$$

Now transform this to new velocity coordinates u, v, w, V, θ', ϕ', where the first three are Cartesian components of the velocity of the center of mass of the two molecules and the latter three are spherical coordinates for the relative velocity of the two molecules.

$$
\begin{aligned}
u &= (m_A\dot{x}_A + m_B\dot{x}_B)/(m_A + m_B)\\
v &= (m_A\dot{y}_A + m_B\dot{y}_B)/(m_A + m_B)\\
w &= (m_A\dot{z}_A + m_B\dot{z}_B)/(m_A + m_B)\\
V &= \sqrt{(\dot{x}_B - \dot{x}_A)^2 + (\dot{y}_B - \dot{y}_A)^2 + (\dot{z}_B - \dot{z}_A)^2}\\
\theta' &= \cos^{-1}[(\dot{z}_B - \dot{z}_A)/V]\\
\phi' &= \tan^{-1}[(\dot{y}_B - \dot{y}_A)/(\dot{x}_B - \dot{x}_A)]
\end{aligned}
\quad (19)
$$

The volume element in (18) becomes

$$d\dot{x}_A\cdots d\dot{z}_B = V^2\sin\theta'\,du\,dv\,dw\,dV\,d\theta'\,d\phi'$$

Letting $c^2 = u^2 + v^2 + w^2$, where c is the magnitude of relative velocity, and introducing $\mu = (m_Am_B)/(m_A + m_B)$, the reduced mass, there results

$$dn_A\,dn_B = n_An_B\left[\frac{\mu(m_A + m_B)}{4\pi^2k^2T^2}\right]^{3/2}e^{-([(m_A+m_B)c^2+\mu V^2]/2kT)}$$
$$\cdot V^2\sin\theta'\,du\,dv\,dw\,dV\,d\theta'\,d\phi' \quad (20)$$

Integrating over the complete range of all variables except V gives

$$dn_{AB} = 4\pi n_An_B(\mu/2\pi kT)^{3/2}e^{-\mu V^2/2kT}V^2\,dV \quad (21)$$

as the number of pairs of molecules in 1 cc with the magnitude of relative velocity in range V to $V + dV$.

Now multiplying (21) by the volume swept out yields dZ', the number of collisions per cubic centimeter per second with magnitude of relative velocity in the range V to $V + dV$, and the angle between line of centers at time of contact from θ to $\theta + d\theta$

$$dZ' = 8\pi^2n_An_B(\mu/2\pi kT)^{3/2}e^{-\mu V^2/2kT}V^3\sigma_{AB}^2\sin\theta\cos\theta\,dV\,d\theta \quad (22)$$

The component of relative velocity along the line of centers is $V \cos \theta$. The kinetic energy associated with this component of relative velocity is $\frac{1}{2}\mu V^2 \cos^2 \theta$. Assuming that for successful reaction this must be greater than a certain value E, the activation energy, the expression dZ is to be integrated for all values of θ and V such that $\frac{1}{2}\mu V^2 \cos^2 \theta > E$.

In particular integrate θ from 0 to $\cos^{-1}(2E/\mu V^2)^{1/2}$ and the V from $(2E/\mu)^{1/2}$ to ∞.

The integral involving θ is

$$\int_0^{\cos^{-1}(2E/\mu V^2)^{1/2}} \sin \theta \cos \theta \, d\theta = \frac{1}{2}\left(1 - \frac{2E}{\mu V^2}\right) \tag{23}$$

Using this result and then integrating over V gives

$$Z_{AB}'(E) = \int dZ'$$

$$= 4\pi^2 n_A n_B \left(\frac{\mu}{2\pi kT}\right)^{3/2} \sigma_{AB}{}^2 \int_{(2E/\mu)^{1/2}}^{\infty}\left(1 - \frac{2E}{\mu V^2}\right) V^3 e^{-\mu V^2/2kT} dV$$

and

$$Z_{AB}'(E) = (8\pi kT/\mu)^{1/2}\sigma_{AB}{}^2 n_A n_B e^{-E/kT} \tag{24}$$

If E is set equal to zero, the formula would give Z_{AB}', the total number of collisions without restriction on the energy. Therefore,

$$Z_{AB}'(E) = Z_{AB}' e^{-E/kT} \tag{25}$$

where

$$Z_{AB}' = (8\pi kT/\mu)^{1/2}\sigma_{AB}{}^2 n_A n_B \tag{26}$$

Equation 26 agrees with (12), which was obtained by more elementary considerations. Equation 25 provides the surprisingly simple result that the fraction of all collisions having the desired energy is given by $e^{-E/kT}$ or $e^{-E/RT}$, where in the latter formula E is the activation energy per "mole" of collisions.

The efficiency factor q, introduced earlier, may be written

$$q = pe^{-E/RT} \tag{27}$$

where the steric factor, p, really includes all limitations on the collision rate except that due to the activation energy. With this result for q the rate constants for bimolecular reactions, equations 16 and 17 become for *unlike molecules*

$$k = (8\pi RT/N\mu)^{1/2}\sigma_{AB}{}^2 pe^{-E/RT} \tag{28}$$

or

$$k = pZ_{AB}e^{-E/RT} \tag{29}$$

where Z_{AB} is the collision number at unit concentration, and for *like molecules*

$$k = 2(\pi RT/M)^{1/2}\sigma^2 p e^{-E/RT} \tag{30}$$

or

$$k = pZe^{-E/RT} \tag{31}$$

Because in general p cannot be evaluated *a priori* and σ can only be estimated, it is perhaps more sensible to lump together p and σ^2, defining the combination as the square of an effective molecular diameter, σ_e, where

$$\sigma_e = \sigma p^{1/2}$$

or

$$\sigma_e = \sigma_{AB} p^{1/2} \tag{32}$$

σ_e^2 is then somewhat equivalent to a collision cross section as used in connection with nuclear experiments. In terms of σ_e, equations 28 and 30 become for *unlike molecules*

$$k = (8\pi RT/N\mu)^{1/2}\sigma_e^2 e^{-E/RT} \tag{33}$$

and for *like molecules*

$$k = 2(\pi RT/M)^{1/2}\sigma_e^2 e^{-E/RT} \tag{34}$$

Comparison of these predictions of the kinetic theory of gases with experimental results will be given in detail in Chapter 6 after the statistical mechanical transition-state theory is developed. It is of interest to notice here that the kinetic theory predicts not only that a bimolecular reaction is second-order, but also that the temperature dependence of its rate constant will be approximately of the Arrhenius type inasmuch as the $T^{1/2}$ factor is of minor importance compared with the exponential.

Termolecular Reactions

In discussing termolecular collisions, the rigid sphere model presents a fundamental difficulty. If a collision is defined as simultaneous contact of the spherical surfaces of the three molecules, the chance of having such a collision is zero, since two spheres would be in contact for an instant and there is only an infinitesimal probability that a third molecule would make contact with the other two at that instant. In order to have a finite number of termolecular collisions, it is necessary to consider either that the spheres are not rigid or that the approach of rigid spheres to within an arbitrary distance of each other constitutes a collision. This latter technique was used by Tolman.[5] The result is that for collisions between molecules A,

B, and C such that A and C are to be within a distance δ of B, the collision number is

$$Z'_{ABC} = 8\sqrt{2}\pi^{3/2}\sigma_{AB}{}^2\sigma_{BC}{}^2\delta\sqrt{kT}\left(\frac{1}{\sqrt{\mu_{AB}}} + \frac{1}{\sqrt{\mu_{BC}}}\right)n_A n_B n_C \qquad (35)$$

The symbols have the same or corresponding significance to those used previously. Nothing can be said about δ except that it should be of the order of magnitude of 1 A.

The rate of a termolecular reaction should be equation 35 multiplied by an efficiency factor q. To avoid a detailed calculation such as was carried out for bimolecular reactions, it will simply be assumed that the previous result, $q = pe^{-E/RT}$, applies here too. Although for large activation energies the simple exponential would probably be far from correct, it is a fact that most observed termolecular reactions have small or zero activation energies where the exponential approaches unity as it should. If a possible termolecular reaction has a high activation energy and therefore a very slow rate, both on account of the activation requirement and because termolecular collisions are rare compared with bimolecular collisions, then there is probably some other mechanism of reaction that takes precedence over the termolecular mechanism. Therefore, termolecular reactions with high activation energy are expected to be uncommon except in solution.

With the collision number (35) proportional to three concentration factors, it is seen that a termolecular reaction is third-order. The third-order rate constant is then

$$k = 8\sqrt{2}\pi^{3/2}\sigma_{AB}{}^2\sigma_{BC}{}^2\delta\sqrt{kT}\left(\frac{1}{\sqrt{\mu_{AB}}} + \frac{1}{\sqrt{\mu_{BC}}}\right)pe^{-E/RT} \qquad (36)$$

or

$$k = Z_{ABC}pe^{-E/RT} \qquad (37)$$

where Z_{ABC} is the collision rate at unit concentration.

Unimolecular Reactions

Since a collision involves at least two molecules, it would appear that a unimolecular reaction must be one taking place without collision, perhaps as a spontaneous disruption or transformation of the reacting molecule.

Let the probability of such a spontaneous reaction in time dt be $k\,dt$ where k is a constant characteristic of the molecule and the temperature. For a large number of reacting molecules the fraction that reacts in time

dt can be set equal to the probability $k\,dt$, where $-dn$ is the number reacting

$$-dn/n = k\,dt \qquad (38)$$

or

$$-dn/dt = kn$$

Dividing both sides by the volume, we get

$$-dc/dt = kc \qquad (39)$$

where c is the concentration. This is the familiar first-order rate equation with rate constant k. It shows that a unimolecular reaction would be expected to be first-order and that the rate constant can be interpreted as a probability of reaction per unit time. Simple radioactive disintegrations follow this same law with constants characteristic of the particular disintegrations but independent of the temperature. In first-order chemical reactions suspected of being unimolecular, the temperature is of great importance, the rate constant following an exponential Arrhenius expression. In unimolecular reactions energy must be supplied to activate the molecule, whereas in a radioactive disintegration, the energy is already on hand in the nucleus. The manner in which the molecules become activated was long a mystery, since activation by collision was first ruled out because it was thought that the reaction would have to be bimolecular and second-order in that event rather than unimolecular and first-order. However, the Lindemann-Hinshelwood theory shows that a mechanism can be set up to account for the observed results.[6]

Let the stoichiometric reaction be $A = B + C$. Assume that the mechanism is

I. Activation by collision:

$$A + A \rightarrow A' + A \qquad k_1$$

where A' is an activated molecule, that is, one with sufficient energy for reaction. k_1 is the rate constant of the bimolecular reaction.

2. Deactivation by collision:

$$A' + A \rightarrow A + A \qquad k_2$$

This is the reverse of the activation process and is expected to occur at the first collision of A' after it has been formed if it has not already reacted in the meantime. $k_2 \gg k_1$, since the activation process is limited by the energy requirement.

3. Spontaneous reaction:

$$A' \rightarrow B + C \qquad k_3$$

This is a "true" unimolecular process in the same sense as is a radioactive disintegration. k_3 is a first-order rate constant.

On the basis of this mechanism the following expressions for the rates of formation of A′ and of B apply:

$$d[A']/dt = k_1[A]^2 - k_2[A'][A] - k_3[A'] \tag{40}$$

$$d[B]/dt = k_3[A'] \tag{41}$$

It is difficult to solve these equations accurately, but an excellent simplifying approximation is to assume that there is a *steady state* where the concentration of A′, which is always very small, does not change with time, that is,

$$d[A']/dt = 0$$

This gives from (40)

$$k_1[A]^2 = (k_2[A] + k_3)[A'] \tag{42}$$

from which it follows that

$$[A'] = k[A]^2/(k_2[A] + k_3) \tag{43}$$

Inserting this in (41)

$$d[B]/dt = k_3 k_1[A]^2/(k_2[A] + k_3) \tag{44}$$

Although the gas mixture contains A, A′, B, and C, the concentration of A is practically identical with A + A′, since A′ is very small. Equation 44 shows that the expected rate is neither first- nor second-order with respect to A; however, there are two limiting cases where it does become so. At high pressure where [A] is large so that $k_2[A] \gg k_3$, (44) simplifies to

$$d[B]/dt = (k_3 k_1/k_2)[A] \tag{45}$$

so that under this condition a first-order rate is expected, despite the fact that bimolecular processes occur in activation and deactivation.

At low pressure where [A] is small so that $k_2[A] \ll k_3$, (44) simplifies to

$$d[B]/dt = k_1[A]^2 \tag{46}$$

so that the rate then becomes second-order. This is also the rate of activation, deactivation becoming unimportant, since at sufficiently low pressures there will be a great enough time between collisions so that an A′ once formed will usually react to B and C before it can undergo another collision. In a complex mechanism such as this the reaction rate will often be fixed by the rate of the reaction that tends to be slowest. Such a slow step is called a *rate-determining step*. In this mechanism at low pressure the activation is the rate-determining step, whereas at high pressure the third, unimolecular process is the rate-determining step. At high pressure it is interesting to notice that there is essentially an equilibrium between

A and A'. For, supposing that such an equilibrium exists, the rates of activation and deactivation would be equal, resulting in

$$k_1[A]^2 = k_2[A'][A]$$

or   (47)

$$[A'] = (k_1/k_2)[A]$$

Substituting this in (41) gives

$$d[B]/dt = (k_3k_1/k_2)[A]$$ (48)

which is the same as (45).

It is necessary to generalize this mechanism to allow for activation and deactivation by collisions with other molecules than reactant A. The symbol M will be used to designate any molecule which may be a reactant, a product, or an added non-reactive gas molecule. Assuming that the efficiency of the various molecules in causing activation or deactivation is not very different, all such effects can be designated by the mechanism:

$$A + M \rightarrow A' + M \qquad k_1$$
$$A' + M \rightarrow A + M \qquad k_2$$
$$A' \rightarrow B + C \qquad k_3$$

The steady-state approximation applied here leads to

$$d[B]/dt = k_3k_1[M][A]/(k_2[M] + k_3)$$ (49)

which has high- and low-pressure limiting expressions similar to (45) and (46). If [M] is approximately constant during a run, as when the product compensates for the disappearing reactant as activator and deactivator, or when a foreign gas is present, equation 49 is that of a pseudo-first-order reaction. Define k as

$$k = k_3k_1[M]/(k_2[M] + k_3)$$ (50)

so that

$$d[B]/dt = k[A]$$ (51)

k is the observed first-order rate constant. It also has limiting values for high and low pressure.

High pressure

$$k_2[M] \gg k_3$$
$$k = k_3k_1/k_2$$ (52)

Low pressure

$$k_2[M] \ll k_3$$
$$k = k_1[M]$$ (53)

Thus k has a value which decreases if the pressure is lowered sufficiently. For the purpose of verifying this theory, (50) can be inverted to give

$$1/k = k_2/k_3 k_1 + 1/k_1[\text{M}] \tag{54}$$

A plot of $1/k$ versus $1/[\text{M}]$ should yield a straight line.

A complete theory should predict suitable values for the constants k_1, k_2, and k_3. Both k_1 and k_2 are bimolecular rate constants and might be expected to be given satisfactorily by equations 27 to 31, k_2, in particular, for deactivation should be equal to $Z_{AA'}$, since no activation energy is required in this process and no important steric effect is expected. The value of k_1 may be much larger than would be expected on the simple theory, because of the neglect of internal degrees of freedom, in particular the vibrations, within the molecule. Since the ratio of k_1 to k_2 is the equilibrium constant for formation of A' from A as shown in (47), it is possible to evaluate this by statistical reasoning. Hinshelwood[7] derives on the basis of classical mechanics the chance that a molecule possesses energy greater than E distributed at random among $2f$ square terms, equivalent to f vibrational degrees of freedom. A square term is an energy such as $\frac{1}{2}mv^2$ for kinetic or $\frac{1}{2}kx^2$ for the potential energy of an oscillator. The result, which is a calculation of k_1/k_2, is

$$\frac{k_1}{k_2} = \frac{e^{-E/RT}(E/RT)^{f-1}}{(f-1)!} \tag{55}$$

This is accurate only if $E \gg (f-1)RT$, which may not always be true for large molecules. For $f = 1$ the right side reduces to $e^{-E/RT}$. For polyatomic molecules f may be as large as $3n - 5$, where n is the number of atoms. For typical values of f the expression is much larger than $e^{-E/RT}$, and so k_1 is larger than that calculated from the simple bimolecular collision theory.

The value of k_3 depends on the rapidity with which the energy, once having been given to activate the molecule, can accumulate in the bond to be broken or rearranged. Polanyi and Wigner[8] estimate that $k_1 k_3/k_2$ should be equal to $Ae^{-E/RT}$ where A is approximately equal to the frequency of the bond, typically about 10^{13} sec^{-1}. Sometimes it is naively assumed that k_3 is 10^{13} sec^{-1}. This leads to an anomaly however.[9] For a reaction to appear unimolecular it is necessary that $k_2[\text{A}] \gg k_3$ (see equation 45). However, since k_2 cannot be greater than the collision number $Z_{AA'}$, it can be calculated that, for a gas under ordinary conditions of temperature and pressure, $k_2[\text{A}]$ is no greater than 10^{10} sec^{-1}. Thus the inequality above cannot be obtained. Using the estimates of $k_1 k_3/k_2$ and of k_1/k_2 given above it is found that k_3 is 10^{13} only if f is equal to 1, that is, if a diatomic molecule is involved. For values of f equal to 5 or 10, and for reasonable values of E/RT, k_3 becomes 10^9 to 10^6 sec^{-1}. Rice and Ramsperger, Kassel, Marcus and Hinshelwood[10] have developed the theory of unimolecular reactions in more detail, considering the effect of quantization of vibration and making the more reasonable assumption that the rate of the final step depends upon how much activation energy is given to the molecule in excess of the critical amount. Thus account is taken of the fact that there is a whole series of activated molecules A' with different energies. The result for the rate of decomposition of a molecule with energy E is

$$k_{(E)} = 0 \qquad E < E_0 \tag{56}$$

$$k_{(E)} = A\left(\frac{E - E_0}{E}\right)^s \qquad E > E_0$$

Here E_0 is the minimum energy needed for dissociation or reaction and s is an integer related to the number of vibrational or other degrees of freedom, but differing slightly according to the various theories as to its precise meaning.

The most elaborate theory of unimolecular decomposition is due to Slater.[11] The molecule is treated as a collection of uncoupled harmonic oscillators and the time required for one coordinate to reach a critical value is calculated. All the vibrational properties of the normal molecule are assumed to be known and the minimum energy E_0 is taken as empirical. No other arbitrary parameters are needed.

The detailed calculations are rather involved but $k_1 k_3/k_2$ is shown to be equal to $Ae^{-E_0/RT}$ where A is a weighted root-mean-square average of all the vibrational frequencies of the molecule and hence of the order of 10^{13} sec^{-1}. The value of A must always be between the least and the greatest vibrational frequency of the molecule. More important, the theory predicts the behavior of the observed rate constant k as a function of total pressure. The rate constant falls off with decreasing pressure and in the limit becomes a second-order constant as in the Lindeman-Hinshelwood theory. However, the fall-off is more complex than given by the simple equation 50. This is because the contribution of molecules with different excitation energies to the overall rate changes with the pressure. Thus at high pressures, low energy molecules contribute little to the rate because they are deactivated by collision before decomposing (k_3 small). At low pressures the low energy molecules will contribute proportionately more because of the longer interval before deactivation.

The results of equation 56 are a natural consequence of the theory and do not need to be assumed. It is predicted that the observed activation energy will decrease at low pressures. The high-pressure value will be E_0 and the low-pressure limit will be $E_0 - sRT$. There are not many molecules for which all of the information to apply the Slater theory is available. Also an assumption about the critical coordinate must be made.

Numerical Summary of Rate-Constant Predictions

Because of the uncertainty of the theory developed so far, not much stress can be laid on exact quantitative comparison with experiment.

Table I

FREQUENCY FACTORS

Concentration Unit	Molecules/cc	Moles/cc	Moles/liter
Uni-	10^{13}–10^{14} sec^{-1}	10^{13}–10^{14} sec^{-1}	10^{13}–10^{14} sec^{-1}
Bi-	10^{-10}–10^{-9} cc molecules^{-1} sec^{-1}	10^{14}–10^{15} cc moles^{-1} sec^{-1}	10^{11}–10^{12} l. moles^{-1} sec^{-1}
Ter-	10^{-32}–10^{-31} cc^2 molecules^{-2} sec^{-1}	10^{15}–10^{16} cc^2 moles^{-2} sec^{-1}	10^{9}–10^{10} l^2 moles^{-2} sec^{-1}

Rather than to attempt precise calculations, it is more useful to compare orders of magnitude. Table 1 gives typical powers of 10 for the pre-exponential or frequency factors of the rate constants for uni-, bi-, and termolecular reactions as developed in this chapter, making substitutions such as follow in the collision formulas: for example, M = 40; μ, 20; T, 300; σ, 3×10^{-8}; δ, 1×10^{-8}; p, 1. The formulas give results directly in terms of molecules per cubic centimeter as the concentration unit. The results are also converted into other units for convenient reference.

PROBLEMS

1. Derive a formula for the rate of effusion, dn/dt, of molecules of molecular weight M, at temperature T, through a pinhole of area A, if there are n molecules in volume V. [The rate of effusion of gas mixtures has been used by F. E. Harris and L. K. Nash, *Anal. Chem.*, 22, 1552 (1950), as an analytical method.]

2. Compare the data of Problem 4, Chapter 2, on the decomposition of nitrogen dioxide with equation 34. Is the resulting value of σ_e reasonable?

3. Verify formulas 4 and 5 for α and $\bar{c}$.

REFERENCES

1. For general reference see L. B. Loeb, *Kinetic Theory of Gases*, McGraw-Hill Book Co., New York, 1934.
2. R. C. Tolman, *Statistical Mechanics*, Chemical Catalog Co., New York, 1927; R. H. Fowler and E. A. Guggenheim, *Statistical Thermodynamics*, Cambridge University Press, 1939, p. 491 ff.
3. S. Glasstone, K. J. Laidler, and H. Eyring, *Theory of Rate Processes*, McGraw-Hill Book Co., New York, 1940, pp. 148, 298 ff.
4. O. Oldenberg and A. A. Frost, *Chem. Rev.*, *20*, 99 (1937); H. S. W. Massey and E. H. S. Burhop, *Electronic and Ionic Impact Phenomena*, Oxford University Press, 1952, Chapter 7.
5. (*a*) R. C. Tolman, *Statistical Mechanics*, Chemical Catalog Co., New York, 1927; (*b*) L. S. Kassel, *Kinetics of Homogeneous Gas Reactions*, Chemical Catalog Co., New York, 1932.
6. F. A. Lindemann, *Trans. Faraday Soc.*, *17*, 598 (1922); C. N. Hinshelwood, *Proc. Roy Soc.*, *A113*, 230 (1927).
7. C. N. Hinshelwood, *Kinetics of Chemical Change*, Oxford University Press, New York, 1940, pp. 26–29.
8. M. Polanyi and E. Wigner, *Z. physik. Chem.*, Haber-Band, 439 (1928).
9. S. W. Benson, *J. Chem. Phys.*, *19*, 802 (1951).
10. See references 5(*b*), 7, and 11. Also R. A. Marcus, *J. Chem. Phys.*, *20*, 359 (1952).
11. N. B. Slater, *Theory of Unimolecular Reactions*, Cornell University Press, Ithaca, 1959.

5.

TRANSITION-STATE THEORY

More modern than the kinetic theory of gases and in many respects more satisfactory and simpler is the transition-state or activated complex theory,† first used for a specific reaction by Pelzer and Wigner in 1932 and later developed in general form particularly by Eyring and co-workers.[1-5] It has the great advantage of taking into consideration, at least in principle, all the internal motions of the reacting molecules. Briefly, what is done is first to suppose that the potential energy of the interacting molecules at the time of collision is known as a function of the relative positions of the various nuclei. There will be a configuration of nuclei of minimum potential energy, related to the activation energy, through which or near which the system would be expected to pass in going from reactants to products. This region of configuration space is called the *transition state*. A system in the transition state is called an *activated complex*. It is assumed that the rate of reaction is given by the rate of passage through the transition state (passage over the potential energy barrier), the number of activated complexes at any instant being determined by an equilibrium with the reactant molecules. The theory naturally divides into three parts which will be discussed in turn: the potential-energy surfaces, the rate of passage of activated complexes over the barrier, and the statistical mechanics of the equilibrium between reactants and activated complex.

Potential-Energy Surfaces

The potential energy of configurations of nuclei of molecules in collision, or of stable molecules for that matter, may be represented by an f-dimensional surface in an $(f + 1)$-dimensional space, where f is the number of

† Also called the theory of absolute reaction rates.

independent variables necessary to specify completely the relative positions of n nuclei. The potential-energy surface is the generalization of the familiar diatomic molecule potential-energy curve in which case only one variable, the internuclear distance, is needed to fix the relative positions of the two nuclei.

The existence of such a potential-energy curve or surface is based on the "adiabatic" assumption: the electronic motion being rapid compared with nuclear motion, the electronic energy can be evaluated with the nuclei being considered at rest. Strictly speaking, in quantum mechanics the electronic motion cannot be separated from nuclear motion except through an approximation method such as that developed by Born and Oppenheimer.[6] However, because of the large masses of nuclei as compared with electrons, the total electronic energy so calculated is probably good to a small fraction of a per cent.

London[7] was the first to apply this idea to a chemical reaction; he developed on the basis of the Heitler-London valence theory approximation formulas for the energy of systems of three and four univalent atoms in interaction. For three atoms X, Y, and Z that might be involved in a simple displacement reaction of the type

$$X + YZ \rightarrow XY + Z$$

the energy E is given by

$$E = A + B + C - \sqrt{\tfrac{1}{2}[(\alpha - \beta)^2 + (\beta - \gamma)^2 + (\gamma - \alpha)^2]}$$

where A, B, and C are so-called "coulombic" energies and α, β, and γ are "exchange" energies, all of which are functions of the internuclear distances. The various coulombic and exchange energies appear in the theory as certain integrals which are evaluated with difficulty and which, even when evaluated accurately, do not necessarily result in a satisfactory calculation of E because of the crude approximations of the Heitler-London theory. Eyring and Polanyi[8] conceived a clever semi-empirical method for obtaining numerical values for these integrals on the basis of spectroscopic data for diatomic molecules XY, YZ, and XZ. At the same time the use of experimental data leads to a cancellation of some of the errors of approximation. The energies of the three possible diatomic molecules are given by $A + \alpha$, $B + \beta$, and $C + \gamma$, respectively. Eyring and Polanyi evaluate the separate terms by supposing the diatomic molecule energy is composed of 10–20 per cent coulombic and the remainder exchange energy.

Although a large number of potential-energy surfaces have been investigated theoretically, especially by Eyring and co-workers, using this semi-empirical method, the results are open to serious question as to their quantitative validity. Nevertheless, from a qualitative standpoint these efforts have been of great value in focusing attention on the idea of a potential-energy surface which serves as a basis for the transition-state theory.

Consider the potential-energy surface for the interaction of three atoms X, Y, and Z. There are three coordinates for each nucleus and, therefore, a total of nine, three of which may be taken as determining the center of mass and three determining the orientation in space. Therefore, three coordinates remain to determine the relative positions of the nuclei. These three coordinates can be chosen as the three inter-nuclear distances between members of each possible pair of atoms. Since it is inconvenient to plot the energy as a function of three coordinates, it is useful for visualization to limit the system arbitrarily to two coordinates by supposing that the atoms are always on a straight line. A linear configuration has been shown by London to be of lower potential energy and therefore, more probable than a corresponding angular configuration, at least if the valence electrons are in s states. With this limitation the two remaining coordinates may be chosen as the X-Y distance, R_{XY}, and the Y-Z distance, R_{YZ}, supposing Y is the intermediate atom. The electronic energy or potential energy, E, of the system can then be plotted in space as a function of the two coordinates, or a contour map can be made.

Such a contour representation is shown in Fig. 1. point a represents a configuration where X is far removed from a molecule YZ; d corresponds to atom Z far removed from molecule XY. point b is where all three atoms are well separated, and c where they are close together as in a collision of X with YZ or XY with Z. A cross section through the potential-energy surface at a to b would look like the diatomic potential energy curve I of Fig. 2, since, with X far away, the energy should depend primarily on R_{YZ}. Curve II of Fig. 2 would be the appearance of a cross section at b to d of Fig. 1, it being supposed arbitrarily that the dissociation energy of XY is greater and the normal internuclear distance smaller than for YZ. The energy will not change rapidly with R_{XY} in the neighborhood of a-b nor with R_{YZ} in the neighborhood of b-d, since both such changes correspond to motion of an atom Z or X already far removed from a molecule. Thus there will be "valleys" in the neighborhood of a and d and a "plateau" at b. The valleys do not join except by way of a "pass" or saddle point near c, since as X approaches YZ or XY approaches Z a van der Waals' repulsive force is expected to set in, thus raising the energy.

This van der Waals' repulsion is of the utmost importance because it is responsible for activation energies in general. Without it all exothermic reactions would have zero or low activation energies and be immeasurably fast. The repulsion is due to the interaction of filled inner shells of electrons and the internuclear repulsion in the molecules as they approach each other.

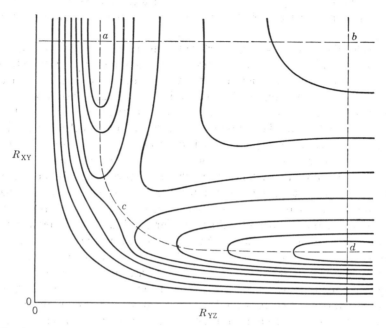

Fig. 1. Potential-energy contour diagram for linear XYZ system as function of internuclear distances. a and d are potential energy minima; b, a maximum, and c, a saddle point.

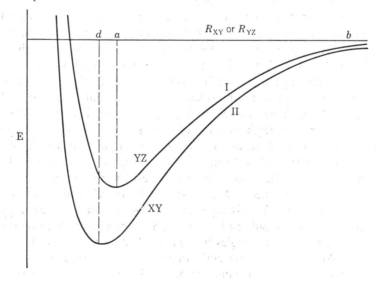

Fig. 2. Potential energy as function of internuclear distance.

The effect is quantum-mechanical in nature and is due to the operation of the Pauli exclusion principle.†

Highly accurate quantum-mechanical calculations are expected to result in a potential-energy surface of the form shown. Hirschfelder and co-workers[13] made the early calculations on the H_3 system. A higher, but still insufficient, accuracy for this difficult problem has been attained by Boys and Shavitt.[14] Eyring's semi-empirical method is qualitatively satisfactory except that there results in some cases near the saddle point a small potential depression, the real existence of which may be questioned.

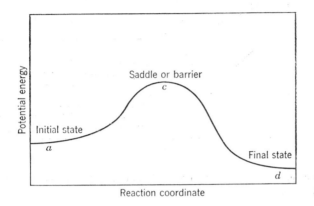

Fig. 3. Potential energy along the reaction coordinate.

The reaction $X + YZ \rightarrow XY + Z$ will correspond to a motion in Fig. 1 from the a valley to the d valley. This motion would take place most probably with the least requirement of energy, and so the point representing the configurations of the system would follow the dashed line moving up the a valley, over the saddle point near c and down into the d valley. The energy as a function of the distance along the dashed line, which may be called the *reaction coordinate*, would appear as in Fig. 3. In going along this coordinate from the initial state to the final state, the energy first rises, reaches a maximum at the saddle point, and then decreases to a level below the original, in this case which happens to correspond to an exothermic reaction. The height of the saddle point above the initial state is a measure of the activation energy.

A system composed of an atom and molecule undergoing reaction does not necessarily follow the line of lowest energy but may also have motion in other directions as, for example, is represented in Fig. 4, where the zigzag line represents an oscillation of the diatomic molecule in addition to

† For a discussion of repulsive forces, see references 9, 10, 11, and 12.

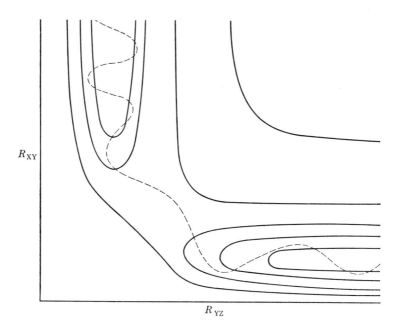

Fig. 4. Potential-energy contour diagram to illustrate vibration of diatomic molecules before and after reaction.

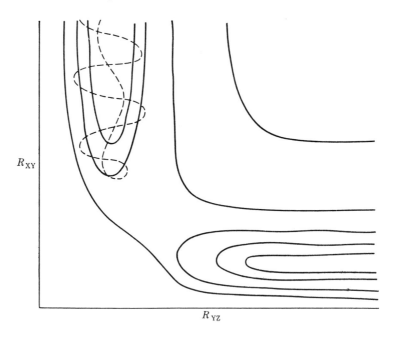

Fig. 5. Potential-energy contour diagram to illustrate transfer between translational and vibrational energy at a collision.

translational motion of approach of the atom and molecule. Also, after passing over the potential energy barrier, the resulting molecule may be vibrating.

Furthermore, as shown by the line in Fig. 5, a collision of X and YZ does not necessarily result in reaction but may cause merely a transfer of energy from translation to vibration, or vice versa.

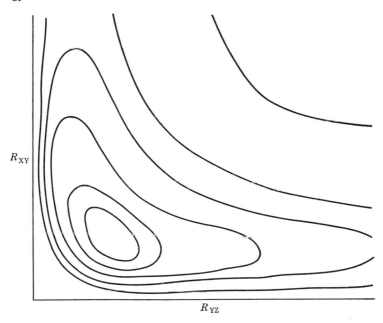

Fig. 6. Potential-energy contour diagram showing potential hole for a stable XYZ molecule.

Another type of potential-energy surface would be that for a linear triatomic molecule XYZ, which might dissociate if sufficiently activated into X and YZ or into XY and Z. This would be expected to have a deep potential hole corresponding to the stable molecule as shown in Fig. 6. A unimolecular reaction would be represented by an oscillating motion of the system in the potential hole followed by a relatively unhindered flight out of one of the valleys.

For a reaction involving more than three atoms, the number of coordinates needed to fix their relative configurations immediately becomes too large to visualize the potential-energy surface. Generally, for n atoms $3n - 6$ coordinates are needed. This is the same as the number of vibrational degrees of freedom of a corresponding nonlinear n-atomic molecule. As with three atoms above, the reaction can be thought of as motion along or near some reaction coordinate and passing "over" a saddle point or potential energy barrier. Of the $3n - 6$ coordinates, one can then be taken as the reaction coordinate, the other $3n - 7$ coordinates being taken normal to the reaction coordinate are of the same nature as the vibrational coordinates of a polyatomic molecule.

In the discussion so far a unique potential-energy surface has been implied. Actually, each system has a large number of such surfaces corresponding to different electronic states, but the lowest such state would presumably be of most importance. For a consideration of excited states, see the review by Laidler and Shuler.[15]

Statistical Mechanical Summary†

Before deriving an expression for the rate of an elementary reaction on the transition-state theory, it will be necessary to review some of the results and terminology of statistical mechanics. Let us start with the Boltzmann law which states that the probability of a molecule being in a state of energy, ϵ_i, and statistical weight, g_i, is proportional to $g_i e^{-\epsilon_i/kT}$. The probability that the molecule will exist in any one or another of its possible states, given by a series of values of the index i, will be proportional to

$$\sum_i g_i e^{-\epsilon_i/kT} = Q \tag{1}$$

This sum is called the *partition function* for the molecule and will be designated by Q.

To show the importance and usefulness of partition functions consider an equilibrium between two isomers A and B

$$A \rightleftharpoons B$$

There will be a set of energy levels for A and another set for B. Corresponding to each will be partition functions, Q_A and Q_B. Assuming ideal gases or dilute solutions, the equilibrium constant for the reaction will be just the ratio of the concentrations; but the ratio of concentrations will be equal to the ratio of probabilities of the two species; therefore,

$$K = [B]/[A] = Q_B/Q_A \tag{2}$$

This is based on the supposition that the same zero of energy is used in expressing the energy levels of A and of B. However, it is more convenient to take the lowest energy level of each molecule as the zero for that molecule. In this case, if $\Delta\epsilon_0$ is the increase in energy from the lowest levels of A to B, or what amounts to the $\Delta\epsilon$ of reaction at the absolute zero

† For general reviews, see references 16, 17, and 18.

of temperature, then Q_B must be replaced by $Q_B e^{-\Delta\epsilon_0/kT}$, since $e^{-\Delta\epsilon_0/kT}$ will factor out of each term in the partition-function summation. With this notation the equilibrium constant becomes

$$K = (Q_B/Q_A)e^{-\Delta\epsilon_0/kT} \tag{3}$$

The generalization of this equation for the equilibrium constant of any arbitrary reaction is found by expressing the free-energy change in terms of partition functions and then relating them to the equilibrium constant through the thermodynamic equation $-\Delta F^0 = RT \ln K$, where ΔF^0 is the standard free energy change for the reaction at the temperature T. The result is that for a reaction

$$a\mathrm{A} + b\mathrm{B} = g\mathrm{G} + h\mathrm{H}$$

the equilibrium constant is given by

$$K = \frac{Q_G^{0g} Q_H^{0h}}{Q_A^{0a} Q_B^{0b}} e^{-\Delta E_0/RT} \tag{4}$$

where K is a concentration equilibrium constant with the concentration unit as 1 molecule/cc and ΔE_0 is as before the ΔE of reaction at the absolute zero but for later convenience put on a mole basis so that the gas constant R replaces Boltzmann's constant. The partition functions used here are specifically for a volume of 1 cc, or stated otherwise:

$$Q_A^0 = Q_A/V \tag{5}$$

It is evident from equation 4 that, if partition functions can be calculated, equilibrium constants can be evaluated theoretically.

Partition Functions for Translation, Rotation, and Vibration

A partition function for a molecule can be written approximately as a product of partition functions for each kind of energy, especially translation, rotation, vibrational, and electronic.

$$Q = Q_t Q_r Q_v Q_e \tag{6}$$

This results from the fact that the energy levels can be represented approximately as a sum of energies of translation, rotation, vibration, and electronic

$$\epsilon = \epsilon_t + \epsilon_r + \epsilon_v + \epsilon_e \tag{7}$$

and each exponential $e^{-\epsilon_i/kT}$ can then be written as a product such as $e^{-\epsilon_t/kT} \cdot e^{-\epsilon_r/kT} \cdot e^{-\epsilon_v/kT} \cdots$.

For *translation* discrete energy levels are not usually considered to exist.

although on the basis of quantum mechanics they can be calculated for molecules moving in a container of a given volume and for certain shapes with the result that they are exceedingly close together. It is more convenient to assume classical motion with continuous energy levels to replace the summation such as in (1) by an appropriate integral which is the phase integral of classical statistics. For 1 degree of freedom of translation (one coordinate, x) the integral is

$$\frac{1}{h} \int_{-\infty}^{+\infty} \int_{0}^{l} e^{-m\dot{x}^2/2kT} m \, d\dot{x} \, dx = \frac{(2\pi m kT)^{\frac{1}{2}}}{h} l \tag{8}$$

where h is Planck's constant and l is the distance in the x direction through which the molecule of mass m is permitted to move.

For 3 degrees of freedom of translation the partition function is a product of three such integrals, and

$$Q_t = \frac{(2\pi m kT)^{\frac{3}{2}}}{h^3} V \tag{9}$$

where V, the volume, appears as the product of three distances, say l_1, l_2, and l_3, where these are the edges of a rectangular parallelepiped, supposing that to be the shape of the container. For other shapes (9) still holds.

For *rotation* the energy levels are discrete but sufficiently close together so that at ordinary temperatures rotation is fully excited. For a *linear* molecule the rotational partition function for 2 degrees of freedom is

$$Q_r = \frac{8\pi^2 IkT}{h^2 \sigma} \tag{10}$$

where I is the moment of inertia and σ is the symmetry number, or the number of equivalent orientations in space. For a *nonlinear* molecule there are 3 rotational degrees of freedom and, in general, three different principal moments of inertia A, B, and C. In this case

$$Q_r = \frac{8\pi^2 (8\pi^3 ABC)^{\frac{1}{2}} (kT)^{\frac{3}{2}}}{h^3 \sigma} \tag{11}$$

For *vibration* the energy levels are far enough apart so that usually only a very few are occupied to any extent. The partition function Q_v must be evaluated strictly as a summation rather than classically as an integral. For the case of an harmonic oscillator with 1 degree of freedom, as in a diatomic molecule, the result is

$$Q_v = (1 - e^{-h\nu/kT})^{-1} \tag{12}$$

with ν as the fundamental vibration frequency. This formula is based upon the lowest vibrational level as defining the zero of energy even though it is a half-quantum of vibration above the minimum of the potential-energy curve.

For several (s) vibrational degrees of freedom the Q_v is approximately a product of terms like (12)

$$Q_v = \prod_{i=1}^{s} (1 - e^{-h\nu_i/kT})^{-1} \tag{13}$$

where the ν_i are the various fundamental vibrational frequencies.

Electronic energy does not usually contribute to a partition function, since most often only the lowest electronic level is occupied at thermal

Table I

APPROXIMATE VALUES OF PARTITION FUNCTIONS PER DEGREE OF FREEDOM

	Designation	Order of Magnitude	Temperature Dependence
Translation	f_t	10^8–10^9	$T^{1/2}$
Rotation	f_r	10^1–10^2	$T^{1/2}$
Vibration	f_v	10^0–10^1	T^0–T^1

equilibrium and that is usually a singlet state. The total partition function may have factors due to electron and nuclear spins, but in the applications of interest here these are of little importance, thus $Q_e = 1$.

The formulas such as (6) to (13) are necessary if a relatively accurate calculation of chemical equilibrium is desired. However, since there are occasions when only order of magnitude can be obtained practically, it is of value to have rough approximations for the various kinds of partition functions. Such values are conveniently given for single degrees of freedom of each type and designated by f_t, f_r, f_v, respectively. f_t is estimated from (8) with l set equal to 1 cm. f_r is obtained from (10) or (11) by setting these Q_r's equal to f_r^2 or f_r^3, respectively, depending upon the number of degrees of freedom. f_v is the same as Q_v in (12). By substitution of typical values of the molecular constants and using a temperature of 300–500° K, the results in Table 1 are obtained. It is to be noticed that $f_t \gg f_r > f_v$. This is because the densities of energy levels are related in the same way. The temperature dependence is also shown. This is of importance in understanding the temperature dependence of an equilibrium constant and later the rate constant, which involves not only the T in the exponential factor $e^{-\Delta E_0/RT}$ but also temperature dependence in the Q's. For vibration f_v is nearly independent of T, that is, roughly proportional to T to the zero power, or constant, since vibrational levels are usually only slightly excited. If, however, the temperature is very high or if the vibration frequency ν is very small, then the vibration is fully excited and f_v is proportional to T. This is T rather than $T^{1/2}$, as in translation and rotation, because there are both kinetic and potential energies associated with the one vibrational degree of freedom.

Derivation of the Rate Equation

Consider the potential energy of the reacting system as a function of the reaction coordinate, the energy being a minimum with respect to variation

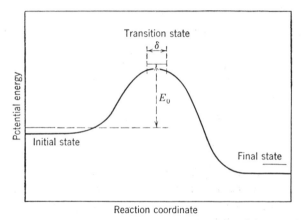

Fig. 7. Transition state at the top of the reaction potential-energy barrier.

in all other coordinates (see Fig. 7). The transition state is defined as including those configurations corresponding to the top of the potential-energy barrier within a small but otherwise arbitrary distance δ along the reaction coordinate and including all possible variation in other co-ordinates, that is, vibrations normal to the reaction coordinate and translations and rotations of the center of mass of the system. Any system with a configuration within the transition state is called an activated complex. The distance δ might be considered of the order of magnitude of 1 A and similar in nature to the δ introduced in the formula for the termolecular collision number of the last chapter, except as that in this case it will turn out that the δ cancels out in the final equation. The E_0 shown in the figure is the difference in energy between the lowest energy levels in the initial and transition states. These levels are above the potential-energy curve because of half-quanta of vibration due to motion in other coordinates. E_0 is analogous to ΔE_0 introduced above and may be called the activation energy at the absolute zero.

If the reaction in question is at equilibrium, let the concentration of activated complexes be $C^{\ddagger}$ (the double dagger will be used exclusively to designate quantities referring to the activated complex or transition state). Of these activated complexes half will be moving to the right according to the forward reaction, and half to the left for the reverse reaction. Let us now calculate the mean velocity $\bar{v}$ of those activated complexes that are going to the right. Using the distribution law for a single velocity component, equation 1 of Chapter 4,

$$\bar{v} = \frac{\int_0^\infty e^{-m^{\ddagger}\dot{x}^2/2kT}\dot{x}\,d\dot{x}}{\int_0^\infty e^{-m^{\ddagger}\dot{x}^2/2kT}d\dot{x}} \tag{14}$$

where $m^{\ddagger}$ is the effective mass of the activated complex for motion along the reaction coordinate. Integration gives

$$\bar{v} = (2kT/\pi m^{\ddagger})^{\frac{1}{2}} \tag{15}$$

The average time to move through the transition state to the right will then be

$$\delta/\bar{v} = \delta(\pi m^{\ddagger}/2kT)^{\frac{1}{2}} \tag{16}$$

and the rate of reaction in terms of number of complexes crossing to the right per unit volume per unit time will be

$$\text{rate} = (C^{\ddagger}/2)/\delta(\pi m^{\ddagger}/2kT)^{\frac{1}{2}}$$
$$= (C^{\ddagger}/\delta)(kT/2\pi m^{\ddagger})^{\frac{1}{2}} \tag{17}$$

For the reaction at equilibrium, as is being discussed at present, the reverse reaction rate is also given by (17) with the net rate zero as is necessary for equilibrium.

It is desired to have an equation for the rate of reaction when far removed from chemical equilibrium and, in particular, at the beginning when only reactants are present. It will be assumed that under these conditions there will still be an equilibrium between reactants and the activated complexes that are moving to the right, in which case equation 17 is still applicable. Of course, there cannot be a complete equilibrium with the activated complexes, since those moving to the left will be missing at the earliest stage of reaction as they are formed from product molecules. But the complexes of importance are only those moving in the desired direction, and, since they are formed by collisions of reactant molecules, there should be just as many of them present in this situation as at a true chemical equilibrium. There is still the possibility, however, that the rate of reaction may be comparable to the rate at which energy can be redistributed among various degrees of freedom at collisions between reactant molecules. If such is the case, the derived rate (17) would be too great. This effect is analogous to that already discussed for unimolecular reactions in Chapter 4, except that there seems to be no simple way of getting experimental evidence for the effect in other than unimolecular reactions.[19],[20]

To continue the derivation, let us now apply statistical-mechanical theory to evaluate $C^{\ddagger}$ in terms of experimental reactant concentrations. Suppose that the chemical equilibrium between reactants and activated complex is of the form

$$a\mathrm{A} + b\mathrm{B} + \cdots \rightleftharpoons \mathrm{M}^{\ddagger} \tag{18}$$

where $\mathrm{M}^{\ddagger}$ represents the complex. This equation implies that the molecularity of the reaction is $a + b + \cdots$. Let the equilibrium constant be

$$K = \frac{C^{\ddagger}}{C_{\mathrm{A}}{}^{a}C_{\mathrm{B}}{}^{b} \cdots} \tag{19}$$

Then solving for $C^{\ddagger}$

$$C^{\ddagger} = KC_{\mathrm{A}}{}^{a}C_{\mathrm{B}}{}^{b} \cdots \tag{20}$$

and substituting in (17)

$$\text{rate} = (K/\delta)(kT/2\pi m^{\ddagger})^{\frac{1}{2}}C_{\mathrm{A}}{}^{a}C_{\mathrm{B}}{}^{b} \cdots \tag{21}$$

This shows immediately that the order of this elementary reaction will be $a + b + \cdots$, the same as the molecularity. Let the rate constant be designated by k_r, using the subscript to avoid confusion with Boltzmann's constant. Then

$$\text{rate} = k_r C_{\mathrm{A}}{}^{a}C_{\mathrm{B}}{}^{b} \cdots \tag{22}$$

and

$$k_r = (K/\delta)(kT/2\pi m^{\ddagger})^{\frac{1}{2}} \tag{23}$$

The next step is to express K in terms of partition functions. For the activated complex the partition function will differ in form from that for a stable molecule in that 1 degree of freedom, that along the reaction coordinate, is unusual. This degree of freedom for motion across the barrier and confined within the distance δ, since that defines the transition state, is very similar to a translational degree of freedom, where the potential energy remains constant. If δ is chosen small enough, the change of potential energy in the transition state may be made as small as is desired. Therefore, it is a good assumption that this special degree of freedom behaves as a translation and contributes to the total partition function a factor

$$(2\pi m^{\ddagger}kT)^{\frac{1}{2}}(\delta/h) \tag{24}$$

Let the total partition function per unit volume for the activated complex be

$$Q_{\ddagger}^{0}(2\pi m^{\ddagger}kT)^{\frac{1}{2}}(\delta/h) \tag{25}$$

where $Q_{\ddagger}^{0}$ includes the contribution of all degrees of freedom except that for the reaction coordinate. Then the equilibrium constant K may be expressed

$$K = \frac{Q_{\ddagger}^{0}(2\pi m^{\ddagger}kT)^{\frac{1}{2}}(\delta/h)}{Q_{A}^{0a}Q_{B}^{0b}\cdots} e^{-E_0/RT} \tag{26}$$

Substitution of (26) in (23) gives

$$k_r = \frac{(2\pi m^{\ddagger}kT)^{\frac{1}{2}}(\delta/h)}{\delta} \left(\frac{kT}{2\pi m^{\ddagger}}\right)^{\frac{1}{2}} \frac{Q_{\ddagger}^{0}e^{-E_0/RT}}{Q_{A}^{0a}Q_{B}^{0b}\cdots}$$

or

$$k_r = \frac{kT}{h} \cdot \frac{Q_{\ddagger}^{0}}{Q_{A}^{0a}Q_{B}^{0b}\cdots} e^{-E_0/RT} \tag{27}$$

where fortunately the unknown quantities δ and $m^{\ddagger}$ have canceled out.

In (27) the group of factors other than kT/h has the form of an equilibrium constant. Define

$$K^{\ddagger} = \frac{Q_{\ddagger}^{0}}{Q_{A}^{0a}Q_{B}^{0b}\cdots} e^{-E_0/RT} \tag{28}$$

and notice from (26) and (28) that

$$K = K^{\ddagger}(2\pi m^{\ddagger}kT)^{\frac{1}{2}}\delta/h \tag{29}$$

Then

$$k_r = (kT/h)K^{\ddagger} \tag{30}$$

This simple and general equation predicts that for any elementary reaction the rate constant is an equilibrium constant multiplied by a universal frequency factor kT/h, which varies only with the temperature and which has dimensions of reciprocal time. It must be admitted, however, that $K^{\ddagger}$ is not strictly an equilibrium constant but is only similar to one.

To develop the theory further, detailed expressions for $K^{\ddagger}$ or for the Q's must be given in terms of experimental quantities that can be derived from other than rate data. Q_{A}^{0} for a stable molecule such as A can be successfully calculated, using equations 5 to 13, the molecular constants such as moments of inertia and vibration frequencies being obtained generally from spectroscopic data. However, although $Q_{\ddagger}^{0}$ can be written explicitly, the required moments of inertia and vibration frequencies are unknown experimentally, but in principle they can be determined theoretically if the potential-energy surface is calculated with sufficient accuracy for the transition state. The moments of inertia depend on the distances between nuclei of the activated complex, and the vibration frequencies are functions of the curvature of the potential-energy surface in directions normal to the reaction coordinate.

Two of the examples that have been worked out in detail are for the bimolecular reaction $H_2 + I_2 \rightarrow 2HI$ by Wheeler, Topley, and Eyring and the termolecular reaction $2NO + O_2 \rightarrow 2NO_2$ by Gershinowitz and Eyring.[21] Reasonable estimates of the five vibration frequencies in the first activated complex and the ten or eleven in the second and the use of empirical activation energies give agreement between theory and experiment to within a factor of 2.

Several approximate methods have been worked out for calculating frequency factors, particularly for reactions involving the abstraction of an atom from a molecule by another atom or small radical.[22] These depend on simplification, such as assuming that certain parts of the partition function of the reactants do not change in forming the transition state, and by using classical partition functions. The methods are very successful in simple cases.

Because of the general lack of sufficient details about potential-energy surfaces for arbitrary reactions, it is of value to make estimates of rate constants, using mere order of magnitude values of partition functions as shown in Table 1. Consider a bimolecular reaction between two polyatomic molecules A and B of n_A and n_B atoms, respectively, and assume that these molecules as well as the activated complex are nonlinear so that each has 3 rotational degrees of freedom. The necessary partition functions can be represented by

$$Q_A^0 \simeq f_t^3 f_r^3 f_v^{3n_A - 6}$$
$$Q_B^0 \simeq f_t^3 f_r^3 f_v^{3n_B - 6} \qquad (31)$$
$$Q_{\ddagger}^0 \simeq f_t^3 f_r^3 f_v^{3n_A + 3n_B - 7}$$

where for $Q_{\ddagger}^0$ there is one less vibrational degree of freedom than normal because of the reaction coordinate. Substituting (31) in (27) and canceling f_t's, etc., between different species inasmuch as they represent only order of magnitude:

$$k_r \simeq \frac{kT}{h} \frac{f_r^5}{f_t^3 f_r^3} e^{-E_0/RT} \qquad (32)$$

Substitution of $T \simeq 300–500°$ K and the values of the f's from Table 1 results to within one or two powers of 10 in

$$k_r \simeq 10^{-15} e^{-E_0/RT} \qquad (33)$$

where the units are cubic centimeter molecules^{-1} sec^{-1}.

A bimolecular reaction between two diatomic molecules forming a nonlinear complex, as probably is the case for the $H_2 + I_2$ reaction, would have a rate constant

$$k_r \simeq \frac{kT}{h} \frac{f_v^3}{f_t^3 f_r} e^{-E_0/RT}$$

or

$$k_r \simeq 10^{-13} e^{-E_0/RT} \tag{34}$$

except that, if such a light molecule as H_2 is participating, the f_t for it is perhaps 10-fold smaller than that for an I_2 molecule or H_2-I_2 complex. This would increase k_r to

$$k_r \simeq 10^{-10} e^{-E_0/RT} \tag{35}$$

which it is interesting to note is the same order of magnitude as that predicted by the collision theory (see Table 1 in Chapter 4).

Comparison of Collision and Transition-State Theories

Although the transition-state theory is capable of handling more difficult situations than the collision theory can, it is important to verify whether the two theories agree when applied to the same situation. That they can agree at least as to order of magnitude is shown in the last paragraph. Consider the rigid sphere model for a bimolecular reaction from the standpoint of the present theory. As only translational motion of the rigid spheres was considered, it is apparent that they would correspond to monatomic molecules here. The activated complex would be equivalent to a diatomic molecule with σ_{AB} for the internuclear distance but with no vibrational degree of freedom, since instead there is motion along the reaction coordinate. If m_A and m_B are the masses of the two reactant molecules, the moment of inertia I works out to be

$$I = \sigma_{AB}^2 \frac{m_A m_B}{m_A + m_B}$$

and k_r written in detail is

$$k_r = \frac{kT}{h} \frac{\dfrac{[2\pi(m_A + m_B)kT]^{3/2}}{h^3} \dfrac{8\pi^2 \sigma_{AB}^2 [m_A m_B/(m_A + m_B)]kT}{h^2}}{\dfrac{(2\pi m_A kT)^{3/2}}{h^3} \cdot \dfrac{(2\pi m_B kT)^{3/2}}{h^3}} e^{-E_0/RT}$$

or after cancellation

$$k_r = \left[8\pi kT \left(\frac{m_A + m_B}{m_A m_B} \right) \right]^{1/2} \sigma_{AB}^2 e^{-E_0/RT} \tag{36}$$

This is identical with equation 28 of Chapter 4 for the collision theory if the steric factor is taken as unity.

Table 2

APPROXIMATE EXPRESSIONS AND VALUES FOR BIMOLECULAR RATE CONSTANTS
FOR DIFFERENT TYPES OF REACTANTS AND COMPLEXES

Frequency Factor			Steric Factor	
Formula	Value, cc/molecule-sec	T Exponent	Formula	Value
Two atoms				
$\dfrac{kT}{h}\dfrac{f_r^2}{f_t^3}$	10^{-10}–10^{-9}	$\frac{1}{2}$	1	1
Atom + linear molecule, linear complex				
$\dfrac{kT}{h}\dfrac{f_v^2}{f_t^3}$	10^{-12}–10^{-11}	$-\frac{1}{2}$ to $\frac{1}{2}$	$\left(\dfrac{f_v}{f_r}\right)^2$	10^{-2}
Atom + linear molecule, nonlinear complex				
$\dfrac{kT}{h}\dfrac{f_v f_r}{f_t^3}$	10^{-11}–10^{-10}	0 to $\frac{1}{2}$	$\dfrac{f_v}{f_r}$	10^{-1}
Atom + nonlinear molecule, nonlinear complex				
$\dfrac{kT}{h}\dfrac{f_v^2}{f_t^3}$	10^{-12}–10^{-11}	$-\frac{1}{2}$ to $\frac{1}{2}$	$\left(\dfrac{f_v}{f_r}\right)^2$	10^{-2}
Two linear molecules, linear complex				
$\dfrac{kT}{h}\dfrac{f_v^4}{f_t^3 f_r^2}$	10^{-14}–10^{-13}	$-\frac{3}{2}$ to $\frac{1}{2}$	$\left(\dfrac{f_v}{f_r}\right)^4$	10^{-4}
Two linear molecules, nonlinear complex				
$\dfrac{kT}{h}\dfrac{f_v^3}{f_t^3 f_r}$	10^{-13}–10^{-12}	-1 to $\frac{1}{2}$	$\left(\dfrac{f_v}{f_r}\right)^3$	10^{-3}
One linear + one nonlinear molecule, nonlinear complex				
$\dfrac{kT}{h}\dfrac{f_v^4}{f_t^3 f_r^2}$	10^{-14}–10^{-13}	$-\frac{3}{2}$ to $\frac{1}{2}$	$\left(\dfrac{f_v}{f_r}\right)^4$	10^{-4}
Two nonlinear molecules, nonlinear complex				
$\dfrac{kT}{h}\dfrac{f_v^5}{f_t^3 f_r^3}$	10^{-15}–10^{-14}	-2 to $\frac{1}{2}$	$\left(\dfrac{f_v}{f_r}\right)^5$	10^{-5}

For comparison with the approximate expressions of transition-state theory such as equations 32 to 34, it is desirable to have this rigid sphere model also expressed in the same way.

$$Q_A{}^0 \simeq f_t^3$$

$$Q_B{}^0 \simeq f_t^3$$

$$Q_{\ddagger}{}^0 \simeq f_t^3 f_r^2$$

and so (37) results, which is, of course, also in agreement with simple

$$k_r \simeq \frac{kT}{h} \frac{f_r^2}{f_t^3} e^{-E_0/RT} \tag{37}$$

$$\simeq 10^{-10} e^{-E_0/RT}$$

collision theory for $p \simeq 1$. Now, for a reaction between polyatomic molecules or diatomic molecules, transition-state theory gives a result (equations 32 to 34 above) that is somewhat smaller than this by as much as 10^{-5}. This factor may be likened to a steric factor, p, which may then be defined simply as a ratio of the actual rate such as (32) divided by the rate on the rigid sphere model (37). In this case

$$p \simeq \left(\frac{f_v}{f_r}\right)^5 \tag{38}$$

and $p \ll 1$, since generally $f_r > f_v$. The conclusion is that reactions between polyatomic molecules are slower than expected on the rigid sphere model, since several rotational degrees of freedom are lost, being replaced by vibrational degrees of freedom, which have a smaller probability associated with them.

Table 2 summarizes the results of these approximate considerations, showing formulas and values for the steric factor as defined above and also for the "frequency factor," which is everything in the rate constant except the exponential containing the activation energy. Also included is the exponent of T in the frequency factor, which is calculated using the information of Table 1 with f_v taken in the range T^0 to $T^{1/2}$. This would be the n of equation 38 of Chapter 2.

The general conclusion from this table is that, the more complicated the molecules, up to a certain point, the smaller is the steric factor. The pre-exponential temperature dependence appears to decrease perhaps to negative exponents. This effect would be difficult to test experimentally because of the usually large effect of the exponential.

For *termolecular* reactions, a similar treatment can be given, but only two cases will be mentioned. For reaction of three atoms with a linear complex,

$$k_r \simeq \frac{kT}{h} \frac{f_t^3 f_r^2 f_v^3}{f_t^9} e^{-E_0/RT}$$

$$\simeq \frac{kT}{h} \frac{f_r^2 f_v^3}{f_t^6} e^{-E_0/RT} \tag{39}$$

or

$$k_r \simeq 10^{-33} e^{-E_0/RT}$$

This being for molecules per cubic centimeter as the concentration unit, it is in fair agreement with the collision formula, although because of the high exponent on f_t the estimate is very crude. For three diatomic molecules in a nonlinear complex,

$$k_r \simeq \frac{kT}{h} \frac{f_v^8}{f_t^6 f_r^3} e^{-E_0/RT} \tag{40}$$

In this case it is interesting that the temperature dependence would be as $T^{-7/2} e^{-E_0/RT}$ if vibration is not excited. For a reaction with E_0 small or zero, such a dependence can lead to a rate which decreases as T increases. This constitutes a possible explanation of the peculiar temperature dependence of the nitric oxide-oxygen reaction.[21]

For *unimolecular* reaction, the overall first-order rate constant should be as in (41). The rate might vary from this either way by one or two

$$k_r \simeq \frac{kT}{h} \frac{1}{f_v} e^{-E_0/RT}$$

$$\simeq 10^{13} e^{-E_0/RT} \tag{41}$$

factors of 10, depending upon slight changes in the moment of inertia or large changes in vibration frequencies. In the Lindemann-Hinshelwood mechanism the bimolecular activation process was found to have a rate that was much larger than expected on the simple collision theory. Since the present transition-state theory indicates that bimolecular reactions between polyatomic molecules should be comparatively slow, there seems to be disagreement. However, the transition-state theory was developed on the supposition that the system goes over a potential-energy barrier, whereas in the bimolecular activation process the system merely goes up and returns back down the same potential valley (see Fig. 5). Therefore, the transition-state theory does not apply to the activation process.

There is a fundamental difficulty in the transition-state theory in that it is assumed that the equilibrium concentration of activated complexes is maintained. But, if the rate of decomposition of the activated molecule is comparable to, or greater than, the rate of deactivation, then equilibrium cannot be maintained. Thus the theory does not apply to the low-pressure region of unimolecular reactions. The rate of decomposition of the activated complex, which takes place in a time of the order of 10^{-13} seconds, must not be confused with the rate of decomposition of the activated molecule, which is much slower. Thus the activated molecule must undergo some rearrangement before it becomes the activated complex. At high pressures where the transition-state theory is applicable, it becomes equivalent to Slater's theory.[23]

Transmission Coefficient

So far it has been supposed that any activated complex that crosses the potential-energy barrier continues on to form the products of reaction. However, there is the possibility that because of some peculiarity in the shape of the potential-energy surface, the system will be reflected back across with the result that reaction does not take place. Let κ be the fraction of the crossings that are successful in leading to the final products. This is called the transmission coefficient. The corrected rate constant will then be

$$k_1 = \kappa(kT/h)K^{\ddagger} \qquad (42)$$

The value of κ could presumably be calculated if the potential-energy surface were known well enough. It is believed that κ ordinarily is rather close to unity, say within 5 or 10 per cent, and so because of the much greater uncertainties elsewhere in the theory the κ will usually be considered unity, with one important exception. Consider a possible bimolecular association reaction of two atoms to form a diatomic molecule. The potential-energy surface for such a reaction is nothing other than the common potential-energy curve for a diatomic molecule. As the two atoms approach each other the potential energy drops and the kinetic energy increases, causing the system to move past the point of equilibrium until at a small internuclear distance the motion reverses and the atoms fly apart. There will be perfect reflection, and κ is essentially zero. Therefore, atoms are not likely to combine in such a bimolecular process, but must do so in a termolecular collision where the third molecule removes energy in the form of translation or vibration. Figure 8 shows the difference between the possible bimolecular and termolecular processes for a system of two atoms A, which can combine to A_2, and an inert atom M, all supposed to be in a linear configuration.

The separated atoms are represented by a point such as a in Fig. 8. The path $a \rightarrow b$ would represent a bimolecular collision of A with A but is not successful because of reflection. The path $a \to c \to b$ corresponds to a termolecular collision and can result in combination as represented by point b, with the molecule A_2 left in one of its vibrational levels and the atom M and the molecule A_2 having an excess relative kinetic energy.

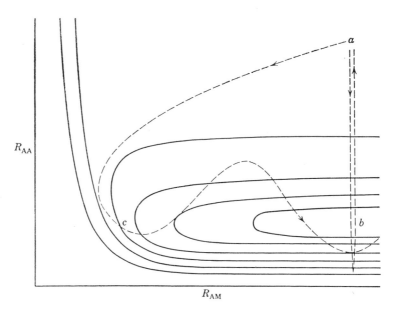

Fig. 8. Potential-energy contour diagram to illustrate the action of a third molecule M in stabilizing a colliding pair of atoms AA.

"Thermodynamic" Treatment of Reaction Rate

Inasmuch as $K^{\ddagger}$ of equations 28 to 30 is similar to an equilibrium constant, it is possible to define quantities analogous to the thermodynamic functions used in connection with ordinary equilibrium constants. Define *free energy of activation* $\Delta F^{\ddagger}$ by the equation

$$\Delta F^{\ddagger} = -RT \ln K^{\ddagger} \tag{43}$$

Define *heat of activation* $\Delta H^{\ddagger}$ by

$$\Delta H^{\ddagger} = RT^2 \frac{d \ln K_p^{\ddagger}}{dT} \tag{44}$$

where $K_p^{\ddagger}$ is $K^{\ddagger}$ in pressure units. Since concentration units are more usual and since

$$K_p^{\ddagger} = K_c^{\ddagger}(RT)^{1-n}$$

with n the molecularity of the gaseous reaction, it follows that

$$\Delta H^{\ddagger} = RT^2 \frac{d \ln K_c^{\ddagger}}{dT} - (n-1)RT \tag{45}$$

Define *entropy of activation* $\Delta S^{\ddagger}$ by

$$\Delta S^{\ddagger} = \frac{\Delta H^{\ddagger} - \Delta F^{\ddagger}}{T} \tag{46}$$

It follows from these definitions that

$$K^{\ddagger} = e^{-\Delta F^{\ddagger}/RT} = e^{\Delta S^{\ddagger}/R} \cdot e^{-\Delta H^{\ddagger}/RT} \tag{47}$$

and for the rate constant

$$k_r = (kT/h)e^{-\Delta F^{\ddagger}/RT} \tag{48}$$

or

$$k_r = (kT/h)e^{\Delta S^{\ddagger}/R} \cdot e^{-\Delta H^{\ddagger}/RT} \tag{49}$$

These quantities $\Delta F^{\ddagger}$, $\Delta H^{\ddagger}$, and $\Delta S^{\ddagger}$ are in reality values for standard states, as is implied in the equation 43 in particular. The notation does not indicate this explicitly because it is understood that standard-state values are always used. However, it must be realized that different standard states may be used at different times, and we must be careful to show in some way what these are. Generally, the standard states of reactants and activated complex are unit concentrations where the concentration unit corresponds to whatever is used in evaluating the rate constant k_r. $\Delta F^{\ddagger}$, etc., may be considered the difference in free energy, etc., between the activated complex and the reactants from which it is formed, all substances being in their standard states.

97982

Relation between $\Delta H^{\ddagger}$ and Various Kinds of Activation Energy

The Arrhenius, or empirical, activation energy E_a is defined in Chapter 2, equation 36.

$$\frac{d \ln k_r}{dT} = \frac{E_a}{RT^2}$$

It follows from (45) and (30) that for a *gaseous reaction*

$$\Delta H^{\ddagger} = RT^2 \frac{d \ln k_r}{dT} - RT - (n-1)RT$$

or

$$\Delta H^{\ddagger} = E_a - nRT \tag{50}$$

where n is the molecularity (and order) of the reaction. For a reaction in *solution* in the liquid state

$$\Delta H^{\ddagger} = RT^2 \frac{d \ln K^{\ddagger}}{dT}$$

$$= RT^2 \frac{d \ln k_r}{dT} - RT$$

or

$$\Delta H^{\ddagger} = E_a - RT \qquad (51)$$

The theoretical activation energy, E_0, introduced earlier in this chapter, cannot be related to $\Delta H^{\ddagger}$ or E_a unless the detailed temperature dependence of the partition functions is known. Suppose that

$$k_r = AT^m e^{-E_0/RT}$$

where A is independent of T and m is as indicated in Table 2 for various types of bimolecular reaction. As shown in equation 40 of Chapter 2,

$$E_0 = E_a - mRT \qquad (52)$$

and therefore, from (50) and (51), for *gases*

$$\Delta H^{\ddagger} = E_0 + (m - n)RT \qquad (53)$$

and for *solutions*

$$\Delta H^{\ddagger} = E_0 + (m - 1)RT \qquad (54)$$

Because m is usually negative, the three kinds of activation energy are related as

$$\Delta H^{\ddagger} < E_a < E_0 \qquad (55)$$

Interpretation of Entropy of Activation

$\Delta S^{\ddagger}$ can be evaluated from rate data, using for example equations 43, 45, and 46. For a bimolecular reaction its value can be correlated with the steric factor of the collision theory in the following way. Writing the rate-constant expression of the two theories

$$k_r = pZ'e^{-E_0/RT}$$

$$k_r = \frac{kT}{h} e^{\Delta S^{\ddagger}/R} \cdot e^{-\Delta H^{\ddagger}/RT}$$

and noting that, since

$$e^{-E_0/RT} \simeq e^{-\Delta H^{\ddagger}/RT}$$

therefore

$$(kT/h)e^{\Delta S^{\ddagger}/R} \simeq pZ$$

More accurately, using (53) or (54) above,

$$(kT/h)e^{\Delta S^{\ddagger}/R}e^{j} = pZ \tag{56}$$

where $j = 2 - m$ or $1 - m$, depending on whether it is a gas reaction or reaction in solution. Since $kT/h \simeq 10^{13}$ sec^{-1} and $Z \simeq 10^{14}$–10^{15} cc moles^{-1} sec^{-1} and $e^{j} \simeq 10$, it follows that

$$e^{\Delta S^{\ddagger}/R} \simeq p \tag{57}$$

when the standard state for $\Delta S^{\ddagger}$ is 1 mole per cc. For 1 mole per liter standard state $Z \simeq 10^{11}$ M^{-1} sec^{-1} and consequently exp $(\Delta S^{\ddagger}/R) \simeq 10^{-3}$ if p were to be unity. This would make $\Delta S^{\ddagger}$ about -13.5 E.U. for a bimolecular reaction with a maximum steric factor. Thus the entropy of activation depends on the choice of standard states, or, what is equivalent, the choice of concentration units. In any event, $\Delta S^{\ddagger}$ is expected to become more negative for a reaction between two polyatomic molecules, than for a reaction between two atoms.

More accurate than the approximate relation (57) is the more exact relation obtained by taking the ratio of two equations (56) written for each of two similar reactions designated by subscripts 1 and 2.

$$e^{(\Delta S_1^{\ddagger} - \Delta S_2^{\ddagger})/R} = (pZ)_1/(pZ)_2$$

or

$$\Delta S_1^{\ddagger} - \Delta S_2^{\ddagger} = R \ln [(pZ)_1/(pZ)_2] \tag{58}$$

or, if Z is the same for the two reactions,

$$\Delta S_1^{\ddagger} - \Delta S_2^{\ddagger} = R \ln (p_1/p_2) \tag{59}$$

This shows that in a series of similar reactions the more negative $\Delta S^{\ddagger}$ corresponds to the smaller steric factor.

For *unimolecular* reactions the corresponding treatment gives

$$k_r = (kT/h)e^{\Delta S^{\ddagger}/R} \cdot e^{-E_0/RT}$$

$\Delta S^{\ddagger}$ is here independent of the standard state. If $10^{13}e^{-E_0/RT}$ is taken as a "normal" value of a unimolecular rate constant, then, since $kT/h \simeq 10^{13}$ sec^{-1}, $e^{\Delta S^{\ddagger}/R}$ is a factor which determines whether the reaction goes faster or slower than normal. If $\Delta S^{\ddagger}$ is positive, corresponding to a more probable activated complex, then the reaction is faster than normal. If $\Delta S^{\ddagger}$ is negative, the activated complex is less probable and the rate slower.

The theoretical equations developed in this chapter will be compared with experimental results in Chapter 6 for gaseous reactions and Chapter 7 for reaction in solution.

REFERENCES

1. H. Pelzer and E. Wigner, *Z. physik. Chem.*, *B15*, 445 (1932).
2. H. Eyring, *J. Chem. Phys.*, *3*, 107 (1935).
3. W. F. K. Wynne-Jones and H. Eyring, *J. Chem. Phys.*, *3*, 492 (1935).
4. H. Eyring, *Chem. Revs.*, *17*, 65 (1935).
5. M. G. Evans and M. Polanyi, *Trans. Faraday Soc.*, *31*, 875 (1935).
6. M. Born and R. Oppenheimer, *Ann. Physik.*, *84*, 457 (1927).
7. F. London, *Probleme der modernen Physik* (*Sommerfeld Festschrift*), Hirzel, Leipzig, 1928, p. 104; c.f. ref 2, 3, Ch. 4, *Z. Elektrochem.*, *35*, 552 (1929).
8. H. Eyring and M. Polanyi, *Z. physik. Chem.*, *B12*, 279 (1931); H. Eyring, *Chem. Revs.*, *10*, 103 (1932).
9. J. C. Slater and J. G. Kirkwood, *Phys. Rev.*, *37*, 682 (1931).
10. M. Born and J. E. Mayer, *Z. Physik*, *75*, 1 (1932).
11. M. L. Huggins, *J. Chem. Phys.*, *5*, 143 (1937).
12. R. A. Buckingham, *Proc. Roy. Soc.*, *A168*, 264 (1938).
13. J. Hirschfelder, H. Eyring, and N. Rosen, *J. Chem. Phys.*, *4*, 121 (1936); J. Hirschfelder, H. Diamond, and H. Eyring, *ibid.*, *5*, 695 (1937); D. Stevenson and J. Hirschfelder, *ibid.*, *5*, 933 (1937); J. Hirschfelder, *ibid.*, *6*, 794 (1938).
14. S. F. Boys and I. Shavitt, preliminary report; I. Shavitt, *J. Chem. Phys.*, *31*, 1359 (1959).
15. K. J. Laidler and K. E. Shuler, *Chem. Revs.*, *48*, 153 (1951); K. J. Laidler, *The Chemical Kinetics of Excited States*, Oxford University Press, London, 1955.
16. J. E. Mayer and M. G. Mayer, *Statistical Mechanics*, John Wiley & Sons, New York, 1940.
17. M. Dole, *Introduction to Statistical Thermodynamics*, Prentice-Hall, New York, 1954.
18. R. H. Fowler and E. A. Guggenheim, *Statistical Thermodynamics*, Cambridge University Press, New York, 1939.
19. H. A. Kramers, *Physica*, *7*, 284 (1940).
20. B. J. Zwolinski and H. Eyring, *J. Am. Chem. Soc.*, *69*, 2702 (1947).
21. H. Gershinowitz and H. Eyring, *J. Am. Chem. Soc.*, *57*, 985 (1935).
22. K. S. Pitzer, *J. A. Chem. Soc.*, *79*, 1804 (1957); K. S. Pitzer and O. Sinanoglu, *J. Chem. Phys.*, *30*, 422 (1959); D. R. Hershbach, H. S. Johnston, and D. Rapp, *ibid.*, *31*, 1652 (1959).
23. N. B. Slater, *Phil. Trans.*, *246A*, 57 (1953).

6.

COMPARISON OF THEORY WITH
EXPERIMENT—SIMPLE GAS REACTIONS

In order to test the theories of the preceding chapters it would be desirable to examine simple gas-phase reactions involving familiar chemical substances. However, simple reactions in the gas phase are uncommon, most mechanisms being complex. Sometimes by a careful analysis of the kinetics of a complex reaction it is possible to get accurate data on the individual steps involved. In this chapter an attempt will be made to present data on rate constants, frequency factors, and activation energies for some simple reactions. The examples are chosen from the recent literature, and the fact must be appreciated that the molecularities may not always be as claimed. This is because the molecularity depends upon an assumed mechanism and is not directly provable. The order of the reaction is experimentally available and is the chief evidence for the molecularity.

Bimolecular Reactions

Table 1 lists some second-order, and presumably bimolecular, gaseous reactions between stable molecules. A and E_a are the values in the expression

$$k = Ae^{-E_a/RT}$$

A is the "frequency factor" and has the same units as k, which are taken to be cc moles^{-1} sec^{-1}. E_a is the Arrhenius activation energy in kilocalories per mole. Since the factor A also may have temperature dependence, the values shown are merely values for an average temperature in

Table I

SECOND-ORDER GASEOUS REACTIONS BETWEEN STABLE MOLECULES

	A, cc/mole-sec	E_a, kcal/mole	$\Delta S^{\ddagger}$, E.U.	E_a Calc., Hirschfelder Rule
$H_2 + I_2 \rightarrow 2HI$	1×10^{14}	40.0	2	(40)
$2HI \rightarrow H_2 + I_2$	6×10^{13}	44.0	0	(44)
$HI + RI \rightarrow RH + I_2$				
$R = CH_3$—	1.6×10^{15}	33.4	8	34
$= C_2H_5$—	4×10^{14}	29.8	5	34
$= n\text{-}C_3H_7$—	1×10^{14}	29.2	2	34
$2NOCl \rightarrow 2NO + Cl_2$	9×10^{12}	24.0	-3	
$NO + O_3 \rightarrow NO_2 + O_2$	8×10^{11}	2.5	-8	
Dimerizations, etc.				
1,3-Butadiene	4.7×10^{10}	25.3	-13	33
2-Isoprene	5.3×10^{11}	28.9	-8	33
1,3-Pentadiene	3.5×10^{10}	26.0	-14	33
2,3-Dimethylbutadiene-1,3	1.4×10^{10}	25.3	-16	33
Cyclopentadiene	8.5×10^7	14.9	-26	33
Butadiene + acrolein	1.5×10^9	19.7	-20	33
Isoprene + acrolein	1.0×10^9	18.7	-21	33
Butadiene + crotonaldehyde	9.0×10^8	22.0	-21	33
Cyclopentadiene + crotonaldehyde	1.0×10^9	15.2	-21	33
Isobutylene + HBr	1.6×10^{10}	22.5	-15	41
Isobutylene + HCl	1.0×10^{11}	28.8	-12	46
Ethylene	7.1×10^{10}	37.7	-12	33
Propylene	1.6×10^{10}	38.0	-15	33
Isobutylene	2.0×10^{12}	43.0	-6	33
Ethylene + hydrogen	4×10^{13}	43.2	0	46

$\Delta S^{\ddagger}$ for standard state of 1 mole/cc.

the range over which the reaction rate is observed experimentally. $\Delta S^{\ddagger}$, the entropy of activation, is calculated as shown in Chapter 5, the standard state being 1 mole per cc, and the units calories per degree mole. The precision with which some of the data are given is misleading. Actually, activation energies are rarely measured to less than kilocalorie accuracy nor frequency factors to within a factor of 4 or 5.

The hydrogen-iodine reaction is the classic example of a bimolecular reaction.[1] The collision theory was successfully used to calculate the frequency factor for this reaction by W. C. McC. Lewis.[2] The transition-state theory has also been applied[3] with good results if the experimental activation energy is used rather than the semi-empirical value calculated from the potential-energy surface. The reaction is reversible, and the reverse reaction also agrees well with the theories.

The reactions listed under dimerizations, etc., such as the Diels-Alder condensations studied by Kistiakowsky and Lacher,[4] involve more complicated molecules. In agreement with the transition-state theory

Table 2

SECOND-ORDER REACTIONS INVOLVING ATOMS OR RADICALS

	E_a	p	E_a Calc., Hirschfelder Rule
$H + H_2 \rightarrow H_2 + H$	7.5	1	6
$Br + H_2 \rightarrow HBr + H$	17.6	10^{-1}	21
$H + HBr \rightarrow H_2 + Br$	1.2	10^{-1}	5
$H + Br_2 \rightarrow HBr + Br$	1.2	1	2.5
$Cl + H_2 \rightarrow HCl + H$	5.5	1	6
$Cl + C_2H_6 \rightarrow HCl + C_2H_5$	1.0	1	5
$CH_3 + (CH_3)_2CO \rightarrow CH_4 + \cdots$	9.2	2×10^{-3}	5
$CH_3 + (CH_3)_2Hg$	9.0	3×10^{-4}	5
$CH_3 + (CH_3)_2O$	9.0	2×10^{-4}	5
$CH_3 + C_2H_6$	10.4	5×10^{-4}	5
$CH_3 + $ neopentane	10.0	5×10^{-4}	5
$CH_3 + n$-butane	8.3	3×10^{-4}	5
$CH_3 + n$-pentane	8.1	2.5×10^{-4}	5
$CH_3 + $ isobutane	7.6	2×10^{-4}	5
$CH_3 + $ 2,3-dimethylbutane	6.9	2×10^{-4}	5
$CF_3 + H_2 \rightarrow CF_3H + H$	8.8	10^{-3}	6
$CF_3 + C_2H_6 \rightarrow CF_3H + C_2H_5$	7.5	10^{-3}	5
$CH_3 + C_2H_4 \rightarrow C_3H_7$	7.0	10^{-3}	3
$CH_3 + C_3H_6 \rightarrow C_4H_9$	6.0	5×10^{-4}	3

these reactions are characterized by low values of the frequency factor and negative entropies of activation. This is because of a conversion of rotational degrees of freedom into vibrations in the activated complex. On the basis of the collision theory these reactions require an assignment of a probability factor of 10^{-4} or 10^{-5}.

Table 2 is a summary of some bimolecular reactions involving atoms or radicals. These reactions are most often found as, or suspected of, being steps in complex reactions, for example, chain reactions. Occasionally such reactions may be observed more or less directly by using some technique such as photochemical or electric discharge for producing atoms or radicals at the time the reaction is to be observed. Information about the rates of numerous atom and radical reactions is available but often the data are of limited accuracy (for references and data see Steacie[5] and Trotman-Dickenson[6]). The chief reason for this limitation is that the concentrations of the free radicals are not observed directly but only estimated from other measurements. The usual technique is to estimate the mean lifetime of the free radical by comparison with some other time-dependent factor of the experiment. Such methods will be discussed further in Chapter 11. The use of electron spin resonance does allow the direct measure of concentration in certain cases.[7]

What are readily available in free radical reactions are ratios of rate constants for competing reactions and differences in activation energies. Thus the reactions involving methyl radical in Table 2 are based on competition experiments between the reactions

$$2CH_3 \cdot \xrightarrow{k_1} C_2H_6 \tag{1}$$

$$CH_3 \cdot + RH \xrightarrow{k_2} CH_4 + R \cdot \tag{2}$$

From the relative amounts of methane and ethane at various temperatures and concentrations of the hydrogen donor RH, it is possible to get values of $k_2/k_1^{1/2}$ and $E_2 - \frac{1}{2}E_1$ where E_2 and E_1 are the activation energies for reactions (2) and (1).[8] The values given in Table 2 for the activation energies and probability factors are based on a knowledge of the absolute values of k_1 and E_1. The ones shown are from an indirect determination by Gomer and Kistiakowsky[9] which leads to $k_1 = 4.5 \times 10^{13}$ cc/mole-sec at 125° and $E_1 = 0$. Thus the rate of recombination of methyl radicals seems to be given by the collision theory with a probability factor of unity and no energy barrier. The rate constant is pressure dependent, the high pressure value being given. At 0.3 mm pressure, the rate constant is only one-tenth that at 100 mm[10]. This is a necessary consequence of the fact that the reverse of reaction (1) is a unimolecular decomposition which would have a first-order rate constant that would also fall off with decreasing pressure and in the same pressure range as for the forward reaction. Such behavior is necessary for equilibrium to be maintained. Table 3 shows most of the available data on rates of simple radical recombinations in the gas phase.[6]

Bimolecular reaction activation energies can be correlated with bond-energy data by the use of the Hirschfelder rules.[11] For a simple displacement reaction involving atoms or radicals, such as $A + BC \rightarrow AB + C$ written in the exothermic direction, the activation energy is about 5.5 per cent of the energy of the bond being broken. For the reverse endothermic reaction, the activation energy would be the 5.5 per cent plus the ΔE of reaction. A rule for reactions of the type $AB + CD \rightarrow AC + BD$, also in the exothermic direction, is that the activation energy is about 28 per cent of the sum of the energies of the two bonds being broken. The value 28 per cent was determined empirically to give agreement with the $H_2 + I_2$ reaction. The 5.5 per cent results from semi-empirical calculations. The comparisons in the tables show the approximate validity of these rules. The rules are not sufficiently sensitive to show differences in a series of related reactions. Thus the same activation energy 33 kcal is predicted for all condensations of the Diels-Alder type in Table 1 (28 per cent of 120 kcal, assuming the second bond of a carbon-carbon double bond to have a strength of 60 kcal). The experimental values range from

Table 3

RATES OF RADICAL RECOMBINATIONS

	E_a	A, $M^{-1} sec^{-1}$
$CH_3 + CH_3 \rightarrow C_2H_6$	0	10.3
$C_2H_5 + C_2H_5 \rightarrow C_4H_{10}$	2	11.2
$CF_3 + CF_3 \rightarrow C_2F_6$	0	10.3
$NO_2 + NO_2 \rightarrow N_2O_4$	0	8.7
$CH_3 + NO \rightarrow CH_3NO$	0	8.3
$COCl + Cl \rightarrow COCl_2$	1	11.6

29 kcal for isoprene to 15 kcal for cyclopentadiene. Also, in most of the reactions of methyl radical shown in Table 2, very nearly the same activation energy would be predicted by the Hirschfelder rule since a carbon-hydrogen bond is broken in each case. The results show that a primary hydrogen atom is more difficult to remove than a secondary hydrogen, which in turn is more difficult to remove than a tertiary hydrogen. These differences in rate are due to differences in activation energies. Even correcting for the known differences in carbon-hydrogen bond energies, depending on whether hydrogen is primary, secondary, or tertiary, would not enable the predicted values to vary sufficiently.

Based on theoretical considerations first given by Evans and Polanyi, Semenov[12] has suggested that for exothermic abstraction and addition reactions of atoms and small radicals, the following approximate equation may be used:

$$E_a = 11.5 - 0.25q \tag{3}$$

Here q is the heat evolved in the reaction. For an endothermic reaction the equation becomes

$$E_a = 11.5 + 0.75q \tag{4}$$

where q is the heat absorbed in the reaction. However, the equation is of limited usefulness and does not seem to apply, for example, to the reactions of halogen atoms.

In Table 1 there is a noticeable tendency for the frequency factor to increase as the activation energy increases. Such a parallelism is a very common observation when a group of related reactions is studied.[13] Although different explanations can be involved in various cases, one factor is probably always operative in helping to bring about this parallelism. This factor has to do with the higher density of energy states as the total energy of a complex molecular system increases. Thus, given that the requisite energy exists in the molecule, or group of molecules, there is a greater number of ways of distributing the energy in the system if the energy is large than if it is small. Alternatively, we may say that a reaction can occur with a minimum of energy and a precisely defined configuration for the transition state, or with an excess of energy and a wide range of configurations for the nuclei of the reacting molecules in the transition state.

Termolecular Reactions

Third-order gaseous reactions are rare. There is a group of such reactions involving nitric oxide and also a group of atom recombination reactions. The recombinations are certainly termolecular, but the nitric oxide reactions are considered by some to be complex, a bimolecular reaction preceded by an equilibrium such as a dimerization equilibrium of NO to $(NO)_2$. Table 4 gives experimental data on these reactions.

Table 4

THIRD-ORDER GASEOUS REACTIONS

	A, cc^2 moles^{-2} sec^{-1}	E_a, kcal
$2NO + O_2 \rightarrow 2NO_2$	8×10^9	0 or negative
$2NO + Br_2 \rightarrow 2NOBr$	$\sim 10^{11}$	~ 4
$2NO + Cl_2 \rightarrow 2NOCl$	$\sim 10^9$	~ 4
$H + H + M \rightarrow H_2 + H$	2.0×10^{16}	0
$D + D + M \rightarrow D_2 + D$	1.5×10^{16}	0
$Br + Br + M \rightarrow Br_2 + M$	10^{16}	0
$I + I + He \rightarrow I_2 + He$	0.34×10^{16}	0
$I + I + Ar \rightarrow I_2 + Ar$	0.72×10^{16}	0
$I + I + H_2 \rightarrow I_2 + H_2$	0.95×10^{16}	0
$I + I + CO_2 \rightarrow I_2 + CO_2$	2.7×10^{16}	0
$I + I + C_6H_6 \rightarrow I_2 + C_6H_6$	17.5×10^{16}	0
$I + I + C_2H_5I \rightarrow I_2 + C_2H_5I$	50×10^{16}	0

The atom recombination rates are about as expected from kinetic theory, but the nitric oxide reactions have small frequency factors. This may correspond to a loss in entropy of activation due to rotations changing into vibrations or may be due to the complex mechanism. The more detailed data given for the iodine recombination reactions[14] show that not all molecules are equally effective in removing energy in a molecular collision. A great deal of experimental work indicates that heavier, and particularly more complex, molecules are more efficient in absorbing energy in a collision. It will be seen that the same molecules must also be more efficient in transferring energy to another molecule (activation) in collision. The activation energies are given as zero for the atom recombinations. Actually, the rates have a negative temperature coefficient corresponding to E_a equal to about -2 kcal. However, the transition state theory predicts a negative temperature coefficient of this magnitude simply from the exponent of the temperature in the frequency factor for the theoretical rate constant (see equation 40, Chapter 5).

Unimolecular Reactions

A number of first-order gaseous decompositions are shown in Table 5. Although some attempt has been made to weed out reactions which are obviously complicated, such as those that show signs of being partially heterogeneous and those that involve definite free redical chains (see Chapter 10), many of the remaining are undoubtedly complex. All of them involve a unimolecular decomposition or rearrangement as the probable first step. Sometimes this may be a dissociation into molecules, as in the thermal decomposition of acetic anhydride into ketene and acetic acid

$$(CH_3CO)_2O \rightarrow CH_3COOH + CH_2{=}C{=}O \tag{5}$$

Szwarc and Murawski[15] have shown that this reaction apparently does not involve free radicals, a cyclic intermediate being proposed

$$\tag{6}$$

Table 5

FIRST-ORDER GASEOUS DECOMPOSITIONS

$\Delta S^{\ddagger}$ calculated at 285°C

Compound	A, sec^{-1}	E_a, kcal	$\Delta S^{\ddagger}$, E.U.
Cyclobutane	4.0×10^{15}	62.5	2
Toluene	2.0×10^{13}	77.5	1
p-Xylene	5.0×10^{13}	76.2	4
Dimethyl mercury	3.1×10^{13}	51.5	3
Propylene oxide	1.2×10^{14}	58.0	6
Ethyl peroxide	5.1×10^{14}	31.5	8
n-Propyl peroxide	2.3×10^{15}	36.5	11
Diacetyl	8.7×10^{15}	63.2	14
Ethyl chlorocarbonate	9.2×10^{8}	29.1	−18
Methyl nitrite	1.8×10^{13}	36.4	2
Ethyl nitrite	1.4×10^{14}	37.7	6
n-Propyl nitrite	2.8×10^{14}	37.7	7
i-Propyl nitrite	1.3×10^{14}	37.0	6
Azomethane	3.5×10^{16}	52.5	17
Methylidene diacetate	1.7×10^{9}	33.0	−17
Methylidene dibutyrate	1.7×10^{9}	33.0	−17
i-Propyl iodide	1.6×10^{13}	42.9	1
t-Butyl bromide	2×10^{13}	40.5	2
Silicon tetramethyl	1.6×10^{15}	78.8	11
Trioxymethylene	1.5×10^{16}	47.4	15
Ethylidene dibutyrate	1.8×10^{10}	33.0	12
Heptylidene diacetate	3.0×10^{10}	33.0	−11
Trichloroethylidene diacetate	1.3×10^{10}	33.0	−13
Furfurylidene diacetate	1.3×10^{11}	33.0	−8
Glyoxal tetraacetate	1.8×10^{12}	39.2	−3
Nitrogen tetroxide	8.0×10^{14}	13.9	9
Trichloromethyl chloroformate	1.4×10^{13}	41.5	1
Methyl azide	3.0×10^{15}	43.5	12
Ethyl azide	2.0×10^{14}	39.7	7
Azopropane	5.7×10^{13}	40.9	4
Methyl iodide	3.9×10^{12}	43.0	−1
Nitromethane	4.1×10^{13}	50.6	2
Ethylidene diacetate	2×10^{10}	32.9	−12
Trimethyl acetic acid	4.8×10^{14}	65.5	8
Dimethylethyl acetic acid	3.3×10^{13}	60.0	3
Paracetaldehyde	1.3×10^{15}	44.2	10
Parabutyraldehyde	2.4×10^{14}	42.0	7
Dicyclopentadiene	1.0×10^{13}	33.7	0

Table 5 (*Continued*)

FIRST-ORDER GASEOUS DECOMPOSITIONS

$\Delta S^{\ddagger}$ calculated at 285° C

Compound	A, sec^{-1}	E_a, kcal	$\Delta S^{\ddagger}$, E.U.
Endomethylene 2,5-tetrahydro-benzaldehyde	2.2×10^{12}	33.6	-3
Acetic anhydride	1.0×10^{12}	34.5	-4
i-Propyl bromide	4.0×10^{13}	47.8	3
Allyl bromide	2.1×10^{12}	45.5	-3
Benzyl bromide	1.0×10^{13}	50.5	1
1-Butene	5.0×10^{12}	63.0	-1
Ethylidene dichloride	1.2×10^{12}	49.5	-4
Ethyl chloride	1.6×10^{14}	59.5	6
t-Butyl chloride	2.5×10^{12}	41.4	-2
Ethyl nitrate	6.3×10^{15}	39.9	14
FIRST-ORDER ISOMERIZATIONS			
trans-Dichloroethylene → *cis*	4.9×10^{12}	41.9	-1
cis-Stilbene → *trans*	6.0×10^{12}	42.8	-1
cis-Methylcinnamic ester → *trans*	3.5×10^{10}	41.6	-11
Methylmaleic acid → methyl-fumaric acid	6.8×10^{5}	26.5	-32
Dimethylmaleic ester → dimethylfumaric ester	1.3×10^{5}	26.5	-36
Vinyl allyl ether → Allylacetaldehyde	5.0×10^{11}	30.6	-5
Cyclopropane → Propylene	1.5×10^{15}	65.0	11

In other cases there is evidence that free radicals or atoms are produced in the primary process, in that substances are added to trap the free radicals (nitric oxide, iodine, toluene, etc.). Thus the pyrolysis of *n*-propyl bromide involves formation of bromine atoms and *n*-propyl radicals,[16] and the pyrolysis of nitromethane involves a split into a methyl radical and a molecule of nitrogen dioxide.[17] Prediction is difficult; for example, *iso*-propyl bromide decomposes by a direct process into hydrogen bromide and propylene (Maccoll and Thomas, *loc. cit.*).

In any event the frequency factors are most commonly in the neighborhood of 10^{13} sec^{-1} in agreement with the transition-state theory if the entropy of activation is zero. Several special cases may be noted. If a cyclic intermediate is formed from a noncyclic reactant, the entropy of activation will be negative and the frequency factor reduced, because internal rotations in the reactant become vibrations in the activated complex with a loss in entropy. This happens, apparently, in the decomposition of acetic anhydride noted above, in the decomposition of methylidene and ethylidene esters into acid anhydride and aldehyde[18] and in the

$$\begin{array}{c} CH_2\!\!=\!\!CH \\ \diagdown \\ CH_2\!\!=\!\!CH\!\!-\!\!CH_2 \end{array}\!\!\!O \rightarrow CH_2\!\!=\!\!CH\!\!-\!\!CH_2\!\!-\!\!CH_2\!\!-\!\!CHO \qquad (7)$$

rearrangement of vinyl allyl ether into allylacetaldehyde.[19] Conversely the decomposition of a ring compound should be accompanied by an increase in entropy and a high frequency factor.[20]

Several *cis-trans* isomerizations have very low values of the frequency factor. This has been attributed[21] to a low value of the transmission coefficient, the assumed mechanism involving a transition to an excited triplet state of the molecule. Such a transition will have a low probability because of the change in electron multiplicity.

If a molecule dissociates into two fragments containing several atoms, and if these two groups rotate freely in the activated complex, then a high frequency factor will be found. Evidence for such free rotation comes from the data of Table 3. The high rates of recombination of some of these radicals can only mean that the orientation of the two methyl groups, for example, is of no importance in collision. There must then be free rotation of the methyl groups in the transition state which is the same for both recombination and dissociation.[22] The reverse reactions of Table 3, either measured directly or calculated from the forward rate and the equilibrium constant, do have high frequency factors (10^{14}–10^{17} sec^{-1}).

An interesting attempt to sort out a simple unimolecular reaction from a complex one appears in the work of Hinshelwood and his co-workers on the pyrolysis of the normal paraffins.[23] The straight-chain hydrocarbons longer than ethane decompose on heating to give initially an olefin and a short-chain paraffin. These primary compounds may then react further, making the total reaction quite complicated. Initial rates may be determined, however, by getting the slope of the pressure-time curve at zero time. For example, *n*-pentane gives the following initial reactions:

$$CH_3\!\!-\!\!CH_2\!\!-\!\!CH_2\!\!-\!\!CH_2\!\!-\!\!CH_3 \rightarrow CH_4 + CH_2\!\!=\!\!CH\!\!-\!\!CH_2\!\!-\!\!CH_3 \qquad (8)$$

$$CH_3\!\!-\!\!CH_2\!\!-\!\!CH_2\!\!-\!\!CH_2\!\!-\!\!CH_3 \rightarrow CH_3\!\!-\!\!CH_3 + CH_2\!\!=\!\!CH\!\!-\!\!CH_3 \qquad (9)$$

$$CH_3\!\!-\!\!CH_2\!\!-\!\!CH_2\!\!-\!\!CH_2\!\!-\!\!CH_3 \rightarrow CH_3\!\!-\!\!CH_2\!\!-\!\!CH_3 + CH_2\!\!=\!\!CH_2 \qquad (10)$$

with the first reaction predominating over the second and the second over the third. In addition to further reactions of these products, it also happens that propylene and the higher olefins inhibit the decomposition of the original paraffin so that the complete pressure-time curve is sigmoid in nature.

If nitric oxide is added, the overall decomposition is also markedly inhibited. This is presumably because the decomposition involves a free-radical chain mechanism which is interrupted by the nitric oxide.[24]

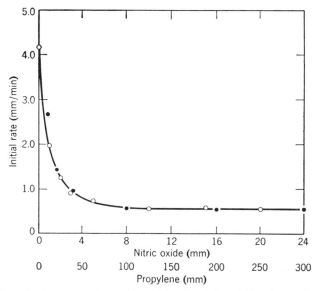

Fig. 1. Quantitative comparison of inhibition by nitric oxide and propylene (100 mm *n*-pentane, 530°C). ● inhibition by propylene; ○ inhibition by nitric oxide (Stubbs and Hinshelwood).

Nitric oxide combines with the alkyl free radicals which are the chain carriers. If sufficient nitric oxide is added, the initial rate of decomposition is reduced to a constant value (for a given pressure of paraffin) and further addition of nitric oxide has no effect. Also propylene will reduce the initial rate and, though less effective, will eventually reduce the rate to the same constant value as nitric oxide.[25] Figure 1 shows the effect of nitric oxide and propylene on the initial rate.

The simplest explanation of this result is that the chain reaction which normally occurs is completely suppressed by large amounts of inhibitor and that a nonchain reaction unaffected by the inhibitors, amounting to $\frac{1}{5}$ to $\frac{1}{3}$ of the total, remains. This residual reaction is believed to be a simple unimolecular decomposition involving a hydrogen atom transfer.

$$R_1 - CH_2 - CH_2 - CH_2 - R_2 \rightarrow R_1 - CH_3 + CH_2 {=} CHR_2 \qquad (11)$$

There are some difficulties with this view however. One is that the products of the reaction, the absolute rates, and the activation energies are very similar for both the uninhibited and completely inhibited reactions. Another is that nitric oxide accelerates the rate of decomposition of acetaldehyde, which is also believed to go by a simultaneous molecular and chain process (see Chapter 10). Several studies in which deuterium isotope labeling is used give ample evidence that the NO inhibited reaction is not a simple unimolecular transformation. Thus pyrolysis of a mixture of C_2H_6 and C_2D_6, and of mixtures of C_3H_8 and D_2, give products containing both light and heavy hydrogen in the same molecule.[26]

The inhibition by NO and propylene presumably involves the following reactions:

$$R\cdot + NO \to R - NO \tag{12}$$

$$R\cdot + C_3H_6 \to RH + C_3H_5\cdot \tag{13}$$

The allyl radical, $C_3H_5\cdot$, is resonance stabilized and hence unreactive. Now, it is known that the relative rates of reactions 12 and 13 are of the order of 400 to 1, with NO the more reactive.[27] But for inhibition purposes, NO is only twelve times as effective as propylene.

An alternate explanation of the results is that the pyrolysis reaction is always of chain type involving free radicals, whether inhibited or not.[28] The initiation and termination steps take place chiefly on the walls of the reaction vessel but the propagation steps occur homogeneously in the gas phase. This is a special example of the Rice-Herzfeld mechanism (see p. 252, Chapter 10).

Both the inhibited and uninhibited reactions have been shown to be sensitive to the nature of the container surface and to surface/volume ratios. A new reaction vessel often gives erratic results until a layer of carbonaceous material is formed. It is assumed that both reversible and irreversible formation of free atoms and radicals occurs at the surface initially, leading to a high rate.

As the concentration of unsaturated products builds up, or if inhibitors are added, the active sites on the surface responsible for irreversible formation of free radicals are blocked out. Eventually only reversible processes occur and steady state concentrations of the chain carrying free radicals exist in the gas phase. This will produce a lower rate of reaction than the initial, uninhibited rate. Further, the rate should now be independent of the surface and of further changes in the nature or concentration of inhibitors.

Because of the dual nature of surfaces in both creating and destroying free radicals it is not easy to prove that a gas phase reaction is partly heterogeneous. The usual techniques of adding glass rods or glass wool to increase the surface area (and hence the rate of a surface reaction) may be ineffective. Free radicals or atoms are produced on surfaces by the adsorption of stable molecules which then dissociate on the surface. The radicals may react on the surface or escape into the gas phase.

Conversely, free radicals in the gas phase will reach the surface by diffusion, become adsorbed, and recombine to form stable molecules. The rate at which they reach the surface depends on the distance over which they must diffuse and the pressure of gas through which they diffuse. Thus in a long cylindrical vessel of narrow diameter, more surface recombination will occur than in a vessel of the same volume but shorter and of greater diameter.[29] This is because the average distance that a radical must travel to reach the wall is less in the first case. At equilibrium the rate of destruction and creation of the radicals by the surface will be equal. There will be an equilibrium surface concentration of adsorbed radicals and an equilibrium volume concentration of radicals in the gas phase. Changes in the amount or nature of the surface cannot change this latter concentration.

Polar Reactions in Gas Phase

A novel interpretation of a series of gas-phase reactions is given by Maccoll for the first-order decomposition of organic halides.[30] An elimination of hydrogen halide occurs and an olefin is formed.

$$
\begin{array}{c}
\text{H} \quad \text{R} \\
| \quad | \\
\text{H—C—C—R} \rightarrow \\
| \quad | \\
\text{H} \quad \text{X}
\end{array}
\qquad
\begin{array}{c}
\text{H} \qquad\qquad \text{R} \\
\diagdown \qquad \diagup \\
\text{C}=\text{C} \\
\diagup \qquad \diagdown \\
\text{H} \qquad\qquad \text{R}
\end{array}
\;+\; \text{H—X}
\qquad (14)
$$

Many of these reactions for substituted halides appear to be simple unimolecular processes not involving free radicals or chain reactions. Table 6 gives some rate data for several bromides and the corresponding acetates in which elimination of acetic acid occurs.

Maccoll calls attention to the large increases in rate caused by α-methyl substitution in the bromides. These increases are very similar to the effect on rates of solvolysis and elimination reactions of those halides in solutions. For example, the activation energies for solvolysis of ethyl, iso-propyl and tertiary-butyl bromides in 80 per cent ethanol-water are 30, 26.7 and 23.3 kcal. For the gas-phase elimination reaction the corresponding energies are 53.9, 47.8 and 42.2 kcal. A number of other correspondences exist in that substituents such as phenyl, vinyl and oxygen increase both the rates of elimination in the gas and solvolysis in solution. The effect of β-methyl substitution is small compared to α-methyl.

The solution reactions are commonly believed to involve ionic intermediates. For the limiting case in which a carbonium ion is formed, it is estimated that an α-methyl substituent on a primary carbon will increase the rate of solvolysis by a factor of 10^4 at room temperature.[31] For the gas-phase decomposition, an extrapolation of the data in Table 6 to 25° C

Table 6

RELATIVE RATES OF ELIMINATION OF HX FROM CH_3CR_2X

(Data from Maccoll, reference 30)

	Bromides, 320° C	Acetates, 400° C
C_2H_5X	1	1
$i\text{-}C_3H_7X$	280	25
$t\text{-}C_4H_9X$	78,000	1,660

by means of the known activation energies will also give a factor of 10^4 for each methyl group. Because of these striking resemblances, Maccoll proposes that an ion-pair structure for the transition state in the gas is most probable.[30]

$$\begin{array}{ccc} \diagup\!\!\!\diagdown & & \\ \text{C---C} & \text{or} & \text{C}=\!\!=\text{C} \\ \diagup \text{H} \quad \text{Br}^- & & \text{H} \quad \text{Br}^- \end{array} \tag{15}$$

As will be discussed in the next chapter, free ions are normally not formed in the gaseous state because of the excessive energy required. However, in this case the ions are not separated and the energy required is much less.

The pyrolysis of acetates. is much less similar to that of the solvolysis reactions. Hence the intermediate is considered to be much less polar. The cyclic mechanism and transition state proposed by Hurd and Blunck[32] seems entirely reasonable, particularly since it explains the stereochemical fact of *cis* elimination in suitably substituted acetates.

$$\begin{array}{ccc} \text{C}\text{-----}\text{C} & & \text{C}\text{------}\text{C} \\ \text{H} \quad \text{O} & \rightarrow & \text{H} \quad \text{O} \\ \text{O}=\!\!=\text{C} & & \text{O---C} \\ \text{R} & & \text{R} \end{array} \tag{16}$$

Effect of Pressure on Unimolecular Reactions

It was shown in Chapter 4 that an expected feature of a true unimolecular reaction would be a decrease in the first-order rate constant with decreasing pressure and an eventual change to second-order kinetics at very low pressure. Such behavior would be noticed at higher and higher pressures as the complexity of the molecule decreases.

The isomerization of cyclopropane to propylene appears to be a

$$\rightarrow \qquad (17)$$

straight-forward unimolecular process in that the usual tests for free radicals do not give any evidence for their existence, nor does the reaction appear heterogeneous in any way. Furthermore, the molecule has the right degree of complexity to give the expected pressure effects in a convenient range of temperatures.

The falling of the rate does occur as predicted and Fig. 2 shows the results of Pritchard, Sowden, and Trotman-Dickenson[33] for the variation of the ratio of the first-order constant at any pressure to the limiting high-pressure constant, k/k_{∞}, as a function of pressure. The rate constant falls by a factor of 10 in going from 100 cm to 0.0067 cm of mercury pressure. The data are presented as a log-log plot for convenience in comparing with the several theories of unimolecular reactions.

The simple Lindeman-Hinshelwood theory does not fit the experimental results at all well. The theories of Kassel, Rice and Ramsperger will fit with an arbitrary choice of the parameter s in equation 56, Chapter 4. The best fit is given by Slater's theory in which no arbitrary parameter occurs.[34] The figure shows the results of Kassel's theory obtained by taking the number of effective oscillators to be 13. In Slater's theory the required number of non-degenerate vibrational modes is known to be 13. The assumption is made by Slater that reaction occurs when a hydrogen atom on one carbon atom approaches to a critical distance of another carbon atom.

The high pressure-rate constant[35] is given by $k = 10^{15.2} e^{-65,000/RT}$ sec^{-1}. This frequency factor is somewhat higher than that predicted by Slater's theory, bearing in mind that, because of the high symmetry of the molecule, there are twelve ways in which the reaction can occur. That is, each of the six equivalent hydrogen atoms can move towards either neighboring carbon atom.

There is a possible complication in that *trans*-cyclopropane-d_2 on partial isomerization (8 %) has been found to give extensive rearrangement

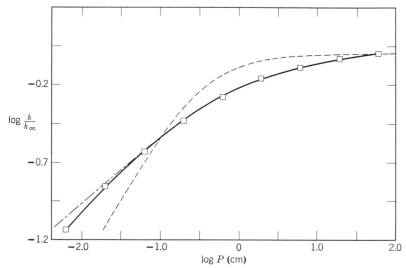

Fig. 2. The dependence of the rate of isomerization of cyclopropane on pressure. The solid line represents the experimental results; - - - -, calculated Lindeman-Hinshelwood curve; —·—·—, calculated Kassel curve; □, points calculated by Slater[34] (after Pritchard, Sowden, and Trotman-Dickenson).

(25 %) to *cis*-cyclopropane-d_2[36].

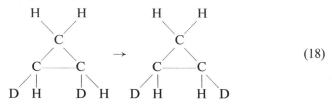

$$\qquad\qquad\qquad\qquad\qquad\qquad\qquad\qquad (18)$$

However, no *asym*-cyclopropane-d_2 was formed in which both deuterium atoms were on the same carbon atom. These results indicate rupture of the cyclopropane ring, internal rotation and then reclosing of the trimethylene radical. The ring opening might be a step which occurs prior to the hydrogen atom shift. There is some evidence that the ring opening is the rate-determining step for the isomerization in that the activation energy of 65 kcal agrees reasonably well with the estimated carbon-carbon bond energy in cyclopropane, 85 kcal for a normal carbon-carbon bond energy minus about 25 kcal strain energy.[37] Slater's theory would not work at all well if ring opening were the rate step. On the other hand, the ring opening and closing may simply be a side reaction to the true isomerization and not a necessary precursor.

An interesting set of experiments can be done in which cyclopropane is formed by the addition of methylene, CH_2, to ethylene.[38]

$$CH_2 + C_2H_4 \rightarrow C_3H_6 \qquad (19)$$

It can be shown by thermochemical calculations that the cyclopropane must be formed initially in a highly energetic state. This "hot" molecule can then either be deactivated by collision or can isomerize to propylene. From the dependence of the yield of the latter on pressure it is possible to get the rate of isomerization of the activated molecules. Furthermore, methylene, or carbene, can be formed itself in various excited states so that the energy content of the hot cyclopropane can be varied.

Table 7

RELATIVE EFFICIENCIES OF ADDED GASES IN THE
ISOMERIZATION OF CYCLOPROPANE
(Data from reference 33)

Cyclopropane	1.00	Water	0.79
Helium	0.06	Toluene	1.59
Argon	0.05	Propylene	1.0
Hydrogen	0.24	Nitrogen	0.06

In this way it has been shown that the rate of reaction (isomerization) of an activated molecule is indeed a function of its excitation energy. The more energetic molecules react faster, as the theories of unimolecular reactions usually postulate. Similar experiments can be performed with methylcyclopropane formed by adding methylene either to propylene or to cyclopropane.[39] The lifetime of "hot" methylcyclopropane molecules can be changed by a factor of 3600 in going from thermally activated molecules to the most energetic photochemically produced methylene adduct.

Another result of the reactions of hot methylcyclopropane is that the relative yields of the various butene isomerization products are very similar, whether formed from methylene plus propylene or methylene plus cyclopropane. This leads to the conclusion[39] that excitation energy in a molecule can flow freely between the various vibrational modes of the molecule. Such a facile energy transfer is not in agreement with Slater's theory of unimolecular reactions, in which the vibrations are treated as uncoupled. It is in agreement with the Rice-Kassel theory.

Adding foreign gases to cyclopropane reacting at low pressure will bring the rate constant back up to its high pressure value. However at equal pressures not all gases are equally effective. Table 7 shows the relative efficiencies, compared to cyclopropane as unity, of various gases in maintaining the rate of isomerization.

These numbers should be related to the efficiencies of the same gases in causing the recombination of atoms and small radicals. A comparison with Table 4 shows some correspondence. However, specific effects are noticeable for some gases in the two different processes.

The unimolecular decompositions of cyclobutane, nitrogen pentoxide (Chapter 12), trioxymethylene, and nitrous oxide also give pressure dependencies in reasonable agreement with the more detailed theories.[40] So also does the *cis-trans* isomerization of *cis*-butene-2.[41]

PROBLEMS

1. Hydrogen gas at room temperature and above is an equilibrium mixture of 25 per cent *para* and 75 per cent *ortho* hydrogen. If almost pure *para* hydrogen is prepared at very low temperatures, the conversion to the equilibrium mixture is a slow process in the absence of a catalyst. The rate of conversion follows a first-order law so that in any one run

$$\ln \frac{(\text{initial } \% \text{ } para - 25)}{(\% \text{ } para - 25)} = kt$$

The first-order k varies with total pressure of hydrogen in the following way at $923°$ K: [A. Farkas, *Z. physik. Chem.*, *B10*, 419 (1930).] Calculate the order

P, mm	50	100	200	400
$k \times 10^3$, sec^{-1}	1.06	1.53	2.17	3.10

with respect to total hydrogen. What mechanism does this suggest? Using thermodynamic data and the above rate data, calculate the rate constant at $923°$ K for the reaction

$$H + p\text{-}H_2 \rightarrow o\text{-}H_2 + H$$

2. The reaction of nitric oxide and hydrogen

$$2NO + 2H_2 \rightarrow N_2 + 2H_2O$$

$P_{H_2}{}^0 = P_{NO}{}^0$	Total Pressure, mm	$t_{1/2}$, sec
	354	81
	288	140
	243	176
	202	224
$P_{H_2}{}^0 = $ constant	$P_{NO}{}^0$, mm	Initial Rate, mm/sec
	359	1.50
	300	1.03
	152	0.25

has been studied by C. N. Hinshelwood and T. E. Green [*J. Chem. Soc.*, *129*, 730 (1926)] who determined the half-life of the reaction and the initial rates as a function of the pressure of the reactants. From the data given calculate the total order of the reaction and the order with respect to nitric oxide.

3. Calculate the rate constant at 500° C for the recombination of methyl radicals, assuming zero activation energy, probability factor of unity, and no third-body restriction.

REFERENCES

1. M. Bodenstein, *Z. physik. Chem.*, *29*, 295 (1899).
2. W. C. McC. Lewis, *J. Chem. Soc.*, *113*, 471 (1918).
3. A. Wheeler, B. Topley, and H. Eyring, *J. Chem. Phys.*, *4*, 178 (1936).
4. G. B. Kistiakowsky and J. R. Lacher, *J. Am. Chem. Soc.*, *58*, 123 (1936).
5. E. W. R. Steacie, *Atomic and Free Radical Reactions*, Reinhold Publishing Corp., New York, 1946.
6. A. F. Trotman-Dickenson, *Gas Kinetics*, Academic Press, New York, 1955.
7. D. J. Ingram, *Free Radicals*, Academic Press, New York, 1958.
8. (a) A. F. Trotman-Dickenson and E. W. R. Steacie, *J. Am. Chem. Soc.*, *72*, 2310 (1950); (b) L. M. Dorfman and R. Gomer, *Chem. Revs.*, *46*, 499 (1950).
9. R. Gomer and G. B. Kistiakowsky, *J. Chem. Phys.*, *19*, 85 (1951).
10. G. B. Kistiakowsky and E. K. Roberts, *J. Chem. Phys.*, *21*, 1637 (1953); R. E. Dodd and E. W. R. Steacie, *Proc. Roy. Soc.*, *223A*, 283 (1954).
11. J. Hirschfelder, *J. Chem. Phys.*, *9*, 645 (1941).
12. N. Semenov, *Some Problems of Chemical Kinetics and Reactivity*, Pergamon Press, New York, 1958, Vol. I, pp. 27 and 58.
13. R. A. Fairclough and C. N. Hinshelwood, *J. Chem. Soc.*, 538, 1573 (1937); J. E. Leffler, *J. Org. Chem.*, *20*, 1202 (1955); C. N. Hinshelwood and D. A. Blackadder, *J. Chem. Soc.*, 2728 (1958).
14. K. E. Russell and J. Simons, *Proc. Roy. Soc.*, *217A*, 271 (1953).
15. M. Szwarc and J. Murawski, *Trans. Faraday Soc.*, *47*, 269 (1951).
16. A. Maccoll and P. T. Thomas, *J. Chem. Phys.*, *19*, 977 (1951).
17. T. L. Cottrell and T. J. Reid, *J. Chem. Phys.*, *18*, 1306 (1950).
18. C. C. Coffin, *Can. J. Res.*, *6*, 417 (1932).
19. F. W. Schuler and G. W. Murphy, *J. Am. Chem. Soc.*, *72*, 3155 (1950).
20. O. K. Rice and H. Gerschinowitz, *J. Chem. Phys.*, *3*, 479 (1935); B. G. Gowenlock, *Quart. Rev.*, *14*, 133 (1960).
21. J. L. Magee, W. Shand, and H. Eyring, *J. Am. Chem. Soc.*, *63*, 677 (1941); see, however, D. Schulte-Frohlinde, *Ann.*, *612*, 138 (1958).
22. R. A. Marcus, *J. Chem. Phys.*, *20*, 359 (1952).
23. F. J. Stubbs and C. N. Hinshelwood, *Disc. Faraday Soc.*, *10*, 129 (1951).
24. J. E. Hobbs and C. N. Hinshelwood, *Proc. Roy. Soc.*, *A167*, 439 (1938).
25. F. J. Stubbs and C. N. Hinshelwood, *Proc. Roy. Soc.*, *200A*, 458 (1950); A. I. Dintses and A. V. Frost, *Zh. Obshch. Khim.*, *3*, 747 (1933).
26. L. A. Wall and W. J. Moore, *J. Am. Chem. Soc.*, *73*, 2840 (1951); F. O. Rice and R. E. Varnerin, *ibid.*, *76*, 324 (1954); V. A. Poltorak and V. V. Voevodskii, *Dokl. Akad. Nauk SSSR*, *91*, 589 (1953).
27. Reference 6, p. 156.

28. (*a*) V. A. Poltorak and V. V. Voevodskii, quoted in reference 12, pp. 248–249; (*b*) F. O. Rice and K. F. Herzfeld, *J. Phys. Colloid Chem.*, *55*, 975 (1951); (*c*) V. V. Voevodskii, *Trans. Faraday Soc.*, *55*, 65 (1959).

29. A. Trifonov, *Z. physikal. Chem.*, *B3*, 195 (1929).

30. A. Maccoll, reviewed in *Theoretical Organic Chemistry*, the Kekulé Symposium, Butterworths Scientific Publications, London, 1959, p. 230 ff.

31. S. Winstein, E. Grunwald, and H. W. Jones, *J. Am. Chem. Soc.*, *73*, 2700 (1951).

32. C. D. Hurd and F. H. Blunck, *J. Am. Chem. Soc.*, *60*, 2421 (1938).

33. H. O. Pritchard, R. G. Sowden, and A. F. Trotman-Dickenson, *Proc. Roy. Soc.*, *217A*, 563 (1953).

34. For a comparison see E. W. Schlag, B. S. Rabinowitch, and F. W. Schneider, *J. Chem. Phys.*, *32*, 1599 (1960).

35. T. S. Chambers and G. B. Kistiakowsky, *J. Am. Chem. Soc.*, *56*, 399 (1934).

36. B. S. Rabinowitch, E. W. Schlag, and K. B. Wiberg, *J. Chem. Phys.*, *28*, 504 (1958).

37. F. H. Seubold, *ibid.*, *21*, 1616 (1953).

38. H. M. Frey and G. B. Kistiakowsky, *J. Am. Chem. Soc.*, *79*, 6373 (1957); the reactions of methylene are reviewed by A. F. Trotman-Dickenson, *Ann. Repts. of Chem. Soc.*, *55*, pp. 47–55; G. Herzberg and J. Shoosmith, *Nature*, *183*, 1801, (1959); see also P. S. Skell and J. Klebe, *J. Am. Chem. Soc.*, *82*, 247 (1960) and earlier papers.

39. J. N. Butler and G. B. Kistiakowsky, *J. Am. Chem. Soc.*, *82*, 759 (1960).

40. R. E. Powell, *J. Chem. Phys.*, *30*, 724 (1959); E. K. Gill and K. J. Laidler, *Proc. Roy. Soc.*, *250A*, 121 (1959).

41. B. S. Rabinowitch and K. W. Michel, *J. Am. Chem. Soc.*, *81*, 5065 (1959).

7.

REACTIONS IN SOLUTION

Although reactions in the gas phase are simpler to deal with theoretically, the fact remains that most reactions encountered in practice occur in solutions of one kind or another. The question then arises whether there is any fundamental difference in the kinetics of reactions occurring in condensed media compared to the gaseous state. The answer is essentially as follows: when a reaction follows the same mechanism in solution and in the gaseous state, the kinetics are not changed appreciably. However, because of the increased interactions in condensed media, the mechanism is frequently changed completely and the kinetics correspondingly altered. In fact, there is a wide variety of reactions which do not occur in the gas phase at all but which go more or less readily in various solvents. The favored mechanisms in solution are ionic ones, involving formation and interaction of charged particles. Such mechanisms are virtually impossible in the vapor state (excluding wall and surface reactions); hence the failure of many reactions to proceed. Polar solvents in general are the best media for ionic reactions.

It will be instructive to consider first some processes which can occur both in the gas phase and in solution to see what differences in rates and equilibria would be expected. For simplicity we shall assume that the solution is ideal, that is, that heat and volume changes are zero when the pure liquid components are mixed and that Raoult's law is obeyed by all components.

Suppose we have the chemical equilibrium

$$A + B \cdots \rightleftharpoons C + D \cdots \tag{1}$$

established in a system consisting of an ideal solution, where the solvent need not be specified and the corresponding vapors are in equilibrium with the various components of the solution. Then the concentrations of A, B, C, and D corresponding to the equilibrium of (1) will be established in both phases. The equilibrium constant is not the same in the two phases, even though the solution is ideal and the vapors may also be treated as behaving ideally. We have for the free energy per mole of substance A

$$F_A = F_A{}^0 + RT \ln P_A \text{ in the vapor} \tag{2}$$

$$F_{A'} = F_A{}^0 + RT \ln N_A \text{ in solution} \tag{2a}$$

Since $P_A = P_A{}^0 N_A$ from Raoult's law ($P_A{}^0$ is the vapor pressure of pure liquid A at the temperature of the system) and since $F_A = F_{A'}$ (equilibrium between the phases), then $F_{A'}{}^0 = F_A{}^0 + RT \ln P_A{}^0$. Since K_p in the gas phase is equal to $e^{-\Delta F^0/RT}$ and K_N in solution is $e^{-\Delta F^{0'}/RT}$ we have

$$K_N \text{ (solution)}/K_p \text{ (gas)} = \frac{P_A{}^0 P_B{}^0 \cdots}{P_C{}^0 P_D{}^0 \cdots} \tag{3}$$

where the equilibrium constants are in terms of mole fractions and partial pressures, respectively. Both of these can be changed to the same units, concentrations in moles per liter being convenient.

$$K_c \text{ (gas)} = K_p(1/RT)^{\Delta n} \quad \text{since } C_A \text{ (gas)} = P_A/RT \tag{4}$$

$$K_c \text{ (solution)} = K_N(1/V_0)^{\Delta n} \quad \text{since } C_A \text{ (solution)} = N_A/V_0 \tag{5}$$

where Δn is the increase in the number of moles for reaction 1 and V_0 is the volume per mole of solution, approximately equal to the molar volume of the pure solvent for dilute solutions. Equation 3 now becomes

$$K_c \text{ (solution)}/K_c \text{ (gas)} = \frac{P_A{}^0 P_B{}^0 \cdots}{P_C{}^0 P_D{}^0 \cdots} \cdot (RT/V_0)^{\Delta n} \tag{6}$$

The immediate deduction from (6) is that reactions in solution are favored where the reactants are volatile and the products are less volatile. In the gas phase the reverse is true; volatile products from less volatile reactants are favored. These deductions are in substantial agreement with experience, though other factors combine to make this so. The difference in the two phases arises chiefly from the difference in the heats of vaporization of the various components, as we can see by writing the vapor pressure as

$$P_A{}^0 = e^{\Delta S_A/R} e^{-\Delta H_A/RT} \text{atm} \tag{7}$$

where ΔS_A and ΔH_A are the entropy and heat of vaporization of substance A, respectively, the standard state in the vapor being 1 atmosphere pressure. The term $e^{\Delta S_A/R}$ can be written also as RT/V_f where V_f is the "free volume" per mole of A, presumably the molar volume less the volume occupied by the molecules themselves.[1] Empirically, V_f turns out to be about 0.5 cc and not different by more than a factor of 2 or 3 for different liquids. Treating it as constant the ratio of equilibrium constants can now be written

$$K_c \text{ (solution)}/K_c \text{ (gas)} = \frac{e^{-\Delta H_A/RT} e^{-\Delta H_B/RT} \cdots}{e^{-\Delta H_C/RT} e^{-\Delta H_D/RT} \cdots} \left(\frac{V_f}{V_0}\right)^{\Delta n} \qquad (8)$$

If Δn is zero, the equilibrium constants differ only because of the heats of vaporization. It can be seen that the heat of reaction in solution is related to the heat of reaction in the gas by

$$\Delta H \text{ (solution)} = \Delta H \text{ (gas)} - \Sigma \Delta H \text{ (vaporization)} \qquad (9)$$

where the summation as usual is over the products less the reactants. Since (V_f/V_0) is about $\frac{1}{100}$, there is an appreciable change in the equilibrium constant when Δn is not equal to zero. A decrease in the number of moles is favored in solution, and an increase in the number of moles is favored in the vapor. This again is in agreement with experience, though other factors enter in. Since reactions in the gas phase are usually run at higher temperatures than reactions in solution, there is an entropy factor working which favors an increase in the number of moles. This factor would also operate for reactions in solution if they were run at the same high temperatures.

For nonideal solution (6) becomes

$$K \text{ (solution)}/K \text{ (gas)} = \frac{\pi_A^0 \pi_B^0 \cdots}{\pi_C^0 \pi_D^0 \cdots} \left(\frac{RT}{V_0}\right)^{\Delta n} \frac{\gamma_A \gamma_B \cdots}{\gamma_C \gamma_D \cdots} \qquad (10)$$

where $\pi_A^0 = P_A/N_A$ is the appropriate Henry's law constant for substance A, Henry's law being obeyed as the solution becomes sufficiently dilute. The activity coefficient γ_A is a measure of the deviation from Henry's law and approaches unity at infinite dilution. Equation 10 may also be written as

$$K \text{ (solution)}/K \text{ (gas)} = \frac{e^{-\Delta H_A'/RT} e^{-\Delta H_B'/RT} e^{\Delta S_A'/R} e^{\Delta S_B'/R} \cdots}{e^{-\Delta H_C'/RT} e^{-\Delta H_D'/RT} e^{\Delta S_C'/R} e^{\Delta S_D'/R} \cdots} \left(\frac{RT}{V_0}\right)^{\Delta n} \qquad (11)$$

where $\Delta H_A'$ and $\Delta S_A'$ are the heat and entropy of vaporization of A not from pure liquid A but from the solution itself.

The selection of suitable reactions to illustrate (6) or (10) is difficult because few equilibria have been studied both in the gas phase and in solution. The dimerization of acetic acid has been measured fairly accurately in benzene solution and quite accurately in the vapor, both studies being carried out over a range of temperatures.

$$2CH_3COOH \rightleftharpoons (CH_3COOH)_2 \qquad (12)$$

The value of K_p in the vapor at $80°$ is 0.0355 mm^{-1}, ΔH^0 is -15.3 kcal and ΔS^0 is -50.0 E.U.[2] Converting these into a standard state of 1 mole per liter, $K_c = 0.0355 \times 760 \times 0.082 \times 353 = 780$ liters/mole and $\Delta S^0 = -50.0 + R \ln (760 \times 0.082 \times 353) = -30.2$ E.U. In benzene solution the corresponding values are $K_c = 139$ liters/mole, $\Delta H^0 = -9.7$ kcal, and $\Delta S^0 = -22.3$ E.U.[3]

For this reaction Δn is -1 and equation 8 can be written

$$K_s/K_g = \frac{e^{-2\Delta H_{\text{mono}}/RT}}{e^{-\Delta H_{\text{dim}}/RT}} \left(\frac{V_0}{V_f}\right) \tag{13}$$

where ΔH_{mono} is the heat of vaporization of the monomer and ΔH_{dim} the heat of vaporization of the dimer. From (9) we have

$$-9.7 = -15.3 - \Delta H_{\text{dim}} + 2\Delta H_{\text{mono}} \tag{14}$$

so that it can be calculated that the heat of vaporization of the monomer is 2.8 kcal greater than half the heat of vaporization of the dimer. This is a reasonable result, since the heat of vaporization of the dimer would be expected to be less than twice that of the monomer because of the inter-molecular cancellation of strong hydrogen bonds.

The ratio (V_0/V_f) can also be recognized as equal to $e^{\Delta S_s/R}/e^{\Delta S_g/R}$. Experimentally this is $e^{8/R} \simeq 50$, whereas the expected ratio of (V_0/V_f) is about 100. The agreement is better than expected, since a solution of acetic acid in benzene is not ideal.

The equilibria between the various halogens has been studied in the vapor phase and in carbon tetrachloride solution by Blair and Yost. Table 1 shows their data. Since there is no change in the number of

Table I

EQUILIBRIA BETWEEN THE HALOGENS IN THE GASEOUS STATE AND IN CARBON TETRACHLORIDE SOLUTION

[C. M. Blair and D. M. Yost, *J. Am. Chem. Soc.*, 55, 4489 (1933)]

Reaction	Medium	ΔH^0, cal	ΔF^0, cal	ΔS^0, E.U.
$I_2 + Cl_2 \rightleftharpoons 2ICl$	Gas	$-6,560$	$-7,370$	2.7
	CCl_4	$-7,940$	$-8,040$	0.4
$I_2 + Br_2 \rightleftharpoons 2IBr$	Gas	$-2,740$	$-3,440$	3.1
	CCl_4	$-3,260$	$-3,360$	1.1
$Br_2 + Cl_2 \rightleftharpoons 2BrCl$	Gas	-600	$-1,260$	2.3
	CCl_4	-760	-710	-0.1

moles and because reactants and products are similar, the agreement between the constants in the two phases is expected. Probably to a good approximation $2\Delta H_{ICl} = \Delta H_{I_2} + \Delta H_{Cl_2}$, that is, the heat of vaporization of iodine monochloride is the arithmetical mean of the heats of vaporization of iodine and chlorine. The same would be true for the other interhalogen compounds.

Kinetics in Solution

Let us turn next to a consideration of the rates of chemical reactions in solution compared to the gas phase. Two different approaches may be made, based on the collision theory of reaction rates and on the transition-state theory. That based on the transition-state theory is simpler to discuss, since it follows directly from the previous discussion of equilibrium constants.

If we have the reaction

$$A + B \cdots \rightleftharpoons M^{\ddagger} \rightarrow \cdots \tag{15}$$

where A and B are reactants and $M^{\ddagger}$ is an activated complex, then the specific rate constant for this reaction in the gas phase is $k_g = (RT/Nh)K_g{}^{\ddagger}$, where $K_g{}^{\ddagger}$ is the "equilibrium constant" between the complex and the reactants, assuming the vapors to be ideal. In solution the analogous equation must be corrected to take into account deviations from ideal behavior. The thermodynamic "equilibrium constant" should be defined as a ratio of activities.

$$K_s{}^{\ddagger} = \frac{a_{\ddagger}}{a_A a_B \cdots} = \frac{C^{\ddagger}}{C_A C_B \cdots} \frac{\gamma_{\ddagger}}{\gamma_A \gamma_B \cdots} \tag{16}$$

Consequently if the *rate* of a reaction is proportional to the *concentration* of the activated complex, the rate constant is dependent on the ratio of activity coefficients.

$$\text{rate} = (RT/Nh)C^{\ddagger} = (RT/Nh)K_s{}^{\ddagger}C_A C_B \cdots \frac{\gamma_A \gamma_B \cdots}{\gamma_{\ddagger}} \tag{17}$$

$$k_s = (RT/Nh)K_s{}^{\ddagger} \frac{\gamma_A \gamma_B \cdots}{\gamma_{\ddagger}} \tag{18}$$

The activity coefficients can be referred to any convenient standard state, the usual one being that of infinite dilution for the solutes. The rate constant in the standard state will then be equal to $(RT/Nh)K_s{}^{\ddagger}$. A comparison of the rates in the gas phase and in solution based on equations 16 and 18 leads to a repetition of equation 10.

For a bimolecular reaction in an ideal solution, the ratio of rate constants becomes

$$k_s/k_g = \frac{P_A^0 P_B^0}{P_{\ddagger}^0} \frac{V_0}{RT} = \frac{V_0 e^{-\Delta H_A/RT} e^{-\Delta H_B/RT}}{V_f e^{-\Delta H_{\ddagger}/RT}} \tag{19}$$

where $\Delta H_{\ddagger}$ is the heat of vaporization of the activated complex. Thus the rate would be expected to increase in solution by a factor of V_0/V_f equal to about 100. This increase is related to the probability factor or the entropy of activation. In addition, the heat of activation in solution is related to the heat of activation in the gas by

$$\Delta H_s^{\ddagger} = \Delta H_g^{\ddagger} - \Sigma\Delta H \text{ (vaporization)} \tag{20}$$

If the activated complex has a high heat of vaporization, which would be true if it were a very polar aggregate, the reaction in solution is favored over that in the gas. On the other hand, if the reactants are very polar and the activated complex is less polar so that its heat of vaporization is reduced, the reaction in the gas is favored. It is true that reactions leading to highly polar products or ions seldom occur in the vapor, whereas they are quite common in solution.

Again it is difficult to find adequate data for comparison, since the rates of few reactions have been measured both in the gaseous state and in solution. The decomposition of chlorine monoxide, the decomposition of ethylene iodide, the condensation of acrolein with cyclopentadiene, the condensation of cyclopentadiene with itself, and the *ortho-para* hydrogen conversion catalyzed by oxygen molecules are second-order reactions which have been found to have the same rates within a factor of 3 or 4 in the gas phase and in solutions of various inert solvents.[4] Furthermore, the activation energies and frequency factors are very nearly the same for any one of these reactions in solution or in the vapor. It appears then that the heat of vaporization of the activated complex is about equal to the sum of the heats of vaporization of the two reactants. There is little evidence for the increased frequency factor of about 100 for the solution reaction predicted. This is partly because the free volume, V_f, is not the same for all liquids as assumed.

For a unimolecular reaction, equation 10 becomes

$$k_s/k_g = \frac{\pi_A^0 \gamma_A}{\pi_{\ddagger}^0 \gamma_{\ddagger}} \tag{21}$$

If the chemical change involved does not lead to a great change in polarity, it will be expected that the properties of the activated complex will resemble those of the reactant and that the ratio k_s/k_g will be approximately unity.

The experimental facts are in agreement with this expectation, the decomposition of nitrogen pentoxide[5] and di-*t*-butyl peroxide[6] having about the same rate constants and activation energies in the gas phase and in a number of solvents. The unimolecular isomerization and racemization of *d*-pinene has very nearly the same rate in the vapor and in several high boiling solvents[7] at 218°. The decomposition of acetyl peroxide also is very similar in the gas and in a variety of solvents.[8]

Although it thus appears that reactions which do occur in both the gas and in solution have about the same velocity, it must be remembered that for the majority of reactions the rates must be quite different in the two phases, since the reactions in general are found in only one phase or the other. This is due partly to the failure of reactions involving ions to occur in the gaseous state because of energetic considerations and partly due to the fact that a set of reactants that reacts one way in the gas phase will frequently react another way in solution.

A consideration of the rates of reaction in solution from the viewpoint of the collision theory is enlightening. In Chapter 4 the number of collisions that a given molecule made with other molecules was found by calculating the volume swept out by the molecule per second and multiplying this by the number of other molecules in unit volume. A correction should now be made for the effective volume occupied by the molecules themselves. This correction does not affect the volume swept out by the central molecule but changes the concentration of other molecules from N/V to $N/(V - Nb)$, where b is similar to van der Waal's constant per molecule but not identical with it, and $(V - Nb)$ is the same as the free volume V_f. Consequently, the number of bimolecular collisions in solution is increased by a factor of V/V_f. This is the same factor that was deduced from the transition-state theory as given by equation 19. The difference in the activation energy by the collision theory is also given by (20). There is an interesting example of a reaction in the gas phase studied over a wide enough range of pressures so that a correction for the free volume becomes important. This is the work of Kistiakowsky[9] on the thermal decomposition of hydrogen iodide. The concentrations ranged from 0.02 mole to 7 moles per liter. The rate constant increased with pressure and was found to obey the equation

$$k/k_0 = V/(V - Nb) = V/V_f \tag{22}$$

in agreement with the deduction made above. Here k_0 is the rate constant at very low concentrations. The agreement may be illusory, since it cannot be definitely stated that the activation energy remained constant over the range of pressures used.

The effect of the solvent on unimolecular reactions was a subject of controversy in the early days of the development of kinetic theory. It was argued that such reactions were slowed down by collisions with the solvent, which tended to rob activated molecules of their energy before they could react. Though it is true that such deactivating collisions do occur, they are always compensated for by an equal number of activating collisions between solvent molecules and unactivated molecules of reactant. This compensation is evident from a consideration of the general theory of unimolecular reactions (Chapter 4). The most important effect that the solvent may be expected to have is perhaps in changing the nature of the products formed by the reaction, since reactive fragments formed in unimolecular decompositions (free radicals) frequently will react with the solvent.

There is one important difference between collisions in solution and in the gaseous state. Collisions in solution often tend to be repeated, so that multiple collisions of the same two molecules occur. This is because the solvent tends to surround the colliding particles in a "cage" which traps them for a certain time before one or both can diffuse out. During this time the probability of the two particles colliding is very high. Such multiple collisions or "encounters" are greatly influenced by the viscosity of the solvent. Thus we have the situation that the total number of collisions in solution compares to that in the gas phase, but that repeated collisions are favored over fresh collisions. For ordinary chemical reactions where most collisions are ineffective, this has no effect on the rate. For some reactions which occur at essentially every collision, such as the coagulation of colloids and the quenching of fluorescence, the effect is very important. For such cases special collision equations must be worked out. The reaction rates are said to be diffusion-controlled. Since such reactions are usually very rapid, they will be discussed in Chapter 11.

For the majority of reactions in solution, it seems that the elementary collision theory is as adequate for reactions in solution (without any correction for free volume) as for gases in that it provides a "normal" or expected value of the frequency factor A in the Arrhenius equation

$$k = Ae^{-E_a/RT}$$

Those reactions which have much larger or much smaller A values then represent in some way "abnormal" reactions for which an explanation must be sought in terms of complex mechanisms, the effect of electrical charges, etc. The justification for this lies in a consideration of the experimental facts for the bimolecular reactions known in solution. The most probable as well as the average value of the frequency factor lies close to the 10^{11} liters/mole-sec predicted by the collision theory. This value can then be considered the norm.

However, since the reactions often involve complicated molecules, the probability factor of unity implied in such a statement cannot easily be reconciled with the theories of kinetics discussed in Chapters 4 and 5. Part of the answer lies in the obvious fact that strong intermolecular forces must exist in any condensed system at room temperature or above. Hence any molecule, solute or solvent, in such a system is a part of a large aggregate which, in some respects may be considered a giant molecule. This aggregation greatly reduces the entropy (freedom of motion) of the individual molecule. When rearrangement occurs to form an activated complex, this simply represents a new aggregate with perhaps not too much difference in lability. Thus the loss of translational entropy characteristic of the gas phase does not occur to the same extent in solution.

The Influence of the Solvent

The preceding arguments have shown that in certain cases there is not much difference between reactions in the gaseous state and in solution. For such reactions the nature of the solvent seems to make little difference. For the great majority of reactions, however, which occur in solution but not in the gas phase, the specific properties of the solvent are important in determining not only the rate but also the equilibrium. A discussion will be presented now of the effect of the properties of the solvent, particularly those leading to deviations from ideal behavior.

Scatchard[10] and particularly Hildebrand[11] have developed expressions for the activity coefficients of nonelectrolytes as solutes in various liquid solvents. On the assumption that the heat of mixing is responsible for all deviations from ideal behavior and that the interaction energy of a solute molecule and a solvent molecule is the geometric mean of solute-solute and solvent-solvent interactions, it is possible to derive the following equation:

$$RT \ln f_2 = V_2 \phi_1^2 [(\Delta E_2/V_2)^{1/2} - (\Delta E_1/V_1)^{1/2}]^2 \tag{23}$$

where V_2 is the molar volume of solute (as a liquid), ΔE_2 is the molar energy of vaporization of the solute (as a pure liquid), and V_1 and ΔE_1 are the same quantities for the solvent. ϕ_1 is the volume fraction of the solvent, equal to unity for a dilute solution, and f_2 is the activity coefficient of the solute referred to a standard state of pure liquid solute (not infinite dilution). For dilute solutions (23) may be written as

$$RT \ln f_2 = V_2 (\delta_2 - \delta_1)^2 \tag{24}$$

where $\delta_2^2 = (\Delta E_2/V_2)$ is a parameter which may be called the internal pressure or the cohesive energy density. Equation 24 has been useful in predicting solubilities of nonelectrolytes as a function of the differences in internal pressures (Hildebrand and Scott[11]).

It may now be applied to a kinetic problem by writing equation 18 in the form

$$k = k_0 \frac{f_A f_B}{f_{\ddagger}} \tag{25}$$

for a bimolecular reaction where k_0 is the rate constant of a given reaction in an ideal solution. Hence k_0 is independent of the solvent whose properties are brought in only as they affect the ratio of activity coefficients. From (24) and (25) we now have

$$\ln k = \ln k_0 + \frac{V_A}{RT}(\delta_1 - \delta_A)^2 + \frac{V_B}{RT}(\delta_1 - \delta_B)^2 - \frac{V_{\ddagger}}{RT}(\delta_1 - \delta_{\ddagger})^2 \tag{26}$$

The reaction rate in a given solvent is thus seen to depend on the difference in internal pressures of the solvent and the reactants and the solvent and the activated complex. If the activated complex is similar to the solvent but the reactants are not, the rate will be large compared to the rate in the ideal solution. If the reactants resemble the solvent, but the transition state does not, the rate will be small compared to the ideal situation. Briefly, reactions producing products more polar than the reactants will go well in polar solvents; reactions giving products less polar than the reactants will go well in nonpolar solvents. This deduction is in agreement with experience.

The concept of internal pressure is useful for reactions involving neutral molecules only; other properties of the solvent such as the solvating power and dielectric constant become important when ions are involved. The effect of such properties will be discussed in the remainder of the chapter.

The Ionization of Neutral Molecules

Since the chief difference between reactions in solution and in the gas phase is the formation or participation of ions, it will be desirable to consider in some detail the process of forming ions from neutral molecules and the effect of the solvent on such a process.[12] Consider the ionization of a simple acid such as HCl in the gas phase:

$$HCl\ (g) \rightarrow H^+\ (g) + Cl^-\ (g) \tag{27}$$

The change in heat content, ΔH^0, for this reaction can readily be calculated. It is essentially equal to the heat of dissociation of the H—Cl bond plus the ionization energy of the hydrogen atom plus the electron affinity of the chlorine atom.

$$\Delta H^0 = 103 + 312 - 88 = 327 \text{ kcal} \tag{28}$$

The change in entropy, ΔS^0, can also be calculated at 25° and a concentration of 1 mole per liter from the absolute entropies of HCl, the hydrogen atom, and the chlorine atom (the two latter being virtually the same as for hydrogen ion and chloride ion).

$$\Delta S^0 = 19.6 + 30.2 - 38.3 = 11.6 \text{ E.U.} \tag{29}$$

The reaction is so unfavorable from an energetic point of view that the entropy increase, though favorable, is not sufficient to cause the reaction to go to any perceptible degree at ordinary temperatures.

The same reaction carried out in aqueous solution

$$\text{HCl (aq)} \rightarrow \text{H}^+ \text{ (aq)} + \text{Cl}^- \text{ (aq)} \tag{30}$$

has a value of ΔH^0 which may be found from that in the gas phase and the heats of solution of the substances involved. The heat of solution of the HCl molecule in water is not known, but it must be of the same order as the heat of vaporization of HCl, say about -6 kcal per mole to allow for the greater polarity of water. The heats of solution of the two ions can be found from tables of ionic heats of hydration. They are -89 kcal for Cl^- and about -270 kcal for H^+. So the value of ΔH^0 for reaction 30 is about -25 kcal ($327 - 89 - 270 + 6$). The change in entropy for the reaction in water can be calculated from a knowledge of the individual entropies of solution. Although the entropy of hydration of the HCl molecule is not known, it is probably not much different from that of the chloride ion, since for a large ion the effect of the charge has little influence on the entropy of hydration.[13] Hence, the change in ΔS^0 over that in the gas phase will be determined by the entropy of hydration of the hydrogen ion which is -30.6 E.U. This makes ΔS^0 in solution about -19 E.U. Although these calculations are too crude to permit a calculation of the degree of ionization, it is shown that the reaction can occur readily in solution. It is of interest to note that the negative entropy change of about 20 units would be predicted on the basis of the previous calculation for the ionization in water of any uncharged acid, strong or weak. Table 2, which gives the thermodynamics of ionization for a number of weak acids, shows that the experimental facts are in agreement. Since the entropy of hydration depends on the size of the ion (as well as its charge), ΔS^0 is not as negative when large ions are produced.

The most important change was in the lowering of ΔH^0 in the gas from a large positive number to a small negative one in solution. Thus the heat of solution of the ions compensates for the energy required to separate the ions from each other. This is a very definite property of the solvent and is related to the orientation of solvent molecule dipoles around the ions in such a way as to lead to electrostatic attraction. The decrease in entropy observed in solution is also associated with the orientation of the solvent molecules around the ions with an attendant loss in freedom of motion.

Table 2

IONIC EQUILIBRIA IN WATER AT $25°$

	ΔH^0, kcal	ΔS^0, E.U.
$H_2O \rightleftharpoons H^+ + OH^-$ [a]	13.52	−19.2
$HCOOH \rightleftharpoons H^+ + HCOO^-$ [a]	−0.02	−17.6
$CH_3COOH \rightleftharpoons H^+ + CH_3COO^-$ [a]	−0.09	−22.1
$C_2H_5COOH \rightleftharpoons H^+ + C_2H_5COO^-$ [a]	−0.16	−22.8
$ClCH_2COOH \rightleftharpoons H^+ + ClCH_2COO^-$ [a]	−1.16	−17.0
$H_3PO_4 \rightleftharpoons H^+ + H_2PO_4^-$ [a]	−1.80	−16.1
$HSO_4^- \rightleftharpoons H^+ + SO_4^{-2}$ [a]	−5.20	−26.5
$NH_3 + H_2O \rightleftharpoons NH_4^+ + OH^-$ [b]	1.22	−18.7
$CO(NH_2)_2 \rightleftharpoons NH_4^+ + CNO^-$ [c]	11.1	16.5
$H_2O + Cr(H_2O)_5Cl^{2+} \rightleftharpoons Cr(H_2O)_6^{3+} + Cl^-$ [d]	−6.1	−17.2

[a] H. S. Harned and B. B. Owen, *The Physical Chemistry of Electrolytic Solutions*, Reinhold Publishing Corp., New York, 2nd edition, 1950, Chapter 15.
[b] D. H. Everett and W. F. K. Wynne-Jones, *Proc. Roy. Soc.*, *A169*, 190 (1938).
[c] NBS circ. 500, 1952.
[d] H. S. Gates and E. L. King, *J. Am. Chem. Soc.*, *80*, 5011 (1958).

A useful analogy has been drawn between such oriented and strongly held water molecules and "frozen" water molecules.[14] According to this view, the loss in entropy in an ionization includes not only the change in entropy of the molecules which ionize but also the change in entropy of the solvent molecules which surround the ions. Since the entropy of freezing water is −5.5 E.U., it may be interpreted that the ions are solvated by the equivalent of four water molecules more than the neutral acid molecule. The actual loss in entropy, of course, may be shared by a larger number of solvent molecules each of which is only partially frozen. There is also a decrease in heat capacity, ΔCp^0, for the ionization in water of all neutral acids equal to about −40 E.U. Since ice has a lower heat capacity than water by 9 E.U., we see again that ionization is accompanied by the freezing of solvent molecules around the ions.

An important question arising next has to do with the effect of a change in solvent from water to something less polar. Table 3 shows the experimental results of progressively replacing water by nonpolar dioxane on the ionization of acetic acid. The change in the equilibrium constant is in the expected direction in that less ionization occurs as the percentage of dioxane in the solvent increases. A rather surprising feature, however, is that the change is almost entirely due to a more negative entropy of ionization. The heat of ionization changes in a direction opposite to that

Table 3

IONIC EQUILIBRIA IN NONAQUEOUS SOLUTIONS AT 25°

	Solvent	ΔH^0, kcal	ΔS^0, E.U.
$CH_3COOH \rightleftharpoons H^+ + CH_3COO^-$ [a]	Water	−0.09	−22.1
	20% dioxane	−0.05	−24.4
	45% dioxane	−0.44	−30.3
	70% dioxane	−0.61	−40.1
	82% dioxane	−1.34	−50.8
p-$BrC_6H_4N(CH_3)_2 + C_3H_5Br \rightleftharpoons$	$\{C_2H_2Cl_4$	−19.1	−59.7
p-$BrC_6H_4N(CH_3)_2C_3H_5^+ + Br^-$ [b]	$\{CHCl_3$	−18.9	−59.5
$C_6H_5N(CH_3)_2 + CH_3I \rightleftharpoons$	$\{C_6H_5NO_2$	−15.0	−36.9
$C_6H_5N(CH_3)_3^+ + I^-$ [c]	$\{90\%$ acetone	−18.5	−47.5

[a] H. S. Harned and B. B. Owen, *loc. cit.*; see Table 2.

[b] H. von Halban, *Z. physik. Chem.*, 67, 129 (1909); W. C. Davies and R. G. Cox, *J. Chem. Soc.*, 614 (1937).

[c] H. Essex and O. Gelormini, *J. Am. Chem. Soc.*, 48, 882 (1920).

which might have been guessed. Table 3 also has some data on quaternary salt formation in nonaqueous solvents, showing also a large negative entropy of ionization.

An exact prediction of the effect of different solvents on ionic equilibria is not possible. However, certain generalizations can be made from a simple electrostatic picture, which views an ionization such as (27) or (30) as the separation of two point charges of opposite sign in a continuous medium. If we separate two such charges from an equilibrium distance r to an infinite distance in a medium of dielectric constant D, then the electrical work is given by Coulomb's law

$$W_{el} = \frac{Z_A Z_B e^2}{Dr} = -\Delta F_{el} \tag{31}$$

and can be equated to the corresponding change in free energy, $-\Delta F_{el}$.

From the thermodynamic equation

$$(\partial \Delta F / \partial T)_p = -\Delta S \tag{32}$$

the change in entropy can also be calculated as

$$\Delta S_{el} = - \frac{Z_A Z_B e^2}{Dr} \left(\frac{\partial \ln D}{\partial T} \right)_p \tag{33}$$

If ΔF and ΔS are known, then ΔH can readily be found from $\Delta H = \Delta F + T\Delta S$. The values of D, ΔF_{el}, ΔS_{el}, and ΔH_{el} for a number of representative solvents are given in Table 4 for the separation of two unit charges, positive and negative, from an equilibrium distance of 1 A. Experimental values of the change in dielectric constant with temperature are used in making the calculations.†

Table 4

CALCULATED ELECTROSTATIC EFFECTS IN VARIOUS SOLVENTS

Solvent	$D_{20°}$	ΔF_{el}, kcal	ΔS_{el}, E.U.	ΔH_{el}, kcal
Vacuum	1.0	312	0	312
Hexane	1.9	164	−117	130
Chlorobenzene	6.65	47	−136	7.1
Ethyl alcohol	25.0	12.5	−75	−9.5
Water	80.0	3.9	−18	−1.5

For $Z_A = +1$, $Z_B = -1$, and $r = 1$ A.

The free-energy change decreases steadily with increasing polarity of the solvent as expected. The entropy change is always negative in a solvent but goes through an apparent minimum for solvents of moderate polarity. It is small for highly polar solvents, very large for solvents of moderate polarity, not so large for nonpolar solvents, and zero for a vacuum.

These calculations are not exact, since the interaction of an ion with a molecular medium is not governed entirely by the macroscopic dielectric constant. They do, however, give some idea of the effect of the solvent on the stability of independent ions in solution. The entropy decrease is again seen to be a property of the solvent, since, when the solvent is completely removed, there is zero entropy of ionization. The variation of ΔS_{el} with polarity is understandable if interpreted as the loss of freedom of solvent molecules when frozen around the ions. Highly polar liquids such as water already exist in a partially frozen state because of strong intermolecular forces, hydrogen bonding, etc. Hence, while strongly bound to the ions, such molecules suffer a relatively small entropy decrease.

† For a similar calculation of solvent effects see reference 15 at close of chapter.

Nonpolar solvents are quite free in the liquid state, and, on freezing, a large decrease in entropy occurs. For example, the entropy of fusion of most organic substances is about 10 E.U. (unless the solid state has free rotation). If such molecules are bound tightly to ions, a large negative entropy change would be expected.

The change in heat content calculated on the basis of electrostatic interactions only is apt to be very inaccurate because it neglects the fact that one or more solvent molecules may form an essentially covalent bond, or at least a bond with some covalent character, with the ions produced. For example, the hydrogen ion in almost any medium forms a covalent bond with one molecule of solvent, giving rise to H_3O^+, $C_2H_5OH_2^+$, etc. The energy of such an interaction is not completely coulombic and must be calculated by quantum-mechanical methods. Also reactions producing ions in solution are frequently the result of the ionizing of two solute molecules with breaking of old covalent bonds and formation of new ones as in the reaction of an alkyl halide with an amine

$$C_2H_5Br + (C_2H_5)_3N \rightarrow (C_2H_5)_4N^+ + Br^- \qquad (34)$$

The total change in heat content will then include the differences in bond energies.

Kinetics of Ionization

Table 5 gives the experimental values of the activation energies and entropies of activation for a number of reactions involving the formation of ions from neutral molecules in a variety of solvents. It appears that two generalizations can be made: the activation energy depends on the type of reaction and does not change rapidly from solvent to solvent; the entropy of activation is always negative and changes with the solvent, becoming more negative as the polarity of the solvent decreases. It follows from this that the rates of reactions producing ions in solution increase with the polarity of the solvent and that the increase is governed largely by the change in the entropy of activation. There is no evidence for the minimum value of the entropy of activation in solvents of moderate polarity indicated by Table 4.

It seems reasonable again to relate the entropy decrease in going from reactants to activated complex to the freezing of solvent molecules around the incipient ions.[16] The activated complex is almost an ion pair at its distance of closest approach or at least an exceedingly polar complex approaching an ion pair. Each end of the polar complex has already accumulated a layer of solvent molecules, whose presence is necessary to allow the process of separating the ions completely to continue. As the

Table 5

ACTIVATION ENERGIES AND ENTROPIES OF ACTIVATION FOR IONIZATIONS

Reaction	Solvent	E_a, kcal	$\Delta S^{\ddagger}$, E.U.
$C_6H_5NH_2 + C_6H_5\overset{\overset{O}{\|\|}}{C}CH_2Br \rightarrow$	Benzene	8.1	-56
	Chloroform	10.8	-46
	Acetone	11.1	-39
$C_6H_5\overset{\overset{O}{\|\|}}{-C}-CH_2-NH_2C_6H_5^+ + Br^{-}$ ^{a}	Nitrobenzene	13.5	-33
	Methanol	12.4	-33
	Ethanol	13.9	-28
$CH_3I + (C_2H_5)_2S \rightarrow (C_2H_5)_2S-CH_3^+$	Acetone	12.2	-39.8
$+ I^{-}$ ^{b}	Methanol	16.7	-25.2
	Ethanol	17.2	-25.3
$p\text{-NO}_2\langle\rangle-CH_2Br + H_2O \rightarrow$	50% dioxane	18.8	-30.1
	70% dioxane	17.2	-36.8
	90% dioxane	15.4	-44.2
$p\text{-NO}_2\langle\rangle-CH_2OH + H^+$ $+ Br^{-}$ ^{c}			
$HA + H_2O \rightarrow H_3O^+ + A^{-}$ ^{d}	Water		
Acetoacetic ester		14.2	-26.2
α-Methylacetylacetone		18.0	-16.0
Nitromethane		22.6	-20.2
Nitroethane		22.9	-18.4
Ethyl nitroacetate		16.0	-15.3
Solvolysis of t-butyl chloride ^{e}	Water	23.8	$+12.2$
	Methanol	25.5	-3.1
	Ethanol	26.7	-3.2
	Acetic acid	26.4	-2.5
	90% dioxane	22.2	-18.5
	90% acetone	22.4	-16.8

^{a} H. E. Cox, *J. Chem. Soc.*, *119*, 142 (1921).
^{b} J. K. Syrkin and I. T. Gladischew, *Acta Physicochim. U.R.S.S.*, *2*, 291 (1935).
^{c} J. W. Hackett and H. C. Thomas, *J. Am. Chem. Soc.*, *72*, 4962 (1950).
^{d} R. G. Pearson and R. L. Dillon, *J. Am. Chem. Soc.*, *75*, 2439 (1953).
^{e} S. Winstein and A. H. Fainberg, *J. Am. Chem. Soc.*, *79*, 5937 (1957).
The standard state for $\Delta S^{\ddagger}$ is one mole per liter.

separation occurs, the layer of solvent molecules is completed so that usually a further decrease in entropy occurs in going from the transition state to the products. This last statement has an important bearing on the rate of reverse reaction, the formation of neutral molecules from ions, which will be discussed shortly. From the same arguments given in regard to equilibria, the less polar solvents have a greater loss in freedom in becoming frozen to the ions that do the more polar solvents. Hence, the lower rates of reaction in the less polar solvents. It should be mentioned at this point that there is considerable difficulty in studying rates of ionization reactions in very nonpolar solutions such as those in hexane or benzene. Very small quantities of polar materials, or salts, can cause large changes in reaction rate.[17] Also the reactions may be reversible and will not occur appreciably unless the product precipitates or a material is added to combine with product irreversibly. The rate may then depend on the impurities present, the added substances, or even the surface of the crystallized salt and is not dependent on the solvating properties of the medium. There is also evidence that a polar reactant in a nonpolar solvent will tend to act as a solvating agent as well.[18] For this reason reactions run in nonpolar solvents must be measured at low concentrations of reactants if the effect of the solvent is being studied.

For a given medium the decrease in entropy is greater for a small ion than for a larger one, since the smaller ions are more highly solvated. The hydrogen ion in particular is not formed without a considerable decrease in entropy. There may be some cases in polar media where large ions are formed with a positive entropy of activation.

The relative insensitivity of the activation energy to the dielectric properties of the medium is not unexpected. Of much greater importance in determining the activation energy for a reaction producing ions, as well as for any other kind of reaction, are the repulsive energies between the reactants as they are brought close together. The forces responsible are of the van der Waals type and result from the interaction of filled electron levels in different molecules. Such interactions are attractive for moderate distances of separation but always become repulsive at the shorter distances needed for making and breaking of chemical bonds. The strengths of such bonds are also of importance in determining the activation energy. Hence, most of the factors that determine the energy barrier to forming the activated complex are properties of the reactants, and the solvent plays only a secondary role. Since the solvent lowers the energy of both the reactants and the activated state by solvation, even this effect tends to cancel. Though the solvation would be expected to be greater for the polar complex, the experimental evidence seems to be that the activation energy is frequently somewhat greater in the more polar solvents.

The explanation of the slowness of reactions producing ions was one of the early stumbling blocks to the application of the collision theory of reaction rates to solutions. The large negative entropies of activation correspond to very low probability factors in the equation

$$k = pZe^{-E_a/RT}$$

Values of p down to 10^{-9} are found experimentally for such reactions. An adequate qualitative explanation of these values can again be given in terms of the solvated, polar activated complex. The collision theory is based upon a p value of unity if reaction occurs every time two reactant molecules with the requisite energy collide. If in addition it is postulated that the collision occurs simultaneously with the presence of several suitably oriented solvent molecules or that the collision be actually an n-body collision instead of a two-body collision, then very low probability factors become reasonable. The variation of the p factor with solvent may be explained, again qualitatively, by the reasonable assumption that a polar molecule will more frequently be coordinated with the reactant molecule in the approximate position necessary for participation in the reaction than will a nonpolar molecule.

A theory for the influence of the dielectric constant of the medium on the free energy of a polar molecule has been given by Kirkwood.[19] By considering electrostatic forces only (neglecting van der Waals' forces), the difference in free energy of a dipole in a medium with dielectric constant D and with a dielectric constant of unity is given by

$$\Delta F = -\frac{\mu^2}{r^3}\frac{(D-1)}{(2D+1)} \tag{35}$$

where μ is the dipole moment and r the radius of the molecule. Applying this to the transition-state theory for the reaction $A + B \rightleftharpoons M^{\ddagger}$ where A, B, and $M^{\ddagger}$ are polar species, and remembering that

$$k = (RT/Nh)e^{-\Delta F^{\ddagger}/RT}$$

we obtain

$$\ln k = \ln k_0 - \frac{N}{RT}\frac{(D-1)}{(2D+1)}\left[\frac{\mu_A^2}{r_A^3} + \frac{\mu_B^2}{r_B^3} - \frac{\mu_{\ddagger}^2}{r_{\ddagger}^3}\right] \tag{36}$$

where k is the rate constant in the medium of dielectric constant D, and k_0 is the rate constant in a condensed medium of dielectric constant unity where the nonelectrostatic forces are the same for the activated complex as for the reactants. Equation 36 predicts that, if the activated complex is more polar than the reactants (as would be true if the products were ions), the rate of the reaction increases with the dielectric constant of the medium. For many such reactions in mixtures of two solvents of such composition that the dielectric constant can be varied, a straight line can be obtained by plotting $\log k$ against $(D - 1)/(2D + 1)$.[20] The equation is not always obeyed, however, and is not valid in general if reaction rates in different solvents of various dielectric constants are compared.

A polar molecule can be taken as a reasonable model of the transition state for many reactions producing ions from neutral molecules. Hence (35) and (36) can be used, even though equation 35 is restricted to point dipoles imbedded in a spherical molecule. Using equation 32, it is also possible to calculate the electrostatic contribution to the entropy of activation.

$$\Delta S^{\ddagger}_{el} = -\left(\frac{\mu_A^2}{r_A^3} + \frac{\mu_B^2}{r_B^3} - \frac{\mu_{\ddagger}^2}{r_{\ddagger}^3}\right)\frac{3D}{(2D + 1)^2}\left(\frac{\partial \ln D}{\partial T}\right)_p \qquad (37)$$

Table 6 shows the results of such calculations for a number of solvents.[21] The electrostatic contributions to the activation enthalpy are also listed.

Table 6

CALCULATED VALUES OF $\Delta F^{\ddagger}_{el}$, $\Delta H^{\ddagger}_{el}$ AND $\Delta S^{\ddagger}_{el}$ FOR A REACTION WHERE $-[(\mu_A^2/r_A^3) + (\mu_A^2/r_A^3) - (\mu_{\ddagger}^2/r_{\ddagger}^3)] = 60$ KCAL

Solvent	$D_{20}°$	$-(\partial \ln D/\partial T)_{Pa}$	$-\Delta F^{\ddagger}_{el}$	$-\Delta H^{\ddagger}_{el}$	$-\Delta S^{\ddagger}_{el}$
Hexane	1.9	0.714×10^{-3}	11.2 kcal	14.4	10.6
Carbon tetrachloride	2.2	0.843	13.7	17.1	11.4
Benzene	2.3	0.876	13.8	17.2	11.5
Toluene	2.3	0.673	14.1	16.7	8.8
Anisole	4.4	5.2	20.9	33.5	42.2
Chloroform	4.5	3.33	21.0	29.1	27.0
Chlorobenzene	6.7	2.89	23.8	28.8	16.8
Ethyl bromide	9.4	4.91	26.6	32.9	21.2
Acetophenone	18.1	4.10	27.4	30.3	9.7
Acetone	20.0	4.63	27.8	30.8	9.9
Ethyl alcohol	25.0	6.02	28.2	31.3	10.4
Methyl alcohol	32.5	5.39	28.6	30.8	7.3
Nitrobenzene	35.8	5.21	28.8	30.7	6.3
Water	80.0	4.63	29.6	30.3	2.4

Reasonable values of the dipole moment-radii differences are assumed and the experimental variation of the dielectric constants with temperature is used. The values of $\Delta H^{\ddagger}_{el}$ are in reasonable agreement with experiment for all the polar molecules in that they show little variation from one solvent to another. For the polar solvents the values of $\Delta S^{\ddagger}_{el}$ are also reasonable in that they are always negative and decrease with decreasing polarity of the solvent. Qualitatively these results imply that the potential energy of the transition state is always lowered about the same amount by either a very polar or only a slightly polar solvent. However, only a few solvent molecules need to be oriented to bring about the reduction in energy for highly polar molecules such as water, whereas many more molecules must be oriented for less polar molecules such as anisole.

For the nonpolar solvents such as benzene or hexane the calculated results do not agree with experimental values for reactions in such solvents. The prediction is that the activation energies would be very large (note that $\Delta H^{\ddagger}_{el}$ is always negative and hence subtracts from the activation energy) and that the entropies of activation would be only moderately negative. The facts for nonpolar solvents are that the activation energies are roughly comparable to those for polar solvents and that the entropies are the most negative of all. In these cases the use of the gross dielectric constant is very misleading. The interaction of the ion or dipole with a nonpolar molecule is much greater than would be expected. This is particularly true for aromatic solvents, which often are surprisingly good for ionic solutes or ionic reactions in spite of a dielectric constant of 2.3 or so.

It must be emphasized that the use of a classical electrostatic model and the dielectric constant is only a very rough guide to predicting the efficiency of a solvent in promoting rates of ionic reactions. There are many examples of specific effects, particularly for hydroxylated solvents, since the ability of a solvent to hydrogen bond to an anion is an important factor. Other solvents, such as nitromethane, although having high dielectric constants, are poor solvents for ionic materials because they cannot effectively coordinate anions. The positive end of the dipole of the solvent molecule is somewhat buried in this case and not so easily available as the positive hydrogen atom of water or alcohol. The specific effect of the benzene ring has already been commented on and is undoubtedly related to the highly polarizable cloud of π electrons above and below the aromatic ring.[22] It is also easy to imagine that certain atoms, or groups of atoms, for example basic nitrogen atoms or oxygen atoms, would greatly influence the interaction of the solvent with certain solutes. Classification of solvents into groups such as aromatic, aliphatic, and hydroxylic, or donor and non-donor have been suggested.[23]

REACTIONS BETWEEN IONS

It is common to think of reactions between ions as being very rapid. This is true, for example, in the combination of hydrogen ion with hydroxyl ion in the neutralization of strong acids with strong bases, and in the combination of simple ions to form an insoluble salt such as silver chloride. In such cases, where two ions simply combine, the rate of reaction is governed by the diffusion of the ions towards each other and the activation energy for the combination is very small. However, there are many reactions between ions which involve the making and breaking of covalent bonds (or in some cases electron transfers) which may be as slow as reactions between neutral molecules and which have normal activation energies. It is with reactions of this type that we shall be mostly concerned.

If the overall process of ionization and the kinetics of ionization are understood, then in principle the reverse process of forming neutral molecules from ions is also understood. This follows from the general relationship between equilibrium constants and the rate constants of forward and reverse reactions, $K_{eq} = k_f/k_r$. This relationship has certain limitations which will be discussed in Chapter 8. For simple processes, however, it is exact and the concept has been used, for example, to calculate the rate constant for ion recombination from measured values of rates of ionization and acid ionization constants for a number of "pseudo" acids.[24] If we represent such an acid by HA, we have in water

$$HA + H_2O \underset{k_r}{\overset{k_f}{\rightleftharpoons}} H_3O^+ + A^- \tag{38}$$

$$K_{eq} = K_a = k_f/k_r \tag{39}$$

Table 7 shows some entropies of activation calculated in this indirect manner and also a number of directly measured values for reactions between ions.

For reactions between ions of unlike sign there is generally an entropy increase going from reactants to activated complex; for ions of like sign there is an entropy decrease. In terms of the collision theory, reactions between oppositely charged ions are more rapid than predicted by theory and the p factor is greater than unity. Reactions between like-charged ions are slower than the collision theory predictions and the p factor is less than unity. In terms of solvation of ions the explanation of the entropy factor is, as previously indicated, that two ions coming together to form a neutral molecule will become partially desolvated in the transition state so that some frozen solvent molecules will be released with an increase of entropy. The same will be true for any two ions of opposite charge, since their union will result in a lowering of the net charge and solvation in general increases with the charge on the ion. For ions of the same sign, however, the transition state will be a more highly charged ion which would be expected to be strongly solvated, so that more solvent molecules might be required than for the separate ions. This would lead to a decrease in entropy in forming the transition state.

Again the elementary electrostatic theory embodied in equation 31 and 33 will be useful in giving the general effect of the charges on the ion reactants and the dielectric constant of the medium. The electrostatic free

Table 7

ENTROPY OF ACTIVATION FOR SOME REACTIONS BETWEEN IONS IN WATER

	pZ, liters/mole-sec	$\Delta S^{\ddagger}$, E.U.
$Cr(H_2O)_6{}^{+3} + CNS^{-}$ [a]	1.3×10^{14}	0.7
$CO(NH_3)_5Br^{++} + OH^{-}$ [b]	4.2×10^{17}	20.1
$ClO^{-} + ClO^{-}$ [c]	9.6×10^{8}	−19.6
$ClO^{-} + ClO_2{}^{-}$ [c]	8.5×10^{8}	−19.8
$CH_2BrCOO^{-} + S_2O_3{}^{=}$ [d]	1.2×10^{7}	−28.3
$CH_2ClCOO^{-} + S_2O_3{}^{=}$ [d]	2.3×10^{9}	−17.7
$Co(NH_3)_5Br^{++} + Hg^{++}$ [b]	1.3×10^{8}	−23.6
$S_2O_4{}^{=} + S_2O_4{}^{=}$ [e]	1.8×10^{4}	−41.2
$S_2O_3{}^{=} + SO_3{}^{=}$ [f]	2.3×10^{6}	−31.4

$H_3O^{+} + A^{-} \to H_2O + HA$ [g]	E_a	$\Delta S^{\ddagger}$, E.U.
Methylacetylacetone	9.0	6
Acetoacetic ester	7.2	2
Nitromethane	15.2	6
Nitroethane	21.2	15

[a] C. Postmus and E. L. King, *J. Phys. Chem.*, **59**, 1216 (1955).
[b] J. N. Brönsted and R. Livingston, *J. Am. Chem. Soc.*, **49**, 435 (1927).
[c] F. Foerster and P. Dolch, *Z. Elektrochem.*, **23**, 137 (1917); F. Giardani, *Gazz. chim. ital.*, **54**, 844 (1929).
[d] A. N. Kappana and H. W. Patwardhan, *J. Indian Chem. Soc.*, **9**, 379 (1932).
[e] K. Jellinek and E. Jellinek, *Z. physik. Chem.* **93**, 325 (1919).
[f] D. P. Ames and J. E. Willard, *J. Am. Chem. Soc.*, **73**, 164 (1951).
[g] R. G. Pearson and R. L. Dillon, *J. Am. Chem. Soc.*, **75**, 2439 (1953).

energy and entropy will now have the opposite sign from that given by the previous equations, since the process under consideration is that of bringing two ions from infinite separation to the equilibrium distance in the activated complex, for example

$$\Delta F^{\ddagger}{}_{el} = \frac{Z_A Z_B e^2}{D r_{\ddagger}} \tag{40}$$

(These equations apply only to the infinitely dilute solution. The effect of ionic strength will be considered later.)

The application of these equations to the collision theory[25] leads to the calculation of an electrostatic contribution to the activation energy equal to $\Delta H^{\ddagger}_{el}$ and to the calculation of a collision number which differs from that for uncharged molecules (simple kinetic molecular theory) by a factor of $e^{\Delta S^{\ddagger}_{el}/R}$. Thus it is found that collisions between like-charged ions are less likely and collisions between unlike-charged ions more likely than collisions between uncharged molecules simply because it is more probable that a negative ion will find itself close to a positive ion rather than to another negative ion.

The transition-state theory leads to exactly the same conclusions: to the heat of activation is added a term equal to $\Delta H^{\ddagger}_{el}$ and to the entropy of activation a term $\Delta S^{\ddagger}_{el}$.[16] For a distance of $r_{\ddagger} = 2$ A in water

$$\Delta S^{\ddagger}_{el} = -10 Z_A Z_B \text{ E.U.} \tag{41}$$

which is seen to be in fair agreement with the experimental entropy factors in Table 7. The mistake must not be made of assuming that equation 41 calculates the experimental entropy of activation. This is not so for several reasons. One is that only a contribution to the entropy is being calculated and other entropy changes may appear. Another reason is that even for a reaction with a probability factor of unity, $\Delta S^{\ddagger}$ would be -9 E.U. or so for a concentration unit of moles/liter as used in Table 7. The electrostatic entropies are independent of the choice of standard state.

The entropy factor often turns out to be the dominant one for ionic reactions. The contribution of the electrostatic effect to the activation energy is small and difficult to measure.[26] For example, in the alkaline hydrolysis of the positively charged ester ethoxycarbonylmethyltriethyl-ammonium ion, $(C_2H_5)_3NCH_2COOC_2H_5{}^+$, it is found that the rate at $25°$ C is some 200 times greater than for the hydrolysis of neutral ethyl acetate.[27] The frequency factor is increased more than 500-fold for the charged ester and the activation energy is also increased from 11.7 kcal to 12.7 kcal. Both changes are in good agreement with equation 40 if $r_{\ddagger}$ is taken as 1.3 A. Nevertheless, the total entropy of activation for the hydrolysis of the betaine ester is about -15 E.U., even though oppositely charged ions are involved.

From (40) it is possible to write an expression for the dependence of the rate constant on the dielectric constant:[28]

$$\ln k = \ln k_0' - \frac{N Z_A Z_B e^2}{D R T r_{\ddagger}} \tag{42}$$

where k_0' is the specific rate constant in a medium of infinite dielectric constant. This equation predicts a linear plot of log k against $1/D$ with a negative slope if the charges of the ions are the same sign and a positive slope if the charges are of opposite sign. Figure 1 shows the data obtained by Amis and La Mer[29] on the alkaline fading of bromphenol blue in mixtures of ethanol and water, and the data of King and Josephs[30] on

Table 8

BROMPHENOL BLUE (TETRABROMPHENOLSULFONPHTHALEIN) + HYDROXIDE ION IN WATER-ETHANOL AT 25°

(Amis and La Mer)

Wt. % Ethanol	D	k ($\mu = 0$)	log k
0.0	78.5	25.2	1.401
10.2	72.5	9.71	0.987
15.4	69.5	5.46	0.737
20.6	66.5	3.01	0.479
31.5	60.0	0.103	−0.987

AZODICARBONATE ION + HYDROGEN ION IN WATER-DIOXANE AT 25°

(King and Josephs)

Vol. % Dioxane	D	log k ($\mu = 0$)	$\Delta S^{\ddagger}$, E.U.	E_a, kcal
0	78.5	10.34	12.9	10.2
10	71.2	10.60	14.1	10.2
20	62.2	10.89	14.9	10.1
30	53.5	11.22	15.3	9.8
40	44.3	11.64	15.7	9.3
50	35.6	12.20	20.2	9.9
60	27.0	12.95	25.3	10.4

the reaction between hydrogen ion and the double negative azodicarbonate ion in mixtures of water and dioxane. The first reaction is between a univalent negative ion and a divalent negative ion

$$OH^- + BPB^{-2} \rightleftharpoons BPBOH^{-3} \tag{43}$$

and the second between a univalent positive ion and a double negative ion

$$N_2(COO)_2^{-2} + H_3O^+ \rightleftharpoons HN_2(COO)_2^- \tag{44}$$

to give an intermediate which decomposes to give nitrogen, hydrazine, and carbonate ion by a complex mechanism. Reaction 44, however, seems to be the rate-determining step and is so rapid that it may be studied only in alkaline solution. In agreement with the predictions of the theory, the slope of log k versus $1/D$ is negative for reaction 43 and positive for reaction 44. The slopes also lead to values of r equal to 2.81 A in the first

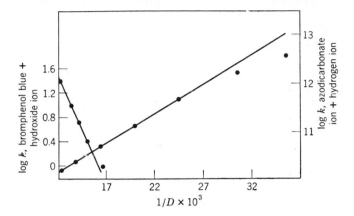

Fig. I. Influence of dielectric constant on rate of reaction between two ions of same sign (left axis) and of opposite sign (right axis) (Amis and La Mer, King and Josephs).

case and 3.42 A in the second. All we can say of these numbers is that they are of the right order of magnitude. Table 8 gives the data from which Fig. 1 was constructed. $\Delta S^{\ddagger}$ has also been given for the data of King and Josephs to show the effect of reducing the dielectric constant on the probability factor. The experimental activation energy remains practically constant, although the rate constant changes by a factor of 140.

Reactions between Ions and Neutral Molecules

Inasmuch as reactions between neutral molecules are often abnormally slow and reactions between ions abnormally fast (or slow, depending on the charge), it may be expected that a reaction between an ion and a neutral molecule will show a normal type of behavior. This is approximately true. Table 9 gives the data for a number of such reactions in various polar solvents. The pZ factor from the experimental data has been calculated and is seen to agree quite well with the collision-theory value of about 10^{11} for most of the examples. Since the small or large values of pZ for reactions forming ions or reactions between ions were explained by the electrostatic effects or by the large changes in polarity accompanying such reaction, it follows that for a reaction between an ion and a neutral molecule the electrostatic effects are small. This can be verified by approximate calculations of the energy of interaction between an ion and

Table 9

REACTIONS BETWEEN IONS AND NEUTRAL MOLECULES

Reaction	Solvent	E_a, kcal	$pZ \times 10^{-11}$, liters/mole-sec
$CH_3I + C_2H_5O^-$ [a]	C_2H_5OH	19.5	2.42
$C_2H_5I + C_2H_5O^-$ [b]	C_2H_5O	20.7	1.49
$C_6H_5CH_2I + C_2H_5O^-$ [b]	C_2H_5OH	19.9	0.15
$n\text{-}C_3H_7I + C_6H_5O^-$ [c]	C_2H_5OH	22.5	3.53
$n\text{-}C_{16}H_{33}I + C_6H_5O^-$ [c]	C_2H_5OH	22.4	2.78
$n\text{-}C_4H_9I + C_6H_5CH_2O^-$ [d]	C_2H_5OH	21.6	2.92
$n\text{-}C_{16}H_{33}I + C_6H_5CH_2O^-$ [d]	C_2H_5OH	21.1	1.26
$C_2H_5I + \beta\text{-}C_{10}H_7O^-$ [e]	CH_3OH	21.0	0.10
$C_2H_5I + \beta\text{-}C_{10}H_7O^-$ [e]	C_2H_5OH	19.8	0.11
$C_2H_5I + \beta\text{-}C_{10}H_7O^-$ [e]	C_3H_7OH	21.3	0.40
$(CH_3)_2SO_4 + CNS^-$ [f]	CH_3OH	17.9	0.57
$CO_2 + OH^-$ [g]	H_2O	9.1	0.15
$CH_2ICOOH + Cl^-$ [h]	H_2O	22.9	7.90
$CH_2ClCOOH + I^-$ [h]	H_2O	19.8	0.13
$n\text{-}C_4H_9Cl + I^-$ [i]	$(CH_3)_2CO$	22.2	2.24
$C_2H_5Br + OH^-$ [j]	C_2H_5OH	21.4	4.30
$(CH_3)_2COHCH_2COCH_3 + OH^-$ [k]	H_2O	18.0	1.31
$CH_3COOC_2H_5 + OH^-$ [l]	H_2O	11.3	1.66×10^{-4}
$CH_3COOC_4H_9 + OH^-$ [l]	H_2O	11.4	2.13×10^{-4}
$n\text{-}C_3H_7Cl + I^-$ [m]	$(CH_3)_2CO$	18.6	1.2×10^{-3}
$n\text{-}C_5H_{11}Cl + I^-$ [m]	$(CH_3)_2CO$	18.4	1.0×10^{-3}
$i\text{-}C_3H_7Br + OH^-$ [n]	$80\% \ C_2H_5OH$	21.7	0.139
$CH_3Br + S_2O_3^{-2}$ [o]	H_2O	19.5	68.2
$CH_3I + S_2O_3^{-2}$ [o]	H_2O	18.8	21.9
$p\text{-}NO_2C_6H_5CH_2Br + Cl^-$ [p]	$50\% \ \text{dioxane-}H_2O$	17.8	1.95×10^{-2}
	$70\% \ \text{dioxane-}H_2O$	17.0	6.5×10^{-3}
	$90\% \ \text{dioxane-}H_2O$	17.1	5.0×10^{-3}

[a] W. Hecht and M. Conrad, *Z. physik. Chem.*, *3*, 450 (1889).
[b] W. Hecht, M. Conrad, and C. Bruckner, *ibid.*, *4*, 273 (1889).
[c] D. Segaller, *J. Chem. Soc.*, *105*, 106 (1914).
[d] P. C. Haywood, *ibid.*, *121*, 1904 (1922).
[e] H. E. Cox, *ibid.*, *119*, 142 (1921).
[f] P. Walden and M. Centnerszwer, *Z. Elektrochem.*, *15*, 310 (1909).
[g] R. Brinkman, R. Margaria, and F. J. W. Roughton, *Phil. Trans.*, *232*, 65 (1933).
[h] C. Wagner, *Z. physik. Chem.*, *115*, 121 (1925).
[i] J. B. Conant and W. R. Kirner, *J. Am. Chem. Soc.*, *46*, 232 (1924).
[j] G. H. Grant and C. N. Hinshelwood, *J. Chem. Soc.*, 258 (*1933*).
[k] G. M. Murphy, *J. Am. Chem. Soc.*, *53*, 977 (1931).
[l] L. Smith and H. Olsson, *Z. physik. Chem.*, *118*, 99 (1925).
[m] J. B. Conant and R. E. Hussey, *J. Am. Chem. Soc.*, 47, 476 (1925).
[n] E. D. Hughes, C. K. Ingold, and U. G. Shapiro, *J. Chem. Soc.*, 225 (*1936*).
[o] A. Slator, *ibid.*, *85*, 1286 (1904).
[p] J. W. Hackett and H. C. Thomas, *J. Am. Chem. Soc.*, *72*, 4962 (1950).

the dipole of a polar, but neutral, molecule. The energy is given (for large values of r) by

$$\Delta F_{\mathrm{el}} = - \frac{|Ze| \mu \cos \theta}{Dr^2} \tag{45}$$

where Ze is the charge on the ion, μ the dipole moment of the molecule, r the distance from the center of the ion to the center of the dipole, and θ the angle of approach of the ion to the line of the dipole (θ is zero when the ion approaches the oppositely charged end of the dipole head-on and $180°$ when it approaches the end of the dipole with the same charge). This energy is small compared to the energy of two ions. Furthermore, we can expect that the total solvation of an ion and a molecule will be not much more than the solvation of the activated complex formed by their union. This activated complex will have a charge equal, of course, to the charge of the ion.

We may estimate the effect of the solvent on the rate of reaction between an ion and a neutral molecule from equation 45. Since the attraction (assuming correct orientation for the dipole) will be somewhat greater, the rate of the reaction will be larger in a medium of lower dielectric constant. Another method that has been used for discussing the effect of the solvent involves the Born equation for the charging of an ion in a continuous dielectric:[31]

$$\Delta F_{\mathrm{el}} = \frac{Z^2 e^2}{2Dr} \tag{46}$$

Because of the difference in the radius r for the reactant ion and the activated complex, there is a difference in free energies which adds to the free energy of activation:

$$\Delta F^{\ddagger}{}_{\mathrm{el}} = \frac{Z^2 e^2}{2D} \left(\frac{1}{r_{\ddagger}} - \frac{1}{r} \right) \tag{47}$$

Accordingly, the rate constant may be written

$$\ln k = \ln k_0' + \frac{NZ^2 e^2}{2DRT} \left(\frac{1}{r} - \frac{1}{r_{\ddagger}} \right) \tag{48}$$

where k_0' is again the rate constant in a medium of infinite dielectric constant. Since $r_{\ddagger}$ will be larger than r, the rate again should be somewhat greater in a medium of lower dielectric constant.[20] This seems to be true if the rate of a reaction is studied in mixtures of two solvents so that the dielectric constant can be varied by changing the proportions of each solvent. However, agreement would not necessarily be expected if the rate were measured in several completely different solvents. One reason is that the ionic radius is not constant but varies from solvent to solvent.

Influence of Ionic Strength

The previous discussions have been based on the assumption that the properties of the solution were those of infinite dilution, that is, that the presence of solute molecules or ions did not affect the properties of the reactants or activated complex. In practice real solutions have deviations from ideality at moderate concentration. If ions are present, the deviations become apparent even at low concentrations. The effect on the rate is given by equation 18, which may be written as

$$k = k_0 \frac{\gamma_A \gamma_B \cdots}{\gamma_{\ddagger}} \qquad (49)$$

where k_0 is the rate constant at infinite dilution in the given solvent. This choice selects the standard state as a hypothetical state of unit concentration with a partial molal heat content, volume and heat capacity for each component being equal to that at infinite dilution. The value of k_0 is experimentally available by measuring k at a series of finite concentrations and extrapolating the results to infinite dilution.

The most important application of (49) occurs when one or more of the reactants are ions. According to the Debye-Hückel theory, the relation between the activity coefficient of an ion and the ionic strength is given for dilute solutions (less than $0.01 M$) by

$$-\ln \gamma_i = \frac{Z_i^2 \alpha \sqrt{\mu}}{1 + \beta a_i \sqrt{\mu}} \qquad (50)$$

where μ is the ionic strength, a_i the distance of closest approach of another ion to the ith ion, and α and β are constants for a given solvent and temperature. The numerical values are $\alpha = 0.509 \times 2.303$ and $\beta = 0.329 \times 10^8$ in water at 25°. Equation 49 can be written, using (50), as

$$\ln k = \ln k_0 - \frac{Z_A^2 \alpha \sqrt{\mu}}{1 + \beta a_A \sqrt{\mu}} - \frac{Z_B^2 \alpha \sqrt{\mu}}{1 + \beta a_B \sqrt{\mu}} + \frac{(Z_A + Z_B)^2 \alpha \sqrt{\mu}}{1 + \beta a_{\ddagger} \sqrt{\mu}} \qquad (51)$$

or, if we adopt a mean value a for the distance of closest approach,

$$\ln k = \ln k_0 + \frac{2 Z_A Z_B \alpha \sqrt{\mu}}{1 + \beta a \sqrt{\mu}} \simeq \ln k_0 + 2 Z_A Z_B \alpha \sqrt{\mu} \qquad (52)$$

The approximation in (52) is valid only for dilute solution where μ is small. Equations 49 and 52 were first derived by Brönsted[32] and by

Bjerrum.[33] They follow directly from the transition-state theory, although that theory was not used as such in the original derivation. The assumption must be made that the activated complex has its equilibrium ionic atmosphere in spite of its short lifetime ($\sim 10^{-13}$ second), which is much smaller than the times of relaxation of ionic atmospheres. However, since reacting ions approach each other relatively slowly, their individual ionic atmospheres have time to adjust themselves to very nearly the equilibrium distribution for the activated complex.

An equation identical to (52) can be derived on the basis of the collision theory.[34] This is done simply by considering the perturbing effect of the ionic atmosphere on the calculated values of the electrostatic free energy for the bringing together of two charged reactants. Thus collisions between oppositely charged ions are reduced and collisions between similarly charged ions are enhanced by the presence of the atmospheres, which tend to reduce the electrostatic attraction in the first case and the repulsion in the second.

Equation 52 predicts a linear relationship if log k is plotted against the square root of the ionic strength, with a slope proportional to the product $Z_A Z_B$. This is found to be true qualitatively at least for a number of reactions between ions. The quantitative agreement is excellent in a number of samples, the observed slope agreeing with the Debye-Hückel theoretical slope.[35] However, the relationship has been much abused, experimental data being plotted against the square root of ionic strength for concentrated solutions. Since the limiting law is only valid for solutions below 0.01 molal for 1–1 electrolytes, and for lower concentrations still for higher charged ions, such procedure has no theoretical justification.

Furthermore, even at quite low concentrations there is often evidence for complex formation between an ion reactant and an added ion of opposite sign which renders equation 52 invalid.[36] Particularly for multiply charged ions and for ions of the transition metals which form strong complexes, it is found that addition products are formed.[37] Such complexes may be reasonably stable or exist only in small concentration. In either case they usually have an influence on the reaction rate, in some cases so great as to be classified as catalytic. Even if complexes are not formed, at higher concentrations specific effects of added ions are certain to be found.

If one of the reactants is a neutral molecule so that $Z_B = 0$, then (50) predicts no effect of the ionic strength. This appears true for very dilute solutions. However, at higher ionic concentration the rate constant may change because of changes in the activity coefficients of the ions not given by the Debye-Hückel theory and because the activity coefficients of neutral molecules are affected by higher ionic strengths. For example,

Hückel[38] proposed a term in the first power of the ionic strength to reproduce the activity coefficient of an ion at higher concentrations:

$$-\ln \gamma_i = \frac{Z_i^2 \alpha \sqrt{\mu}}{1 + \beta a_i \sqrt{\mu}} - b_i \mu \tag{53}$$

Also the activity coefficient of a neutral molecule may be expressed in terms of[39]

$$\ln \gamma_0 = b_0 \mu \tag{54}$$

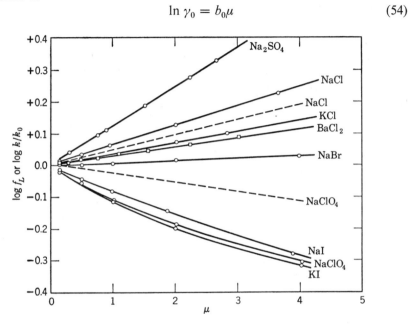

Fig. 2. Plot of logarithm of the activity coefficient (full lines) and of the relative rate of hydrolysis (dotted lines) of γ-butyrolactone in various salt solutions [Long, McDevit, and Dunkle, *J. Phys. & Colloid Chem.*, **55**, 819 (1951)].

We may then write

$$\ln k = \ln k_0 + (b_0 + b_A - b_+)\mu \tag{55}$$

so that the logarithm of the rate constant becomes a linear function of the first power of the ionic strength. This seems to be approximately true for a number of reactions involving two neutral molecules or a neutral molecule and an ion. Unfortunately, the b_0, b_A, and b_+ coefficients are largely empirical, and it is difficult to predict even the sign of the effect on the rate constants. Laidler and Eyring[20] and Amis and Jaffe[40] have attempted a more detailed analysis of the problem.

Figure 2 gives some representative results for a reaction between a neutral molecule and an ion. The reaction is the acid-catalyzed hydrolysis of γ-butyrolactone to γ-hydroxybutyric acid. The rate is proportional to the concentration of hydrogen ion and of lactone. The dotted lines show $\log k/k_0$ plotted against the ionic strength for sodium chloride and sodium perchlorate solutions. The value of k_0 is determined by extrapolation to zero ionic strength. The salt effects are seen to be in opposite directions. For comparison the logarithm of the activity coefficient of the lactone is plotted against the ionic strength for a number of salts. These were determined independently from distribution experiments. It is seen that the opposite salt effects of sodium chloride and perchlorate on the rates are related to the opposite effects of these salts on the activity coefficient of the lactone.

In the event that a reaction involves only neutral molecules it might be anticipated that ionic strength effects would be small except for high ionic concentrations where equation 54 and hence (55) would be applicable. There is the particular case of ionization, however, where a further prediction of salt effects can be made.[41] If one or more neutral molecules are reacting to form oppositely charged ions, as in the hydrolysis of an alkyl halide,

$$RX + H_2O \rightarrow ROH_2^+ + X^- \tag{56}$$

the transition state may be considered a strong dipole. Each end of the dipole will surround itself by an ion atmosphere of opposite charge, which will tend to stabilize the dipole just as an ion is stabilized by its ionic atmosphere. The distribution of such a dipolar ionic atmosphere can be calculated by an extension of the Debye-Hückel theory so as to apply to nonspherical symmetry, such as two opposite charges forming a dipole. Kirkwood[42] gives a complete treatment for a general charge distribution on a spheroid of internal dielectric constant different from that of the solvent.

Considering the simple case of two point charges in a continuous dielectric, the equation for the activity coefficient of the dipole becomes[41]

$$-\ln \gamma_0 = \left(\frac{8\pi N\epsilon^2}{1000\,DkT}\right)^{\frac{1}{2}} Z^2\, d\alpha\mu \tag{57}$$

where α has the same value as in equation 50, Z is the fractional charge of each pole, and d is the distance between them. Now, if it is permissible to neglect the dipoles of the reactants in comparison with the large dipole of the transition state, the rate constant for a reaction such as (56) can be expressed as

$$\ln k = \ln k_0 + \left(\frac{8\pi N\epsilon^2}{1000\,DkT}\right)^{\frac{1}{2}} Z^2\, d\alpha\mu \tag{58}$$

which is similar to (55) in that it predicts a linear dependence of the logarithm of the rate constant on the ionic strength. The magnitude of this effect can be estimated, however, since Z will be approximately one-half and d a little larger than the normal bond distance. Also since Z^2 is positive, the rate always increases with increasing ionic strength.

In the hydrolysis of a 0.1 molar t-butyl bromide solution in 90 per cent acetone-water solvent at 50°, the first-order rate constant is found to increase some 30 per cent as the reaction goes to completion.[41] This is because ions are produced in the process and the ionic strength increases.

Table 10

ACCELERATIONS PRODUCED BY NEUTRAL SALTS IN THE HYDROLYSIS
OF ORGANIC BROMIDES

90 per cent acetone-water solvent, 50° C

(Bateman, Church, Hughes, Ingold, and Taher)

Halide	Added Salt	Molarity	k/k_0 Found	$Z^2 d$, Angstroms
t-Butyl bromide	LiBr	0.1065	1.44	0.82
	LiCl	0.1005	1.39	0.78
	NaN_3	0.1000	1.40	0.80
Benzhydryl chloride	NaN_3	0.1000	2.13	1.74
p-t-Butyl benzhydryl	NaN_3	0.0501	1.70	2.22
chloride	NaN_3	0.1000	2.88	2.22

k_0 is the initial rate constant in the absence of added salts.

The rate-determining step is probably $(CH_3)_3CBr \rightarrow (CH_3)_3C^+ + Br^-$ (see Chapter 12). The observed effect is in good agreement with equation 58 with $Z^2 d$ equal to 0.74 A, which is a reasonable value, the normal carbon bromine bond distance being 2.05 A. Added inert salts have the correct influence of increasing the first-order rate constant as is brought out by Table 10. Also, in the hydrolysis of larger organic halides, the value of $Z^2 d$ needed to give agreement with the experimentally determined accelerations is larger than for t-butyl bromide. This is what would be expected because of the greater spreading of the charge in the activated complex due, for example, to resonance involving structures such as

Equation 58 again, like the Debye-Hückel theory, can only be expected to work in dilute solution and also only in solvents of high dielectric constant. In nonpolar solvents, added ions will interact with reactants and with transition states in a direct way and the concept of an ion atmosphere is invalid. In such solutions the effect of added salts can be quite large. Thus it has been found[43] that the addition of $0.1M$ $LiClO_4$ to ethyl ether will increase the rate of ionization of an organic sulfonate ester by a factor of 10^5. The empirical equation

$$k = k_0 + b \text{ [salt]} \qquad (59)$$

where k_0 is the rate constant in the absence of added salt and b is a constant characteristic of the salt, has been shown to work quite well in solvents such as glacial acetic acid.[44] The reactions studied were solvolysis reactions of organic halides and esters. In such solvents ion-pair formation is extensive.[45]

The Secondary Salt Effect

The effect, described in the previous paragraphs, of the ionic strength on the activity coefficients is called the primary salt effect. For reactions involving catalysis by acids or bases there is a secondary salt effect which has to do with the effect of ionic strength on the dissociation of weak acids or bases. Thus the actual degree of dissociation of a weak acid such as acetic acid can be changed by changing the concentration of salts in solution. If the rate of a reaction depends on the concentration of hydrogen ion coming from the dissociation, it is clear that the rate of reaction will depend on the salt concentration. This phenomenon is quite independent of any influence of the salt on the activity coefficients of the reactants and activated complex which must also be considered.

The theory of the secondary salt effect can be easily worked out for the range of ionic concentrations where the Debye-Hückel theory is valid. If we have an acid of charge Z and its conjugate base of charge $Z - 1$, the equilibrium in question may be written

$$K_a = \frac{[H^+][A^{Z-1}]}{[HA^Z]} \frac{f_{H^+} f_{A^{Z-1}}}{f_{HA^Z}} \qquad (60)$$

Solving for the concentration of hydrogen ion and taking the logarithm

$$\log [H^+] = \log K_a + \log \frac{[HA^Z]}{[A^{Z-1}]} + \log \frac{f_{HA^Z}}{f_{A^{Z-1}} f_{H^+}} \qquad (61)$$

If the assumption is made that the activity coefficient of each of the species is given by equation 50, we may write

$$\log [H^+] = \log K_a + \log \frac{[HA^Z]}{[A^{Z-1}]} - 2\alpha(Z-1)\sqrt{\mu} \qquad (62)$$

which differs from the classical expression by the last term. In the most common case that Z is zero, the effect is such that the concentration of hydrogen ion is increased with increasing ionic strength. Table 11 shows

Table II

DECOMPOSITION OF DIAZOACETIC ESTER IN 0.05M ACETIC ACID

$$CHN_2COOC_2H_5 + H_2O \rightarrow CH_2OHCOOC_2H_5 + N_2$$

Temperature 15° C

[J. N. Brönsted and C. E. Teeter, *J. Phys. Chem.*, 28, 579 (1924)]

Potassium Nitrate, moles/liter	k, min^{-1}	$[H^+] \times 10^4$
0.000	1.27	9.52
0.005	1.31	9.77
0.010	1.35	10.1
0.020	1.37	10.3
0.050	1.42	10.6
0.100	1.46	10.9

the magnitude of the effect on the ionization of acetic acid. The concentration of hydrogen ion is measured by noting the rate of decomposition of diazoacetic ester which is specifically catalyzed by hydrogen ion. An analogous effect would be found for a reaction catalyzed by hydroxide ion, and equation 62 can be used also to solve for the concentration of hydroxide ion by using the ion product of water and the appropriate activity coefficients.

The existence of all the salt effects mentioned indicates the necessity for adequate control of the ionic strength in a kinetic investigation. Either the ionic strength must be kept low so that the effects are small, or a series of measurements must be made and extrapolated to zero ionic concentrations. A different technique, which is often useful, is to keep the ionic strength constant at some large value which does not change sensibly during the course of an investigation. This "swamping" method is particularly useful in determining the order of a reaction which proceeds with a change in ionic strength. However, the rate constants obtained in this way generally are quite different from those in very dilute solution. It is good practice to add small quantities of electrolytes in studying any reaction which may involve ions to see what the influence of ionic strength is. Some care must be exercised to select substances whose ions may be regarded as innocuous in themselves and are acting only because of their charges.

PROBLEMS

1. The rate of dimerization of cyclopentadiene and the reverse rate of dissociation of the dimer have been studied in the gas phase and in the pure liquid [B. S. Khambata and A. Wasserman, *Nature, 137*, 496 (1936); A. Wasserman, *Trans. Faraday Soc., 34*, 128 (1938)]. The values of the frequency factors and activation energies are as follows (concentrations in moles/liter, time in seconds):

	Dimerization Second-Order		Dissociation First-Order	
Gas phase	$10^{6,1}$	16.7 kcal	$10^{13,1}$	35.0 kcal
Pure liquid	$10^{5,7}$	16.0	$10^{13,0}$	34.5

Calculate the equilibrium constant in each case and ΔH^0 and ΔS^0. What properties of the activated complex can be deduced, assuming the mechanism to be the same in both cases?

2. The hydrolysis of benzoyl chloride has been studied in mixtures of water and acetone [B. L. Archer and R. F. Hudson, *J. Chem. Soc.*, 3259 (1950)].

$$C_6H_5COCl + H_2O \rightarrow C_6H_5COOH + H^+ + Cl^-$$

The reaction was found to be first-order in all cases and the rate constants and activation energies determined. From the data calculate the entropy of activation at 25° C for each of the mixtures investigated, and give a reasonable interpretation of the variations. Discuss the possibility of studying this reaction in the gas phase.

Volume % Water	E_a, kcal	$k \times 10^4$, sec^{-1} at 25° C
40.0	16.9	25.25
33.3	16.3	14.11
25.0	14.9	6.94
15.0	13.6	3.00
5.0	11.0	0.55

3. Predict the influence of (*a*) increasing dielectric constant and (*b*) increasing ionic strength on the rates of the following reactions:

$$CH_3Br + H_2O \rightarrow CH_3OH + H^+ + Br^-$$

$$Co(NH_3)_5Br^{+2} + Hg^{+2} + H_2O \rightarrow Co(NH_3)_5H_2O^{+3} + HgBr^+$$

$$Co(NH_3)_5Br^{+2} + NO_2^- \rightarrow Co(NH_3)_5NO_2^{+2} + Br^-$$

$$OH^- + CH_3Br \rightarrow CH_3OH + Br^-$$

$$OH^- + BrCH_2CH_2CH_2CH_2CH_2CO_2^- \rightarrow HOCH_2CH_2CH_2CH_2CH_2CO_2^- + Br^-$$

4. The reaction between persulfate ion and iodide ion is as follows:

$$S_2O_8^{-2} + 2I^- \rightarrow 2SO_4^{-2} + I_2$$

Kinetically the reaction is second-order, first-order each in iodide ion and persulfate ion. The variation of the second-order rate constant with the total concentration of reactants is considerable as the following data shows. Plot the data and compare with the theoretical predictions of the Brönsted-Bjerrum-Christiansen equation. Initial concentration of $K_2S_2O_8$, $0.00015M$. Temperature $25°$ C. [C. V. King and M. B. Jacobs, *J. Am. Chem. Soc.*, **53**, 1704 (1931).]

Potassium Iodide, moles/liter	log k (k in liters/mole-min)
0.0016	$\bar{1}.013$
0.0020	$\bar{1}.021$
0.0032	$\bar{1}.049$
0.0040	$\bar{1}.065$
0.0060	$\bar{1}.072$
0.0080	$\bar{1}.100$
0.0100	$\bar{1}.124$
0.0120	$\bar{1}.143$
0.0180	$\bar{1}.199$
0.0240	$\bar{1}.228$

REFERENCES

1. J. F. Kincaid, H. Eyring, and A. E. Stearn, *Chem. Revs.*, **28**, 301 (1941).
2. M. D. Taylor, *J. Am. Chem. Soc.*, **73**, 315 (1951).
3. E. A. Moelwyn-Hughes, *J. Chem. Soc.*, 850 (1940).
4. R. P. Bell, *Ann. Reports on Progress Chem.*, *Chem. Soc. London*, **36**, 82 (1939).
5. H. Eyring and F. Daniels, *J. Am. Chem. Soc.*, **52**, 1472 (1930).
6. J. H. Raley, F. F. Rust, W. E. Vaughan, and F. H. Seubold, *J. Am. Chem. Soc.*, **70**, 88, 95, 1336 (1948).
7. D. F. Smith, *J. Am. Chem. Soc.*, **49**, 43 (1927).
8. A. Rembaum and M. Szwarc, *J. Am. Chem. Soc.*, **76**, 5975 (1954); M. Levy, M. Steinberg, and M. Szwarc, *ibid.*, 5978.

9. G. B. Kistiakowsky, *J. Am. Chem. Soc.*, *50*, 2315 (1928).
10. G. Scatchard, *Chem. Revs.*, *8*, 321 (1931).
11. J. H. Hildebrand and R. L. Scott, *Solubility of Non-Electrolytes*, Reinhold Publishing Corp., New York, 3rd edition, 1950.
12. R. A. Ogg and M. Polanyi, *Trans. Faraday Soc.*, *31*, 604 (1935).
13. H. S. Frank and M. W. Evans, *J. Chem. Phys.*, *13*, 507 (1945).
14. J. L. Magee, T. Ri, and H. Eyring, *J. Chem. Phys.*, *9*, 419 (1941).
15. E. A. Moelwyn-Hughes, *Proc. Roy. Soc.*, *A155*, 308 (1936).
16. W. F. K. Wynne-Jones and H. Eyring, *J. Chem. Phys.*, *3*, 492 (1935).
17. (*a*) R. G. Pearson and D. C. Vogelsong, *J. Am. Chem. Soc.*, *80*, 1038, 1048 (1958); (*b*) C. C. Swain and E. E. Pegues, *ibid.*, *80*, 1012 (1958); (*c*) E. D. Hughes, C. K. Ingold, S. Patai, S. F. Mok, and Y. Pocker, *J. Chem. Soc.*, 1206 ff. (1957).
18. C. G. Swain, *ibid.*, *70*, 1119 (1948); L. C. King and D. L. Brebner, *ibid.*, *75*, 2330 (1953).
19. J. G. Kirkwood, *J. Chem. Phys.*, *2*, 351 (1934).
20. K. J. Laidler and H. Eyring, *Ann. N.Y. Acad. Sci.*, *39*, 303 (1940).
21. R. G. Pearson, *J. Chem. Phys.*, *20*, 1478 (1952).
22. H. C. Brown and J. D. Brady, *J. Am. Chem. Soc.*, *74*, 3570 (1952).
23. E. F. Caldin and J. Peacock, *Trans. Faraday Soc.*, *51*, 1217 (1955); reference 17(*a*) and R. F. Hudson and B. Saville, *J. Chem. Soc.*, 4121 (1955).
24. S. H. Maron and V. K. La Mer, *J. Am. Chem. Soc.*, *61*, 2018 (1939); R. G. Pearson and J. M. Mills, *J. Am. Chem. Soc.*, *72*, 1692 (1950).
25. J. A. Christiansen, *Z. physik. Chem.*, *113*, 35 (1924).
26. V. K. La Mer, *J. Franklin Inst.*, *225*, 709 (1938); E. S. Amis, *J. Am. Chem. Soc.*, *63*, 1606 (1941).
27. R. P. Bell and F. J. Lindars, *J. Chem. Soc.*, 4601 (1954).
28. G. Scatchard, *Chem. Revs.*, *10*, 229 (1932).
29. E. S. Amis and V. K. La Mer, *J. Am. Chem. Soc.*, *61*, 905 (1939).
30. C. V. King and J. J. Josephs, *J. Am. Chem. Soc.*, *66*, 767 (1944).
31. M. Born, *Z. Physik*, *1*, 45 (1920).
32. J. N. Brönsted, *Z. physik. Chem.*, *102*, 169 (1922); *115*, 337 (1925).
33. N. Bjerrum, *Z. Physik. Chem.*, *108*, 82 (1924); *118*, 251 (1925).
34. J. A. Christiansen, *Z. physik. Chem.*, *113*, 35 (1924); G. Scatchard, *Chem. Revs.*, *10*, 229 (1932).
35. V. K. La Mer, *Chem. Revs.*, *10*, 179 (1932); *J. Franklin Inst.*, *225*, 709 (1938).
36. A. R. Olson and T. R. Simonson, *J. Chem. Phys.*, *17*, 1167 (1949).
37. R. P. Bell and J. E. Prue, *J. Chem. Soc.*, 362 (1949); R. P. Bell and G. M. Waind, *ibid.*, 1979 (1950); J. I. Hoppé and J. E. Prue, *ibid.*, 1775 (1957).
38. E. Hückel, *Physik. Z.*, *26*, 93 (1925).
39. P. Debye and J. McAulay, *Physik. Z.*, *26*, 22 (1925).
40. E. S. Amis and G. Jaffe, *J. Chem. Phys.*, *10*, 598 (1942).
41. L. C. Bateman, M. G. Church, E. D. Hughes, C. K. Ingold, and N. A. Taher, *J. Chem. Soc.*, 979 (1940).
42. J. G. Kirkwood, *Chem. Res.*, *24*, 233 (1939).
43. S. Winstein, S. Smith, and D. Darwish, *J. Am. Chem. Soc.*, *81*, 5511 (1959).
44. A. H. Fainberg and S. Winstein, *J. Am. Chem. Soc.*, *78*, 2763 (1956).
45. S. Bruckenstein and I. M. Kolthoff, *J. Am. Chem. Soc.*, *79*, 1, 5915 (1957).

8.

COMPLEX REACTIONS

Most reactions occurring in nature or in the laboratory do not take place at a single collision between reactant molecules, but have a mechanism which involves several such *elementary processes* or *reaction steps*. Such reactions are called *complex reactions*.

If a reaction mechanism involves a reactant which can undergo two or more reactions independently and concurrently, the reactions are called *parallel* or *side* reactions. If, on the other hand, there is a set of reactions such that the product of one reaction is the reactant for the next, the reactions are called *series* or *consecutive* reactions.

Complex reaction mechanisms may involve various combinations of series and parallel reactions. These give rise to special forms, some of which are generally recognized with special names such as competitive reactions, reversible reactions, coupled reactions, chain reactions, and some types of catalytic reactions.

In dealing with complex reactions we shall treat many general types, discussing the more important ones in detail and giving examples of actual reactions as observed in the laboratory.

Parallel First-Order Reactions

Let the mechanism be given by the following steps with corresponding rate constants

$$A \xrightarrow{k_1} U$$
$$A \xrightarrow{k_2} V \quad\quad (1)$$
$$A \xrightarrow{k_3} W$$

and so on. Then, using A, U, V, and W to represent the corresponding concentrations, and A_0, etc., for the initial concentrations

$$-dA/dt = k_1A + k_2A + k_3A = (k_1 + k_2 + k_3)A$$
$$= kA \qquad k = k_1 + k_2 + k_3 \qquad (2)$$

and

$$\ln (A_0/A) = kt$$

or

$$A = A_0e^{-kt} \qquad (3)$$

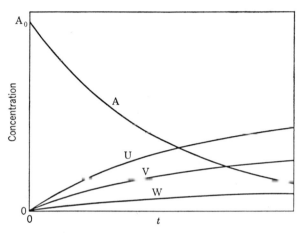

Fig. I. Concentration-time curves for parallel first-order reactions. A, reactant; U, V, W, products.

The reaction is simple first-order as far as A is concerned. Also

$$dU/dt = k_1A = k_1A_0e^{-kt}$$

and

$$U = \frac{-k_1A_0}{k} e^{-kt} + \text{constant}$$

or

$$U = U_0 + (k_1A_0/k)(1 - e^{-kt})$$
$$V = V_0 + (k_2A_0/k)(1 - e^{-kt}) \qquad (4)$$
$$W = W_0 + (k_3A_0/k)(1 - e^{-kt})$$

If $U_0 = V_0 = W_0 = 0$ the equations simplify and it is readily seen that

$$V/U = k_2/k_1 \qquad W/U = k_3/k_1 \qquad (5)$$

or

$$U:V:W = k_1:k_2:k_3$$

The products are in constant ratio to each other, independent of time and initial concentration of the reactant.

Graphically the results appear as in Figs. 1 and 2 for the particular conditions indicated. By using as abscissa $1 - e^{-kt}$ in Fig. 2 the curves become straight lines. This is because all the concentrations vary exponentially with the same exponential constant k. All have the same half-life (or half-growth time for U, V, and W) $t_{1/2} = (1/k) \ln 2$, despite having

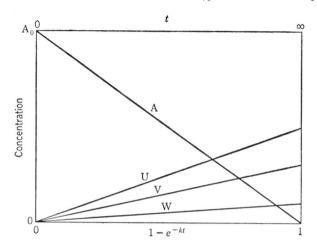

Fig. 2. Linear plot for parallel first-order reactions.

different rate constants. For an example see the isomerization of α-pinene in Chapter 12.

If the rate of a set of reactions of this type is immeasurably fast, it is still possible to determine relative rate constants by measuring the relative concentrations of the products and using equations 5.[1]

Two Parallel First-Order Reactions Producing a Common Product

$$A \xrightarrow{k_1} C + \cdots$$

$$B \xrightarrow{k_2} C + \cdots$$

Such reactions, of course, take place independently as far as the reactants are concerned. But, if the concentration of the common product C or the sum of the reactants is measured experimentally, the task of finding k_1 and k_2 becomes more complex. This situation commonly arises in radioactivity when the radiation from mixed radioactive substances is being observed.

A chemical example is provided by the observations of Brown and Fletcher[2] on the hydrolysis of mixed tertiary aliphatic chlorides.

The rate equations $-dA/dt = k_1 A$ and $-dB/dt = k_2 B$ integrate to

$$A = A_0 e^{-k_1 t} \quad \text{and} \quad B = B_0 e^{-k_2 t}$$

But

$$C = A_0 - A + B_0 - B$$

$$= C_\infty - A_0 e^{-k_1 t} - B_0 e^{-k_2 t}$$

where $C_\infty = A_0 + B_0$ and

$$\log (C_\infty - C) = \log (A_0 e^{-k_1 t} + B_0 e^{-k_2 t})$$

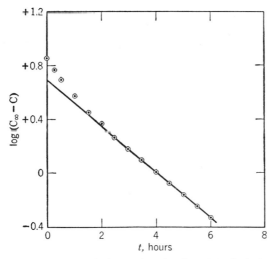

Fig. 3. Deviation from a simple first-order plot for the hydrolysis of diethyl-*t*-butyl-carbinyl chloride (Brown and Fletcher).

If there were just one reactant A or B, or if $k_1 = k_2$, a plot of $\log (C_\infty - C)$ versus t would be linear. If, on the other hand, both are present and $\log (A + B)$ or $\log (C_\infty - C)$ is plotted versus t, there will in general be a curve (k_1 and k_2 assumed unequal). Consider the results of Brown and Fletcher on the hydrolysis of what was supposed to be diethyl-*t*-butyl-carbinyl chloride. Figure 3 shows the curvature just described. After some time the curve becomes linear because the more reactive component, say A, has disappeared and the expression for $\log (C_\infty - C)$ becomes

$$\log B = \log (C_\infty - C) = \log B_0 - (k_2 t / 2.303)$$

Therefore, from the slope and intercept of the line, B_0 and k_2 may be determined and B found at any time. With this information, A may be calculated by difference

$$A = C_\infty - C - B$$

and then log A versus t results in values of A_0 and k_1. Figure 4 shows plots for A and B separately. These compounds are interpreted as being isomers

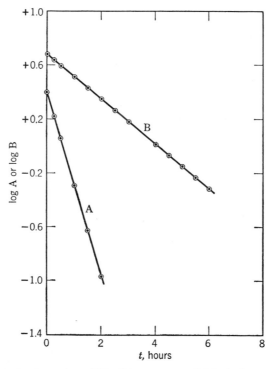

Fig. 4. Analysis of reaction of Fig. 3 into two parallel hydrolyses of isomers (Brown and Fletcher).

which arise in the synthesis of the tertiary chloride. The relative values of A_0 and B_0 indicate that A is 35 per cent and B 65 per cent of the original mixture.

Parallel Higher-Order Reactions, All of the Same Order

Suppose that the mechanism is:

$$aA + bB \xrightarrow{k_1} U + \cdots$$
$$aA + bB \xrightarrow{k_2} V + \cdots$$
$$aA + bB \xrightarrow{k_3} W + \cdots$$

Then

$$-dA/dt = (k_1 + k_2 + k_3)A^a B^b \tag{6}$$

As far as reactants A and B are concerned, the complex reaction behaves kinetically as each of its component parts but with an effective rate constant $k = k_1 + k_2 + k_3$. Although the integrated expressions for the concentrations as a function of the time may be complicated, a simple relation holds between concentrations of products as in the previous example of first-order reactions. Here

$$dU/dt = k_1 A^a B^b$$
$$dV/dt = k_2 A^a B^b \tag{7}$$

Then

$$dV/dU = k_2/k_1 \quad \text{and} \quad V/U = k_2/k_1 \tag{8}$$

if $U_0 = V_0 = 0$.

The products will appear in constant ratio despite the order of the reaction, assuming only that the orders are the same. It is therefore possible to determine the relative rate constants, for example, for the formation of isomers, without knowing the order of the reaction. From the temperature dependence of these relative rate constants, differences of heats of activation and entropies of activation can be calculated. This is of interest in connection with the study of the effect of substituents on reaction rate.

Parallel First- and Second-Order Reactions

The hydrolysis of an organic halide may take place as an S_N1 or first-order reaction or by an S_N2 or second-order reaction. In certain cases both mechanisms may occur side by side.[3] Assuming the first-order mechanism

$$A \xrightarrow{k_1} D + E$$
$$E + B \xrightarrow{\text{fast}} C$$

and the second-order mechanism

$$A + B \xrightarrow{k_2} C + D$$

where A is the organic halide and B is hydroxide ion, the rate of reaction may be written

$$dx/dt = k_1(a - x) + k_2(a - x)(b - x)$$

and easily integrated to give

$$\frac{1}{[(k_1/k_2) + b - a]} \ln \left[\frac{a}{[(k_1/k_2) + b]} \cdot \frac{[(k_1/k_2) + b - x]}{(a - x)} \right] = k_2 t$$

However, this equation is not very useful from the standpoint of determining the constants k_1 and k_2 or of testing the fit to experimental data. Young and Andrews have used the differential equation in the form

$$(dx/dt)/(a - x) = k_1 + k_2(b - x)$$

to apply to data on the hydrolysis of primary butenyl chloride by plotting $(dx/dt)/(a - x)$ versus $(b - x)$.[4]

Series First-Order Reactions

Consider the case

$$A \xrightarrow{k_1} B$$
$$B \xrightarrow{k_2} C$$

This is a commonly discussed mechanism and is exemplified by certain hydrolyses, by radioactive series, and by the reaction of potassium permanganate, manganous sulfate, and oxalic acid, as in the original work of Harcourt and Esson. Esson[5] first integrated the differential equations which are as follows:

$$dA/dt = -k_1A$$
$$dB/dt = k_1A - k_2B \tag{9}$$
$$dC/dt = k_2B$$

The equation for A integrates readily into

$$A = A_0 e^{-k_1 t} \tag{10}$$

This substituted into the second equation yields

$$dB/dt = k_1 A_0 e^{-k_1 t} - k_2 B \tag{11}$$

This linear first-order equation may be integrated by the usual integrating factor methods to give (if $B_0 = 0$)

$$B = \frac{A_0 k_1}{k_2 - k_1} (e^{-k_1 t} - e^{-k_2 t}) \tag{12}$$

C may be found most conveniently from the stoichiometric relation obtained by integrating the sum of equations 9

$$(dA/dt) + (dB/dt) + (dC/dt) = 0 \tag{13}$$

Therefore

$$A + B + C = A_0 \tag{14}$$

if neither B nor C is present initially. Then

$$C = A_0 - A - B$$

or

$$C = A_0\left[1 + \frac{1}{k_1 - k_2}(k_2e^{-k_1t} - k_1e^{-k_2t})\right] \qquad (15)$$

These integrations may also be performed by the general method to be discussed in a later section.

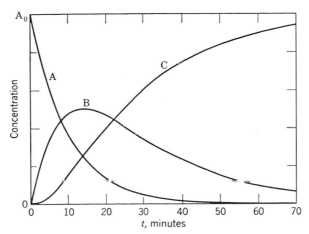

Fig. 5. Concentration-time curves for substances A, B, and C in series first-order reactions.

Figure 5 is a plot of the concentrations A, B, and C as a function of the time for a typical case where $k_1 = 0.1$ min^{-1} and $k_2 = 0.05$ min^{-1}.

The equations 10, 12, and 15 can be put in simpler form by introducing the following dimensionless parameters and variables. Let $\alpha = A/A_0$, $\beta = B/A_0$, $\gamma = C/A_0$. These are essentially concentrations relative to the initial value A_0 and can vary in the range from 0 to 1. Also let $\tau = k_1t$ and $\kappa = k_2/k_1$. Then the equations become

$$\alpha = e^{-\tau} \qquad (16)$$

$$\beta = \frac{1}{\kappa - 1}(e^{-\tau} - e^{-\kappa\tau}) \qquad (17)$$

$$\gamma = 1 + \frac{1}{1 - \kappa}(\kappa e^{-\tau} - e^{-\kappa\tau}) \qquad (18)$$

To visualize the concentration-time relations most effectively it is useful to consider plots of α, β, and γ as functions of τ for various values of

κ as shown in Fig. 6. Instead of using a linear scale for τ, $1 - e^{-\tau}$ or $1 - \alpha$ is plotted linearly. This has the advantage of showing the whole range of time from 0 to ∞ as $1 - \alpha$ goes from 0 to 1, and it also causes certain relations to be linear. It is of interest to note that the concentration of

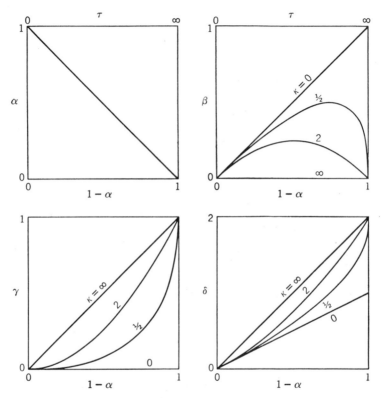

Fig. 6. Plots of relative concentrations α, β, γ, and δ for series first-order reactions as a function of $1 - \alpha$ for various values of κ, the relative rate constant.

the intermediate B, as measured by β, goes through a maximum, the position of which depends on κ, the relative value of the rate constants. By using (17) and setting $d\beta/d\tau = 0$ it is found that

$$\tau_{\max} = \frac{1}{\kappa - 1} \ln \kappa \tag{19}$$

The maximum shifts to smaller τ as κ gets larger and larger. Also the value of β at the maximum is

$$\beta_{\max} = \kappa^{\kappa/(1-\kappa)} \tag{20}$$

which shows that the maximum becomes less pronounced as κ gets larger.

As far as C is concerned its rate of formation is slow at first. This is an example of an *induction period*. The duration of the induction period, arbitrarily taken as the time to reach the point of inflection on the C versus t curve, as in Fig. 5, is easily seen to be equal to the time for B to reach its maximum value since, when $d^2C/dt^2 = 0$, $dB/dt = 0$ from (9).

Included in Fig. 6 is a plot of a function δ which is defined as $\delta = \beta + 2\gamma$. Therefore

$$\delta = 2 - \frac{(1 - 2\kappa)}{(1 - \kappa)} e^{-\tau} - \frac{1}{(1 - \kappa)} e^{-\kappa\tau} \qquad (21)$$

Such a combination of relative concentrations is more closely related to what is usually measured in a reaction than is β or γ alone. In the hydrolysis of a dihalide, δ is a measure of the halide ion produced. In the saponification of a diester with large excess of base, δ measures the amount of base used up relative to the original diester. In general, whenever the same change in some by-product or reactant is produced in each step of the reaction, δ is a useful measure of the extent of reaction. δ varies from 0 to 2 as the reaction proceeds.

The practical problem of the determination of rate constants from experimental data must now be considered. This has been handled in some detail by C. G. Swain.[6] However, two alternate methods which are more convenient than the Swain method will be discussed here.

A Graphical Method

R. E. Powell[7] has shown that for simple, as well as many complex, reactions a plot of relative concentration or per cent reaction versus log t produces a curve the form of which depends only on the type of reaction and possibly some dimensionless parameters. In the present application it is most useful to plot per cent reaction, as given by 50δ, versus log τ as in Fig. 7. There is a family of curves, each identified by a particular value of the parameter κ. Now, since $\tau = k_1 t$ and log $\tau = $ log $k_1 + $ log t, a plot of experimental values of 50δ versus log t will be the same as one of the family but shifted horizontally by an amount $-$log k_1. Therefore, if this experimental plot is made on thin paper, the curve may be matched to one of the family, thus determining the value of κ, and, from the shift of log t with respect to log τ, k_1 may be determined. This technique, besides determining the rate constants, indicates whether the reaction being studied is really of the kind being considered here. For it is unlikely that a similar plot for some other complex mechanism would show just the same form as one of the family of curves calculated in this case from equation 21.

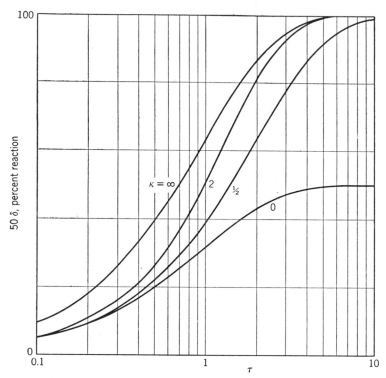

Fig. 7. Per cent reaction versus time parameter τ on logarithmic scale for series first-order reactions.

The Time Ratio Method

The Powell plot method, being graphical, is not so accurate as might be desired. A time ratio method for the accurate determination of the rate constants for competing second-order reactions has been presented.[8,27] This method is of rather general applicability and will now be given as it applies to successive first-order reactions. Let τ_1 and τ_2 be the values of τ corresponding to the values δ_1 and δ_2, which are determined by certain percentages of reactions. The ratio τ_2/τ_1 is equal to the ratio t_2/t_1 of the actual times of reaction for these same percentages. But from equation 21 it is apparent that τ_2/τ_1 and therefore this time ratio are functions only of κ for given fixed δ_1 and δ_2. Therefore a graph or a table of values of κ versus t_2/t_1 can be used to evaluate κ, following which a graph or table of κ versus τ for a certain δ will lead to a value of k_1. Thus the constants are determined. The result may, and should be, verified by using additional pairs of times for certain other combinations of percentage reactions.

Table 1 is from Swain but with certain additions and change in notation. Values of τ and time ratio values are given for various κ's and for 15 per cent, 35 per cent, and 70 per cent reaction, that is, $\delta = 0.30, 0.70$, and 1.40.

Swain has applied his method to the data of Kaufler[9] on the hydrolysis of 2,7-dicyanonaphthalene. The time ratio method gives the same result but with greater ease.

Table I

SERIES FIRST-ORDER REACTIONS. TIME-PERCENTAGE REACTION RELATIONS FOR VARIOUS RELATIVE RATE CONSTANTS

(Swain,[6] modified)

κ	τ_{15}	τ_{35}	τ_{70}	$\log t_{35}/t_{15}$	$\log t_{70}/t_{15}$	$\log t_{70}/t_{35}$
100	0.168	0.436	1.21	0.415	0.858	0.443
50	0.172	0.441	1.21	0.407	0.847	0.440
20	0.188	0.457	1.23	0.385	0.815	0.430
10	0.209	0.484	1.26	0.366	0.781	0.415
5	0.236	0.536	1.32	0.356	0.748	0.392
2	0.277	0.664	1.54	0.367	0.745	0.378
1.5	0.289	0.686	1.65	0.376	0.757	0.381
1.1	0.300	0.734	1.80	0.388	0.779	0.391
0.9	0.308	0.766	1.93	0.395	0.796	0.401
0.7	0.315	0.806	2.11	0.409	0.826	0.417
0.5	0.324	0.863	2.41	0.425	0.871	0.446
0.2	0.342	0.999	3.81	0.465	1.047	0.582
0.1	0.349	1.078	6.19	0.490	1.249	0.759
0.05	0.353	1.132	11.10	0.506	1.497	0.991
0.02	0.355	1.173	26.55	0.519	1.874	1.355
0.01	0.356	1.188	52.09	0.524	2.166	1.642

An interesting application of the theory of consecutive first-order reactions to the problem of measuring the rate of rapid simple first-order reactions has been given by Bell and Clunie.[10] The reaction, such as the hydration of acetaldehyde, is carried out under two conditions: (a) adiabatically, and (b) in a well-stirred reaction vessel which can transfer heat to the surrounding thermostat. The temperature rise due to the heat of reaction is measured in both cases. The ratio of T_m, the maximum rise in (b), to T_0, the total rise on complete reaction in (a), is analogous to β_{max} of equation 20.

$$T_m/T_0 = \kappa^{\kappa/(1-\kappa)} \qquad \kappa = k_2/k_1$$

After determining k_2, the specific rate of heat loss in (b), k_1, the rate constant of the desired reaction, can be determined without any measurement of time. It is said that k_1 can be determined to within 5 per cent with temperature rises less than 0.1°.

The Stationary State

Many reactions more complex than the type just discussed can be handled mathematically only by making certain simplifying assumptions. An assumption of particular value, especially for chain reactions, is the stationary-state or steady-state approximation. It is assumed in such a case that during most of the reaction the concentration of each intermediate, such as B in the present situation, may be considered essentially constant. This approximation is particularly good when the intermediates are very reactive and therefore present at very small concentrations.

Since the reaction just considered can be solved exactly, it is useful to test the stationary-state approximation here and to investigate the conditions under which it is satisfactory.

Assuming

$$dB/dt = 0 = k_1 A - k_2 B \tag{22}$$

there results

$$B/A = k_1/k_2 \tag{23}$$

and since

$$A = A_0 e^{-k_1 t}$$

$$B = A_0 \frac{k_1}{k_2} e^{-k_1 t} \tag{24}$$

Also

$$C = A_0 - A - B = A_0 \left[1 - \left(1 + \frac{k_1}{k_2} \right) e^{-k_1 t} \right] \tag{25}$$

Equations 24 and 25 are to be compared with the exact equations 12 and 15. It will be seen that the equations become equivalent if $k_2 \gg k_1$ and $t \gg 1/k_2$. The first condition states that the intermediate is very reactive compared with the original reactant and also results in the concentration of B being very low. The second condition ensures that the induction period has been passed. Usually where the stationary-state approximation is applied the induction period is immeasurably short, so that the method can be used with confidence. It can also be seen that (24) shows that the concentration of B is not constant, but is decreasing with time. Because of its low value, however, the approximation that $dB/dt = 0$ is still a good one.

Radioactive Series and Radioactive Steady State

Neglecting the branching that may occur in a radioactive series, the successive disintegrations constitute a sequence of first-order processes, such as

$$A \xrightarrow{k_1} B \xrightarrow{k_2} C \xrightarrow{k_3} D \xrightarrow{k_4} \cdots$$

The solution of the differential equations is known, and a variety of cases are discussed by Rutherford, Chadwick, and Ellis.[11]

When a long-lived isotope such as U^{238} is producing a series of daughters, a radioactive "equilibrium" or steady state is reached in which each daughter radioactive isotope attains a nearly constant concentration or amount relative to the parent. The relative amounts follow from the equations

$$dB/dt = 0 = k_1 A - k_2 B$$

$$dC/dt = 0 = k_2 B - k_3 C \quad \text{etc.} \tag{26}$$

As a result

$$B = (k_1/k_2)A$$

$$C = (k_2/k_3)B = (k_1/k_3)A \quad \text{etc.} \tag{27}$$

Because half-life periods are inversely proportional to rate constants, it is seen that each daughter is present in an amount proportional to its half-life period.

General First-Order Series and Parallel Reactions

Consider a set of m substances with concentrations A_1, A_2, $\cdots$, A_m, and suppose that each substance may react by a first-order process to form each of the other substances as in the accompanying diagram.

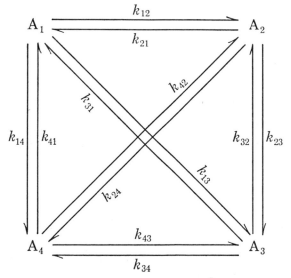

Let the rate constants be in general k_{ij} where the first subscript refers to the substance reacting and the second to the substance formed. Assume that the stoichiometric equations are correspondingly simple, that is, one molecule produced for one reacting.

The m rate equations are then of the form

$$dA_i/dt = -k_{i1}A_i - k_{i2}A_i \cdots + k_{1i}A_1 + k_{2i}A_2 + \cdots$$

or

$$dA_i/dt = (-k_{i1} - k_{i2} \cdots -k_{im})A_i + k_{1i}A_1 + k_{2i}A_2 + \cdots k_{mi}A_m \quad (28)$$

This set of m differential equations in m dependent variables A_i may be integrated in closed form.[12] The notation of Matsen and Franklin will be used here.

Equations 28 may be put into the simpler form

$$dA_i/dt + \sum_{j=1}^{m} K_{ij}A_j = 0 \quad (29)$$

where $K_{ij} = -k_{ji}$ except when $i = j$, and $K_{ii} = \sum_{p=1}^{m} k_{ip}$ except when $p = i$.

Now assume a *particular* solution of the form

$$A_i = B_i e^{-\lambda t} \quad (30)$$

where the B_i are constants and λ is a parameter to be determined. Substitution of (30) into (29) and canceling out the exponential yield the set of simultaneous homogeneous linear equations in the B_i's:

$$-\lambda B_i + \sum_{j=1}^{m} K_{ij}B_j = 0 \quad (31)$$

or

$$\sum_{j=1}^{m}(K_{ij} - \delta_{ij}\lambda)B_j = 0 \qquad i = 1, 2 \cdots m \quad (32)$$

where $\delta_{ij} = 1$ if $i = j$ or 0 if $i \neq j$. The condition for a nontrivial solution for the B_i is that the determinant of the coefficients is zero.

$$|K_{ij} - \delta_{ij}\lambda| = 0 \quad (33)$$

This determinantal equation, or secular equation, as it is often called, is an m-th degree algebraic equation in the parameter λ. It will have as solutions m values of λ. Let these values of λ be λ_r where $r = 1, 2 \cdots m$. Corresponding to each λ_r there will be a set of relative values of B_j which are solutions of equations 32. Let these B_j's be designated B_{jr}.

These particular solutions will not usually be the solutions desired since they do not satisfy the initial conditions of concentration. It is necessary to have a *general* solution which may be written as a linear combination of particular solutions. Let

$$A_i = \sum_{r=1}^{m} B_{ir}Q_r{}^0 e^{-\lambda_r t} \quad (34)$$

where Q_r^0 are coefficients in the linear combinations and may be determined from the initial conditions. Any linear combination of particular solutions is also a solution of the differential equations 29 since the latter are linear equations.

As an example consider the reaction scheme

$$A_1 \underset{k_{21}}{\overset{k_{12}}{\rightleftharpoons}} A_2 \underset{k_{32}}{\overset{k_{23}}{\rightleftharpoons}} A_3$$

that is, two consecutive reversible first-order reactions. This situation has been discussed in detail by Lowry and John.[13] The solution by the present method involves first the secular equation 33 which is explicitly

$$\begin{vmatrix} k_{12} - \lambda & -k_{21} & 0 \\ -k_{12} & k_{21} + k_{23} - \lambda & -k_{32} \\ 0 & -k_{23} & k_{32} - \lambda \end{vmatrix} = 0 \qquad (35)$$

or when expanded

$$-\lambda^3 + \lambda^2(k_{12} + k_{21} + k_{23} + k_{32}) - \lambda(k_{12}k_{23} + k_{21}k_{32} + k_{12}k_{32}) = 0 \qquad (36)$$

The three solutions for λ_r are

$$\lambda_1 = 0$$
$$\lambda_2 = \tfrac{1}{2}(p + q) \qquad (37)$$
$$\lambda_3 = \tfrac{1}{2}(p - q)$$

where $p = (k_{12} + k_{21} + k_{23} + k_{32})$ and $q = [p^2 - 4(k_{12}k_{23} + k_{21}k_{32} + k_{12}k_{32})]^{1/2}$.

Substitution of these λ_r into (32) gives three equations any two of which are independent and which can be solved for B_{2r} and B_{3r}, making the arbitrary assumption that the various B_{1r} are unity. Letting

$$B_{1r} = 1$$
$$B_{2r} = \frac{k_{12} - \lambda_r}{k_{21}} \qquad (38)$$
$$B_{3r} = \frac{k_{23}(k_{12} - \lambda_r)}{k_{21}(k_{32} - \lambda_r)}$$

the general solution is then

$$A_1 = \sum_r Q_r^0 e^{-\lambda_r t}$$
$$A_2 = \sum_r Q_r^0 \frac{(k_{12} - \lambda_r)}{k_{21}} e^{-\lambda_r t} \qquad (39)$$
$$A_3 = \sum_r Q_r^0 \frac{k_{23}(k_{12} - \lambda_r)}{k_{21}(k_{32} - \lambda_r)} e^{-\lambda_r t}$$

Suppose that, at time $t = 0$, $A_1 = A_1{}^0$ and $A_2 = A_3 = 0$, that is, only the first reactant is present initially. Then equations 39 become

$$A_1{}^0 = \sum_r Q_r{}^0$$

$$0 = \sum_r Q_r{}^0 \frac{(k_{12} - \lambda_r)}{k_{21}} \tag{40}$$

$$0 = \sum_r Q_r{}^0 \frac{k_{23}(k_{12} - \lambda_r)}{k_{21}(k_{32} - \lambda_r)}$$

These when solved for $Q_r{}^0$ result in

$$Q_1{}^0 = A_1{}^0 \frac{k_{21}k_{32}}{\lambda_2\lambda_3}$$

$$Q_2{}^0 = A_1{}^0 \frac{k_{12}(\lambda_2 - k_{23} - k_{32})}{\lambda_2(\lambda_2 - \lambda_3)} \tag{41}$$

$$Q_3{}^0 = A_1{}^0 \frac{k_{12}(k_{23} + k_{32} - \lambda_3)}{\lambda_3(\lambda_2 - \lambda_3)}$$

And finally by substitituon into (39) and simplification

$$A_1 = A_1{}^0 \left\{ \frac{k_{21}k_{32}}{\lambda_2\lambda_3} + \frac{k_{12}(\lambda_2 - k_{23} - k_{32})}{\lambda_2(\lambda_2 - \lambda_3)} e^{-\lambda_2 t} \right.$$

$$\left. + \frac{k_{12}(k_{23} + k_{32} - \lambda_3)}{\lambda_3(\lambda_2 - \lambda_3)} e^{-\lambda_3 t} \right\}$$

$$A_2 = A_1{}^0 \left\{ \frac{k_{12}k_{32}}{\lambda_2\lambda_3} + \frac{k_{12}(k_{32} - \lambda_2)}{\lambda_2(\lambda_2 - \lambda_3)} e^{-\lambda_2 t} + \frac{k_{12}(\lambda_3 - k_{32})}{\lambda_3(\lambda_2 - \lambda_3)} e^{-\lambda_3 t} \right\}$$

$$A_3 = A_1{}^0 \left\{ \frac{k_{12}k_{23}}{\lambda_2\lambda_3} + \frac{k_{12}k_{23}}{\lambda_2(\lambda_2 - \lambda_3)} e^{-\lambda_2 t} - \frac{k_{12}k_{23}}{\lambda_3(\lambda_2 - \lambda_3)} e^{-\lambda_3 t} \right\} \tag{42}$$

Lowry and John point out that the concentration of the intermediate (A_2) may or may not go through a maximum, depending on the relative values of the rate constants.

This solution, of course, includes as a special case the irreversible consecutive first-order reactions discussed earlier in this chapter. It may easily be verified that, when we set $k_{12} = k_1$, $k_{23} = k_2$, and $k_{21} = k_{32} = 0$, the expressions for A_1, A_2, and A_3 reduce to equations 10, 12, and 15 for A, B, and C.

Secular equations appear elsewhere in physical chemistry, in particular, in the calculation of the fundamental frequencies in the vibration of polyatomic molecules and also in quantum-mechanical calculations of electronic energies of molecules. Matsen and Franklin have presented an analogy between concentration-time relations in sets of first-order reactions and normal modes of vibration. It is useful to realize that the mathematical methods used in vibration and quantum-mechanical calculations may be applied here. For reactions involving first-order interactions of four or more components the secular equation becomes difficult. However, special methods have been devised to solve them.[14]

Higher-Order Series Reactions

Various possible combinations of consecutive first-order and second-order reactions have been considered in detail by Chien,[15] who has carried out the difficult integrations, expressing the results in terms of special functions such as Bessel's. The cases solved are as follows:

First-order– Second-order	Second-order– First-order	Second-order– Second-order
$A \rightarrow B$	$2A \rightarrow B$	$2A \rightarrow B$
$2B \rightarrow C$	$B \rightarrow C$	$2B \rightarrow C$
and		and
$A \rightarrow B$		$2A \rightarrow B$
$B + D \rightarrow C$		$B + D \rightarrow C$

As a typical solution consider the first case:

$$A \rightarrow B$$

$$2B \rightarrow C$$

Integration of the differential equations results in

$$A = a_0 \tau$$

$$B = a_0 \sqrt{\tau/k} \ \frac{iJ_1(2i\sqrt{\kappa\tau}) - \beta H_0^{(1)}(2i\sqrt{\kappa\tau})}{J_0(2i\sqrt{\kappa\tau}) + \beta H_0^{(1)}(2i\sqrt{\kappa\tau})}$$

$$C = \tfrac{1}{2}(a_0 - A - B)$$

where $\tau = e^{-k_1 t}$, $\kappa = a_0 k_2/k_1$, and $\beta = iJ_1(2k\sqrt{\kappa})/H_1^{(1)}(2i\sqrt{\kappa})$. The J's and H's are Bessel functions.[16]

Chien gives no specific chemical examples of these reactions, and such examples are difficult to find. However, Pearson, King, and Langer[17] have an example only slightly more complicated, arising in the mechanism of formation of cholesterylisothiouronium tosylate from thiourea and cholesteryl tosylate. This involves two concurrent first-order processes, one of which is followed by a second-order reaction.

In more complicated mechanisms the complete solution may be impossible except by approximation. Pshezhetskii and Rubinshtein[18] have used a determinantal method to solve for the relative concentrations.

Competitive, Consecutive Second-Order Reactions

Consider the complex reaction

$$A + B \xrightarrow{k_1} C + E$$

$$A + C \xrightarrow{k_2} D + E$$

with

$$dA/dt = -k_1 AB - k_2 AC \tag{43}$$

$$dB/dt = -k_1 AB \tag{44}$$

$$dC/dt = k_1 AB - k_2 AC \tag{45}$$

$$dD/dt = k_2 AC \tag{46}$$

By material balance there results

$$A - 2B - C = A_0 - 2B_0 \tag{47}$$

$$B + C + D = B_0 \tag{48}$$

where the zero subscripts indicate initial concentrations, it being assumed that only A and B are present at the start.

This situation is more commonly met in practice than are the cases considered by Chien. A common example is the saponification of a symmetrical diester such as ethyl succinate. A would represent the hydroxide ion, B the diester, and C the half-saponified or monoester. The pair of reactions is simultaneously series and parallel, series with respect to B and C and parallel with respect to A. This situation is implied in the name: competitive, consecutive reactions. Such a complex reaction, having been the object of numerous investigations, will be discussed in more than usual detail, as it illustrates a number of methods of handling complex reactions that may be applied to other cases.

The particular difficulty here is that the reaction steps are second-order. Now experimental conditions can be chosen so as to make these pseudo-first-order, as discussed in Chapter 3. For example, a large excess of A may be used. Abel[19] has discussed this situation. The integration for this case is then equivalent to that discussed earlier in this chapter for two successive first-order reactions.

In dealing with the reaction under conditions where the steps are second-order there are various limiting cases that are easily solved. In the first place, suppose that $k_1 \gg k_2$ so that the first step is very rapid compared with the second.[20] The first step then is essentially complete before the second one starts, so that each phase of the reaction can be treated separately as a simple second-order reaction. For example, in the saponification of oxalic acid esters the first step is so fast with excess alkali that it is more convenient to hold the hydroxide ion concentration constant and small with a buffer.[21]

The opposite extreme is when $k_1 \ll k_2$, in which case the second step is a rapid follow-up to the first. The observed rate will then be simply that of a single second-order process but with the provision that two moles of A are reacting for each mole of B. An example is the hydrolysis of an acetal, the hemiacetal reacting much more rapidly. However, in this reaction A is water which is in large excess, causing the kinetics to be pseudo first-order.

A third special case that is easily handled is when $k_1 = 2k_2$. This is approximately true for the saponification of glycerol diacetate.[22] For this combination of k's the reaction appears to be a simple second-order reaction as far as the concentration of A is concerned.

For the more general case of arbitrary k's the situation is more difficult. Ingold[23] was the first to reach a successful result for the determination of k_1 from experimental results. He was interested in the saponification of esters of dibasic acids such as succinic where $k_1 > 2k_2$. To a first approximation the second step can be neglected during the early stages of the reaction. By a method of successive approximations he deduced the formula

$$t = \frac{(a-h)}{k_1 a^2} \left\{ 1 + \frac{(a-h)}{a}\left(1 - \frac{k_2}{2k_1}\right) + \frac{(a-h)^2}{a^2}\left(1 - \frac{7 k_2}{6 k_1} + \frac{2 k_2^2}{3 k_1^2}\right) \right\} \quad (49)$$

where h is the hydroxide concentration at time t and a is the initial concentration of hydroxide and of ester, both equal. Equation 49 amounts to a power series expression for t as a function of $(a - h)$ good to the third degree. Knowing k_2 from the saponification of the corresponding monoester, it was possible to find k_1, using (49). The equation was found to reproduce the data out as far as 30 per cent reaction. However, to extend this to higher-degree terms for greater extents of reaction would be particularly difficult.

Ritchie[24] has used a method of graphical differentiation, whereas Westheimer, Jones, and Lad[25] have developed a method involving a graphical integration. The latter arrived at the approximate equation

$$\frac{1}{h} - \frac{1}{a} +$$

$$\int_a^h \left\{ \frac{c}{1-c} - \left(\frac{a}{h}\right)^{1-c} \bigg/ (1-c) + \frac{a}{h} + \frac{c}{1-c}\left[\left(\frac{a}{h}\right)^{1-c} - 1\right] \right\} \frac{dh}{h^2} = k_1 t$$

$$(50)$$

where a is the initial concentration of hydroxide and of ester and $c = k_2/k_1$. The integral represents only a correction if c is sufficiently small. In actual application the error in using the formula was no more than 4 per cent for reactions up to 50 per cent completion.

An interesting mathematical development has been given by French and applied by French and McIntire[26] to the periodate oxidation of Schardinger dextrins, where up to 40 per cent oxidation the reaction was found to be approximately equivalent to two competitive, consecutive second-order reactions with $k_1 > k_2$. In the notation of equations 43 to 48, a parameter is defined $\theta = \int_0^t A \, dt$. Since $d\theta = A \, dt$ (43), (44), and (45) become, respectively,

$$dA/d\theta = -k_1 B - k_2 C \tag{51}$$

$$dB/d\theta = -k_1 B \tag{52}$$

$$dC/d\theta = k_1 B - k_2 C \tag{53}$$

Equations 52 and 53 are equivalent mathematically to the first two equations of (9) with A replaced by B, B by C, and t by θ. These are "first-order" with the solutions expressed by (10) and (12), with proper substitutions, viz., $B = B_0 e^{-k_1 \theta}$, together with the additional result

$$A = A_0 - B_0[2 - 2e^{-k_1 \theta} - k_1(e^{-k_2 \theta} - e^{-k_1 \theta})/(k_1 - k_2)] \tag{54}$$

Since the resulting equations are in terms of the parameter θ instead of the time t, it is necessary for each run to evaluate θ as a function of t by a graphical integration of the experimental curve of A versus t. As compared with the graphical integration method of Westheimer, Jones, and Lad, the French method would be expected to be less accurate, since in the former the graphical integration involves only a correction term. On the other hand, the present method is of more general value, applying equally well at any stage of the reaction.

Desiring a more general solution of the problem, one that could be used to evaluate both rate constants in a single run as well as one that would avoid graphical integration and also the limitation to only the early phase of the reaction, Frost and Schwemer[27,8] proceeded as follows: for convenience, the initial condition is imposed that $A_0 = 2B_0$, equivalent amounts, rather than equal amounts $A_0 = B_0$, since the second step of the reaction becomes more prominent. It follows then that $C = A - 2B$ and equation 43 becomes

$$dA/dt = (2k_2 - k_1)AB - k_2A^2 \tag{55}$$

The mathematical treatment is somewhat simplified by the introduction of dimensionless variables α, β, and τ and a parameter κ:

$$\alpha = A/A_0 \qquad \beta = B/B_0$$
$$\tau = B_0k_1t \qquad \kappa = k_2/k_1 \tag{56}$$

In terms of these it is possible to solve for $d\alpha/d\beta$, which integrates to give

$$\alpha = \frac{1 - 2\kappa}{2(1 - \kappa)}\beta + \frac{1}{2(1 - \kappa)}\beta^\kappa \tag{57}$$

and then by substitution for α in the equation for $d\beta/d\tau$

$$\tau = \frac{1 - \kappa}{1 - 2\kappa}\int_\beta^1 \frac{d\beta}{\beta^2[1 + (1/1 - 2\kappa)\beta^{\kappa-1}]} \tag{58}$$

The integral in (58) can be evaluated in closed form for any κ which is a rational number, that is, a ratio of integers. In practice only ratios of small integers are conveniently handled, but with κ values sufficiently close interpolation may be used for others. Certain special values of κ give simple integrals. These correspond to the limiting cases mentioned earlier.

For $\kappa = 0$ $k_1 \gg k_2$

$$\tau = \ln\left(\frac{\beta + 1}{2\beta}\right) \quad \text{and} \quad \alpha = \frac{\beta + 1}{2} \tag{59}$$

For $\kappa = \frac{1}{2}$ $k_1 = 2k_2$

$$\tau = \frac{1}{\alpha} - 1 \quad \text{and} \quad \alpha^2 = \beta \tag{60}$$

For $\kappa = \infty$ $k_1 \ll k_2$

$$\tau = \frac{1}{2}\left(\frac{1}{\beta} - 1\right) \quad \text{and} \quad \alpha = \beta \tag{61}$$

All three cases are simple second-order kinetic expressions, at least as far as α is concerned.

Some results of the calculation of τ as a function of α for various κ's are shown in Table 2, and α and β plotted against log τ are shown in Figs. 8 and 9. From Table 2 can be calculated the time ratios of Table 3.

Table 2

τ AS A FUNCTION OF κ AND α

$1/\kappa$	$\alpha = 0.8$ 20% Reaction	$\alpha = 0.7$ 30% Reaction	$\alpha = 0.6$ 40% Reaction	$\alpha = 0.5$ 50% Reaction	$\alpha = 0.4$ 60% Reaction
2.0	0.2500	0.4286	0.6667	1.000	1.500
3.0	0.2599	0.4564	0.7305	1.133	1.770
4.0	0.2656	0.4741	0.7756	1.239	2.011
5.0	0.2693	0.4865	0.8098	1.327	2.235
6.0	0.2720	0.4957	0.8368	1.404	2.449
7.0	0.2740	0.5028	0.8589	1.471	2.657
8.0	0.2755	0.5085	0.8773	1.531	2.862
9.0	0.2768	0.5131	0.8929	1.586	3.066
10.0	0.2778	0.5170	0.9064	1.637	3.270

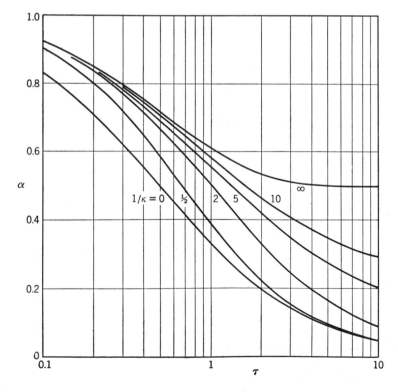

Fig. 8. Competitive, consecutive second-order reactions. α, fraction of reactant A, as a function of time parameter τ and relative rate constant κ (Frost and Schwemer).

Table 3

TIME RATIOS AS A FUNCTION OF κ

(t_{60}/t_{20} is ratio of time for 60 per cent A reacting to time for 20 per cent reaction, or where $\alpha = 0.4$ and 0.8, respectively.)

$1/\kappa$	t_{60}/t_{20}	t_{60}/t_{30}	t_{60}/t_{40}	t_{60}/t_{50}	t_{50}/t_{20}	t_{50}/t_{30}
2.0	6.000	3.500	2.250	1.500	4.000	2.333
3.0	6.812	3.878	2.423	1.562	4.362	2.483
4.0	7.571	4.241	2.592	1.623	4.666	2.614
5.0	8.297	4.593	2.760	1.684	4.928	2.728
6.0	9.003	4.940	2.927	1.745	5.161	2.832
7.0	9.698	5.285	3.094	1.806	5.369	2.925
8.0	10.388	5.629	3.263	1.869	5.558	3.012
9.0	11.078	5.975	3.434	1.933	5.731	3.091
10.0	11.772	6.325	3.607	1.998	5.892	3.166

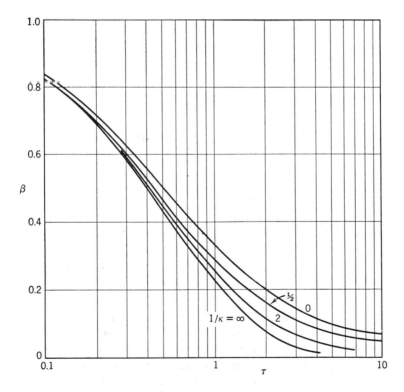

Fig. 9. Competitive, consecutive second-order reactions. β, fraction of reactant B, as a function of τ and κ (Frost and Schwemer).

These can then be used for accurate determination of κ by finding various time ratios in a single run. Once κ is determined, Table 2 provides values of τ which when related to τ and B_0 result in determining k_1. Combination of k_1 with κ then gives k_2 (equation 56). Table 4 shows

Table 4

CALCULATIONS OF RATE CONSTANTS FOR THE SAPONIFICATION
OF ETHYL ADIPATE

% Reaction	t, sec	Percentages Compared	t Ratio	$1/\kappa$
20	605	60/20	6.661	2.808
30	1,060	60/30	3.802	2.795
40	1,690	60/40	2.385	2.777
50	2,595	60/50	1.553	2.850
60	4,030	50/20	4.289	2.785
		50/30	2.448	2.750

Average $1/\kappa = 2.79$.

% Reaction	τ	k_1, liters moles^{-1} sec^{-1}
20	0.2582	0.08570
30	0.4516	0.08555
40	0.7189	0.08542
50	1.108	0.08574
60	1.716	0.08550

Average $k_1 = 0.0856$. $k_2 = 0.0856/2.79 = 0.0307$ liters moles^{-1} sec^{-1}. $B_0 = 0.00996N$. Temperature $= 25.0°$ C.

how this method was used on a single run in the saponification of ethyl adipate. Burkus and Eckert[29] have applied this treatment to data on the reactions of diisocyanates.

Widequist[30] has developed a graphical integration time-variable transformation method[26] so that calculations can be made for non-stoichiometric amounts of reactants. Also he has shown how to treat the data statistically. McMillan[31] has published a relatively simple method for getting the ratio of the rate constants if the simultaneous concentrations of two substances is known. This method is not restricted to stoichiometrical amounts of reactants.

Competitive, Parallel Second-Order Reactions

The saponification of a mixture of simple esters is an example following the general scheme

$$A + B \xrightarrow{k_1} D + F$$

$$A + C \xrightarrow{k_2} E + F$$

B and C, for example, are esters competing for the common reactant A, such as hydroxide ion. This situation is not usually of interest in kinetics because obviously each reaction could be carried out independently of the other as a simple second-order reaction. However, Lee[28] has shown that it is possible to analyze mixtures of reactants such as B and C by observations on the amount of A reacted in time t.

There is an interesting relation between this complex reaction and the one discussed in the previous section. Define the variables S and T so that the B and C of equations 43 to 45 are:

$$B = \left(\frac{k_1 - k_2}{k_1 - 2k_2}\right) S$$

$$C - T = \left(\frac{k_1}{k_1 - 2k_2}\right) S$$

(62)

Substitution in (43) to (45) results in

$$dA/dt = -k_1 AS - k_2 AT \tag{63}$$

$$dS/dt = -k_1 AS \tag{64}$$

$$dT/dt = -k_2 AT \tag{65}$$

which are the rate equations for a hypothetical complex reaction with the steps

$$A + S \xrightarrow{k_1} U + \cdots$$

$$A + T \xrightarrow{k_2} V + \cdots$$

Therefore the solution to the present complex reaction can be obtained as a by-product of the solution of the preceding case.

Van der Corput and Backer[32] have considered this mechanism in detail in connection with the action of lead tetraacetate on isomeric diols.

Reversible Reactions

For a number of simple reversible reactions the rate law is easily integrated.

First-order. Let

$$A \underset{k'}{\overset{k}{\rightleftharpoons}} B$$

Then

$$-dA/dt = kA - k'B \tag{66}$$

If only A is present at the start,

$$A_0 - A = B$$

and therefore

$$-dA/dt = (k + k')A - k'A_0 \tag{67}$$

Integration yields

$$\ln \frac{kA_0}{(k + k')A - k'A_0} = (k + k')t \tag{68}$$

Equations such as these involving approach to equilibrium usually simplify when the equilibrium concentration is explicitly introduced. For this case by setting $dA/dt = 0$

$$kA_e = k'B_e = k'(A_0 - A_e) \tag{69}$$

Where the subscript e refers to equilibrium concentrations, then

$$A_e = \frac{k'}{k + k'} A_0 \tag{70}$$

and (68) can then be written

$$\ln \left(\frac{A_0 - A_e}{A - A_e} \right) = (k + k')t \tag{71}$$

The result is that the approach to equilibrium is a first-order process with effective rate constant the sum of the constants for the forward and reverse directions.

First- and second-order. Suppose that

$$A \underset{k'}{\overset{k}{\rightleftharpoons}} B + C$$

Then

$$-dA/dt = kA - k'BC \tag{72}$$

and, if $B_0 = C_0 = 0$,

$$B = C = A_0 - A$$

Introducing the equilibrium concentration A_e so that

$$kA_e = k'(A_0 - A_e)^2$$

$$k' = k \frac{A_e}{(A_0 - A_e)^2} \tag{73}$$

(72) becomes

$$-dA/dt = k\left[A - \frac{A_e}{(A_0 - A_e)^2}(A_0 - A)^2\right] \tag{74}$$

which can also be written as

$$-\frac{d(A - A_e)}{dt} = k\frac{A_e(A - A_e)}{(A_0 - A_e)^2}\left\{\frac{A_0^2 - A_e^2}{A_e} - (A - A_e)\right\} \tag{75}$$

Now this has the same form as the equation for a simple second-order reaction between two reactants with unequal concentration such as that discussed in Chapter 2. That case being familiar, it is very easy to write the integrated result here by making proper substitutions. Simplification gives the result

$$\ln\frac{A_0^2 - A_eA}{(A - A_e)A_0} = k\left(\frac{A_0 + A_e}{A_0 - A_e}\right)t \tag{76}$$

or, in terms of a reaction variable $x = A_0 - A$, $x_e = A_0 - A_e$ and, using $a = A_0$,

$$\ln\frac{ax_e + x(a - x_e)}{a(x_e - x)} = k\frac{(2a - x_e)}{x_e}t \tag{77}$$

As another example, suppose that the reaction starts from the other direction: $A_0 = 0$ and $B_0 = C_0 = a$ and $x = B_0 - B = C_0 - C$. Integration yields

$$\ln\frac{x_e(a^2 - xx_e)}{(x_e - x)a^2} = k'\frac{(a^2 - x_e^2)}{x_e}t \tag{78}$$

Second-order. Let

$$A + B \underset{k'}{\overset{k}{\rightleftharpoons}} C + D$$

and

$$-dA/dt = kAB - k'CD \tag{79}$$

If only A and B are present initially and at concentrations A_0 and B_0, (79) becomes

$$-dA/dt = kA(B_0 - A_0 + A) - k'(A_0 - A)^2 \tag{80}$$

Integration can be handled by the method used in the last case, or by using partial fractions. The result can be expressed by

$$\ln\frac{(A_0 - A_e)(A - A_e + Q)}{(A - A_e)(A_0 - A_e + Q)} = (k - k')Qt \tag{81}$$

where

$$Q = [1/(K - 1)]\sqrt{K^2(B_0 - A_0)^2 + 4A_0B_0K}$$

and $K = k/k'$. Also A_e, the equilibrium A, is

$$A_e = \frac{-[K(B_0 - A_0) + 2A_0] + Q(K - 1)}{2(K - 1)}$$

If equal concentrations of A and B are taken for the reaction, (81) simplifies to

$$\ln \left\{ \frac{(K - \sqrt{K})[A(K - 1) + A_0(\sqrt{K} + 1)]}{(K + \sqrt{K})[A(K - 1) - A_0(\sqrt{K} - 1)]} \right\} = (k - k') \left(\frac{2A_0 \sqrt{K}}{K - 1} \right) t \quad (82)$$

or in reaction variable notation (79) becomes

$$dx/dt = k(a - x)^2 - k'x^2 \quad (83)$$

and (82) becomes

$$\ln \left(\frac{x(a - 2x_e) + ax_e}{a(x_e - x)} \right) = k \frac{2a(a - x_e)}{x_e} t \quad (84)$$

These relations can be used also for a reversible second-order reaction of the form

$$2A \rightleftharpoons C + D$$

for example, the homogeneous gaseous decomposition of hydrogen iodide

$$2HI \rightleftharpoons H_2 + I_2$$

In reaction variable notation, the rate law corresponding to (83) now is

$$dx/dt = k(a - x)^2 - k'(x/2)^2 \quad (85)$$

where x is the number of moles of hydrogen iodide decomposed per unit volume in time t, a is the initial concentration of hydrogen iodide, and the rate constants are defined in terms of hydrogen iodide reacted or formed. Although the last term of (85) differs from the last term of (83) through a factor of 4, the integrated result is identical with (84) because the latter is explicitly expressed in terms of x_e, the equilibrium value of x, rather than showing k' or K. In particular, writing the equilibrium constant

$$K = \frac{(x_e/2)^2}{(a - x_e)^2} = \frac{k}{k'}$$

therefore

$$\frac{x_e^2}{(a - x_e)^2} = 4K = \frac{k}{(k'/4)}$$

But this is the expression for the equilibrium constant for the reaction $A + B \rightleftharpoons C + D$, and so the same integrated formula (84) still applies but with the new rate constant k' being the old one divided by 4.

Bodenstein,[33] in his classical work on the decomposition of hydrogen iodide, used a formula equivalent to (84) for the special case in which the initial concentration a was unity, the concentration unit being 1 mole per 22.4 liters.

Equilibrium from the Kinetic Viewpoint

In the preceding section the idea has already been used that at equilibrium the net rate of reaction, for example, $-dA/dt$ of equations 66, 72, and 79, is zero. This idea, of course, goes back to the famous work of Guldberg and Waage[31] on the law of mass action and the condition of equilibrium deduced from it. Assuming that forward and reverse reactions have rates given by simple expressions such that each reactant has an order equal to its coefficient in the stoichiometric equation, the equating of such rates, therefore zero net rate, gives the usual equilibrium constant expression. This derivation must be restricted to kinetically simple reactions; otherwise the order with respect to each reactant, or product, cannot be predicted in such a simple way. However, this method is often applied in elementary textbooks indiscriminately to both simple and complex reactions. Although for the latter the kinetic assumptions are false, the derivation still results surprisingly in the correct equilibrium relation. It is of interest to see how equilibrium in a complex reaction can be treated more accurately.[35]

Consider the general case of a complex homogeneous reaction consisting of a number of simple steps. It will be assumed that the reaction involves ideal gases or ideal solutions. Corrections for nonideal behavior can readily be made. Advantage will be taken of the principle of microscopic reversibility or detailed balancing (see Chapter 9), which states that at equilibrium each step will be in equilibrium and for each step the rate of the forward reaction will equal the rate of the reverse reaction.

To simplify the algebra the following notation will be used. A reaction step will be expressed by placing all molecules on the right side of the equation, reacting molecules being then given a negative coefficient. For example, a reaction such as

$$AB + X \rightarrow AX + B$$

will be written as

$$\rightarrow AX + B - AB - X$$

meaning that AX and B are formed from AB and X.

Let the steps in the general reaction be

$$\rightarrow p_a A + p_b B + p_c C + \cdots$$
$$\rightarrow q_a A + q_b B + q_c C + \cdots$$
$$\rightarrow r_a A + r_b B + r_c C + \cdots$$

where A, B, C, etc., are the molecular species involved in the reaction and the p's, q's, r's, etc., represent numerical coefficients showing the number of molecules of each kind which are involved in a particular step, a negative value of the coefficient indicating that the corresponding molecular species is a reactant in that step. All molecular species are indicated in each expression for a reaction step. A zero coefficient would mean that the particular species either does not take part in that reaction step or else if it does take part it remains unchanged, such as when a third body in a collision removes energy.

At equilibrium each step is in equilibrium and the forward rate equals the reverse rate. If k_1 and k_1' are rate constants for the forward and reverse of the first step,

$$k_1[\mathrm{A}]^{-p_a}[\mathrm{B}]^{-p_b} \cdots = k_1'[\mathrm{C}]^{p_c}[\mathrm{D}]^{p_d} \cdots$$

where the molecular species are placed on one side or the other, depending upon the signs of the coefficients, that is, whether they are reactants in the forward step or reverse step. In the above, p_a and p_b were supposed to be negative, meaning that A and B were reactants in the forward step. After rearranging the equation so that all concentrations are on the right and the constants on the left, we have

$$k_1/k_1' = K_p = [\mathrm{A}]^{p_a}[\mathrm{B}]^{p_b}[\mathrm{C}]^{p_c}[\mathrm{D}]^{p_d}$$

which is the equilibrium expression for the first reaction step. It should be noticed that this expression has the same algebraic form, regardless of which substances are reactants or products. Similarly for the other steps

$$K_q = [\mathrm{A}]^{q_a}[\mathrm{B}]^{q_b}[\mathrm{C}]^{q_c}[\mathrm{D}]^{q_d} \cdots, \text{ etc.}$$

All these expressions hold simultaneously and so they may be combined algebraically. Suppose that the chemical equations for the reaction steps are added together in such a way that the stoichiometric equation is obtained. This may require multiplying some of the steps by a simple number. Let the stoichiometric equation be

$$\rightarrow x_a\mathrm{A} + x_b\mathrm{B} + x_c\mathrm{C} \cdots$$

where

$$x_a = n_p p_a + n_q q_a + n_r r_a + \cdots$$

$$x_b = n_p p_b + n_q q_b + n_r r_b + \cdots$$

the coefficients $n_p, n_q, n_r, \cdots$ being small integers, chosen in such a way that the x's for the catalysts and intermediates are zero. (If these were not zero, it would mean the catalyst or intermediate molecules would appear in the stoichiometric equation.) Now combine the equilibrium expressions for each step by multiplying together after raising each to the corresponding power n_p, n_q, $\cdots$

$$K_p{}^{n_p} K_q{}^{n_q} \cdots = [A]^{n_p p_a + n_q q_a +} \cdot [B]^{n_p p_a + n_q q_a +} \cdots$$

but

$$n_p p_a + n_q q_a + \cdots = x_a \quad \text{etc.}$$

Let K replace the product $K_p{}^{n_p} K_q{}^{n_q} \cdots$. Then

$$K = [A]^{x_a} [B]^{x_b} [C]^{x_c} \cdots$$

which is the desired equilibrium expression for the stoichiometric equation of the complex reaction as a whole.

Since at equilibrium the forward and reverse rates must be equal, no matter how complex the reaction, and since these rate expressions must combine to give a relation between concentrations consistent with the thermodynamic equilibrium constant expression, it would appear possible to deduce a reverse rate expression from a measured forward rate expression. This is not necessarily true, however, since the equating of forward and reverse rates can lead to a relation between concentrations which might be a power of the usual equilibrium constant expression, or any function for that matter, and still be consistent. As a very simple example of the difficulty involved, consider the reaction whose stoichiometry is

$$A \rightarrow B$$

and for which two mechanisms are possible

$$2A \rightarrow 2B$$

$$2A \rightarrow A + B$$

In both cases the forward rate would be given by $k[A]^2$. If the first mechanism predominated, the reverse rate would be given by $(k/K^2)[B]^2$. However, if the second mechanism was the one operating, the reverse rate would be $(k/K)[A][B]$.

Thus in the absence of a knowledge of the mechanism it is not possible to state what the reverse rate is. Manes, Hofer, and Weller[36] have considered the problem in some detail and show, in addition to the above conclusion, it is possible, in principle, to determine the rate expressions of forward and reverse rates by measurements of the net reaction rate near equilibrium over a range of conditions. It is also interesting to note that all reactions near equilibrium, no matter how complex, become first-order with respect to any variable that indicates the distance from equilibrium.

It is, of course, possible to observe the gross rate of reaction, that is, forward rate or reverse rate, *at* equilibrium by the use of isotopic tracers. An example of this situation, exchange reactions, will now be considered.

Isotope Exchange Reactions

Consider the exchange reaction

$$AX + BX^* = AX^* + BX$$

where X^* is an isotope, most conveniently radioactive, that can exchange with a normal isotope X. Let the concentrations be designated as follows

$$[AX] + [AX^*] = a$$

$$[BX] + [BX^*] = b$$

$$[AX^*] = x \qquad [AX] = a - x$$

$$[BX^*] = y \qquad [BX] = b - y$$

Let R be the gross rate of exchange, that is, the rate of exchange of all atoms X whether like or different isotopes. It is assumed that R is independent of the varying isotope masses, that there is no isotope effect. This is not quite true, but except for hydrogen isotopes the error is small. Now, regardless of the mechanism of the exchange reaction, the rate at which X^* in BX^* can exchange with X in AX will be proportional to the fraction of BX and BX^* molecules that are BX^* and proportional to the fraction of AX and AX^* that are AX. Similarly for the reverse exchange. The net rate is then

$$dx/dt = -dy/dt = R(y/b)[(a - x)/a] - R[(b - y)/b]x/a \qquad (86)$$

or

$$dx/dt = R[(ay - bx)/ab] \qquad (87)$$

But since

$$y - y_\infty = x_\infty - x$$

and

$$x_\infty/y_\infty = a/b$$

$$dx/dt = (R/ab)[(a + b)(x_\infty - x)] \qquad (88)$$

This integrates into

$$\ln [x_\infty/(x_\infty - x)] = (R/ab)(a + b)t \qquad (89)$$

since R, a, and b are constant during a run. The result is a first-order rate expression, even though no assumptions have been made about the order of the exchange reaction with respect to the chemical constituents. R may be any function of a and b. This relation has been derived by several workers[37-43] essentially in this same form. For radioactive tracers at low concentrations $(a - x)/a$ and $(b - y)/b$ in (86) are close to unity and are often omitted. The final expression (89) is unchanged, however. Alternatively, (89) may be written as

$$-\ln(1 - F) = (R/ab)(a + b)t \tag{90}$$

where F is the fraction of exchange that has occurred. For the more general case

$$AX_n + nBX^* = AX_n^* + nBX$$

the equivalent expression is

$$-\ln(1 - F) = (R/nab)(na + b)t \tag{91}$$

The references cited give illustrations of the use of (89) for getting information about the mechanism of exchange. Harris has discussed the influence of isotope effects and concludes that, if the relative concentration of the distinguishable isotope is small enough, the isotope exchange will always be closely first-order. For large quantities of the differing isotopes, differences in reactivities of the isotopes must be considered. The use of deuterium and its isotope effect in rates of reaction is widespread in studying reaction mechanisms. The isotope effects of the heavier elements, although small, can often give valuable information about mechanisms.[44]

Complex Mechanisms Involving Equilibria

One of the most common types of complex reactions is that in which a simple reaction is preceded by an equilibrium. Consider as an elementary example

$$A \rightleftharpoons B \qquad \text{equilibrium constant } K$$
$$B \xrightarrow{k} C \qquad \text{rate constant } k \tag{92}$$

The rate at which the original reactant A disappears will be determined by the shift of its equilibrium with B, owing to the definite rate at which B reacts to form C. Mathematically this is treated as follows:

The rate of formation of C is

$$dC/dt = kB \tag{93}$$

But B and A are related through the equilibrium constant $B/A = K$ or $B = KA$. By substitution

$$dC/dt = kKA \tag{94}$$

Assuming the concentration B is small compared with A and C, that is, B is a reactive intermediate.

$$dC/dt = -dA/dt \tag{95}$$

Therefore

$$-dA/dt = kKA \qquad (96)$$

and the reaction would appear first-order despite the complexity of the mechanism. The observed rate constant would be a product of an actual rate constant for the rate-determining process and the equilibrium constant of the preceding equilibrium.

It is characteristic of these mechanisms that it is usually possible to explain an observed simple rate expression on the basis of various mechanisms. Consider the well-debated case of the reaction between nitric oxide and oxygen:

$$2NO + O_2 = 2NO_2$$

The rate was found[45] to follow a simple third-order expression

$$-d[NO]/dt = k[NO]^2[O_2]$$

and the reaction was first interpreted as being termolecular. However, because of the negative temperature dependence, that is, the rate decreases as the temperature increases, further consideration was given and an alternate mechanism was devised involving the equilibrium

$$2NO \rightleftharpoons N_2O_2 \qquad (97)$$

preceding the bimolecular rate-determining step

$$N_2O_2 + O_2 \xrightarrow{k'} 2NO_2 \qquad (98)$$

This would result, by a parallel treatment to that above, in the rate expression

$$-d[NO]/dt = k'K[NO]^2[O_2] \qquad (99)$$

which agrees with the observed third-order expression, but in which the third-order rate constant k is the product of a bimolecular second-order constant k' and an equilibrium constant K for formation of the nitric oxide dimer. Because of the heat of dissociation of the dimer, K will decrease as temperature rises. It is only necessary to assume that the normal increase in k' is of a smaller magnitude so that the product $k'K$ decreases.[46]

The alternate explanation of this abnormal temperature effect in terms of transition-state theory[47] where only a simple termolecular mechanism is assumed has already been discussed in Chapter 5. It should be realized that these two mechanisms are not inconsistent from the transition-state standpoint. In the transition-state theory it is assumed that there is an equilibrium between the reactants and the activated complex. Equilibria can, of course, also exist with any number of intermediates without affecting the rate or the rate expression in the least.

Other examples of equilibria preceding rate-determining steps will be given in Chapters 9 and 12.

Since as a reaction of this type proceeds there is a shift of the equilibrium at a finite rate, exact equilibrium does not quite exist. In fact the equilibrium assumption is a limiting case of a more complex reaction. Reverting to the example considered at the beginning of this section, consider the separate rates of the forward and reverse reactions as well as the rate of the succeeding process

$$A \underset{k_2}{\overset{k_1}{\rightleftharpoons}} B$$
$$B \xrightarrow{k_3} C \tag{100}$$

As a set of coupled first-order reactions this could be integrated accurately as discussed earlier in this chapter. However, if the intermediate B is at low concentration, the steady-state approximation will be useful. Applying it to this case, we have

$$dB/dt = k_1A - k_2B - k_3B = 0 \tag{101}$$

Therefore

$$B = k_1A/(k_2 + k_3) \tag{102}$$

and

$$-dA/dt = dC/dt = [k_3k_1/(k_2 + k_3)]A \tag{103}$$

If, on the other hand, equilibrium had been assumed between A and B, the equilibrium constant K would equal k_1/k_2 and the rate expression (96) would have been:

$$-dA/dt = (k_1k_3/k_2)A \tag{104}$$

This would also follow from (101), the steady-state formula, if $k_2 \gg k_3$. In other words, for an equilibrium to occur preceding a rate-determining step, not only must steady-state conditions be satisfied but also the reverse step of the process forming the intermediate must be very rapid compared with the rate at which the intermediate undergoes reaction to the final product.

As a further simple example, consider the Lindemann mechanism of unimolecular reactions as presented in Chapter 4. The limiting expression for high pressure, which gave the simple first-order rate, was equivalent to assuming an equilibrium between normal molecules and activated molecules and also was the condition in which the rate of deactivation was large compared with the rate of reaction.

In what has been considered up to now it has been supposed that the intermediate which is in equilibrium with the reactant is at very low concentration so that it makes no important contribution to the stoichiometry of the reaction. If, on the other hand, there is an appreciable amount, say a few per cent, the rate of disappearance of the reactant would become complicated. However, in such a case the concentration of the intermediate would presumably be an observable amount, and so the kinetics might then be handled more directly. Simplification also occurs if the intermediate is formed almost completely from the reactants.

There also exists the possibility of an equilibrium after a rate process. For example, the decomposition of nitrogen pentoxide gives the overall result

$$2N_2O_5 = 2N_2O_4 + O_2 \tag{105}$$

but N_2O_4 is in equilibrium with NO_2

$$N_2O_4 \rightleftharpoons 2NO_2 \tag{106}$$

Such an equilibrium would not affect the rate in a constant-volume system; however, it can easily affect the method of measurement. In this case the reaction is conveniently followed by observing the increase in pressure of the gaseous reaction mixture. As the products increase in partial pressure the equilibrium shifts, thus necessitating the application of a correction to the pressure increase before the rate of decrease of partial pressure of the reactant can be determined.[48]

Two important types of complex reactions remain to be considered: homogeneous catalytic reactions, which will be presented in the next chapter, and chain reactions, which will be presented in Chapter 10. Heterogeneous catalysis can be discussed from the present viewpoint by making certain suppositions about possible intermediate compounds between the reactants and the catalyst. However, because of the great complexity of the subject it will be omitted from this treatise.

PROBLEMS

1. Apply the time ratio method using Table 1 to evaluate k_1 and k_2 from the following data on the hydrolysis of 2,7-dicyanonaphthalene. (Kaufler[9]; see

% reaction	15	35	70
Time, hr	0.367	1.067	4.200

also Swain.[6])

2. Consider the case of consecutive, competitive first- and second-order reactions:

$$A \xrightarrow{k_1} B$$
$$A + B \xrightarrow{k_2} C$$

(a) Set up the differential rate equations.

(*b*) Obtain the stationary-state solution. Under what conditions will this be a good approximation?

(*c*) Derive a more accurate solution for concentrations A and B that will apply if the relative concentration B/A is very small throughout the reaction.

3. Use the secular equation method to obtain the integrated solution for two successive first-order reactions.

4. An ethyl acetate (E), water (W) and ethyl alcohol (A) mixture, acidified with hydrochloric acid to provide a constant amount of catalyst, undergoes an acid-catalyzed hydrolysis of the ethyl acetate, approaching an equilibrium mixture. The rate may be expressed by

$$dx/dt = k(E - x)(W - x) - k'(A + x)x$$

For an original mixture where $E = 1M$, $W = A = 12.215M$, and where $1/16.35N$ $Ba(OH)_2$ is used to titrate 1-cc samples, the following results were observed [O. Knoblauch, *Z. physik. Chem.*, *22*, 268 (1897)]:

Time, min	cc $Ba(OH)_2$
0	(7.68)
78	8.95
94	9.20
138	9.72
169	9.99
348	11.10
415	11.35
464	11.48
∞	12.00

Obtain a suitable integrated form of the rate equation and evaluate the constants k and k'. (k/k' can be calculated directly from the titer after infinite time.)

REFERENCES

1. A. W. Francis, *J. Am. Chem. Soc.*, *48*, 655 (1926); H. von Halban and H. Eisner, *Helv. Chim. Acta*, *19*, 915 (1936).
2. H. C. Brown and R. S. Fletcher, *J. Am. Chem. Soc.*, *71*, 1845 (1949).
3. W. G. Young and L. J. Andrews, *J. Am. Chem. Soc. 66*, 421 (1944).
4. E. D. Hughes, C. K. Ingold, and U. G. Shapiro, *J. Chem. Soc.*, 225 (1936).
5. W. Esson, *Phil. Trans. Roy. Soc. (London)*, *156*, 220 (1866).
6. C. G. Swain, *J. Am. Chem. Soc.*, *66*, 1696 (1944).
7. R. E. Powell, private communication; D. French, *J. Am. Chem. Soc.*, *72*, 4806 (1950).
8. W. C. Schwemer, Ph.D. thesis, Northwestern University, 1950.
9. F. Kaufler, *Z. physik. Chem.*, *55*, 502 (1906).

10. R. P. Bell and J. C. Clunie, *Nature, 167,* 363 (1951); *Proc. Roy. Soc., 212A,* 16, 33 (1952).

11. E. Rutherford, J. Chadwick, and C. D. Ellis, *Radiations from Radioactive Substances,* Cambridge, 1930, pp. 10–23; H. Bateman, *Proc. Cambridge Phil. Soc., 15,* 423 (1910).

12. A. Skrabal, *Homogenetik,* pp. 190–199; A. Rakowski, *Z. physik. Chem., 57,* 321 (1907); B. J. Zwolinski and H. Eyring, *J. Am. Chem. Soc., 69,* 2702 (1947); F. A. Matsen and J. L. Franklin, *J. Am. Chem. Soc., 72,* 3337 (1950).

13. T. M. Lowry and W. T. John, *J. Chem. Soc., 97,* 2634 (1910); for the solution to a similar problem using matrix methods see E. S. Lewis and M. D. Johnson, *J. Am. Chem. Soc., 82,* 5406 (1960).

14. G. Herzberg, *Infrared and Raman Spectra,* Van Nostrand, New York, 1945, pp. 140, 157–158; A. A. Frost and M. Tamres, *J. Chem. Phys., 15,* 383 (1947); high-speed computers are now commonly used for this purpose.

15. J. Chien, *J. Am. Chem. Soc., 70,* 2256 (1948).

16. E. Jahnke and F. Emde, *Tables of Functions,* Dover Publications, New York, 1945.

17. R. G. Pearson, L. C. King, and S. H. Langer, *J. Am. Chem. Soc., 73,* 4149 (1951).

18. S. Y. Pshezhetskii and R. N. Rubinshtein, *J. Phys. Chem.* (*USSR*), *21,* 659 (1947).

19. E. Abel, *Z. physik. Chem., 56,* 558 (1906).

20. O. Knoblauch, *Z. physik. Chem., 26,* 96 (1898).

21. A. Skrabal, *Monatsh., 38,* 29, 159 (1917).

22. J. Meyer, *Z. physik. Chem., 67,* 272 (1909).

23. C. K. Ingold, *J. Chem. Soc.,* 2170 (1931).

24. M. Ritchie, *J. Chem. Soc.,* 3112 (1931).

25. F. H. Westheimer, W. A. Jones, and R. A. Lad, *J. Chem. Phys., 10,* 478 (1942).

26. D. French, *J. Am. Chem. Soc., 72,* 4806 (1950); D. French and R. L. McIntire, *J. Am. Chem. Soc., 72,* 5148 (1950); S. Widequist, *Acta Chem. Scand., 4,* 1216 (1950).

27. A. A. Frost and W. C. Schwemer, *J. Am. Chem. Soc., 74,* 1268 (1952); W. C. Schwemer and A. A. Frost, *J. Am. Chem. Soc., 73,* 4541 (1951). The numerical tables have been extended by C. A. Burkhard, *Ind. Eng. Chem., 52,* 678 (1960).

28. T. S. Lee, *Anal. Chem., 21,* 537 (1949); T. S. Lee and I. M. Kolthoff, *Ann. N. Y. Acad. Sci.,* 53, 1093 (1951).

29. J. Burkus and C. F. Eckert, *J. Am. Chem. Soc., 80,* 5948 (1958).

30. S. Widequist, *Arkiv. Kemi, 8,* 545 (1955).

31. W. G. McMillan, *J. Am. Chem. Soc., 79,* 4838 (1957).

32. J. G. van der Corput and H. S. Backer, *Proc. Acad. Sci. Amsterdam, 41,* 1058 (1938).

33. M. Bodenstein, *Z. physik. Chem., 13,* 56 (1894); *22,* 1 (1897).

34. C. M. Guldberg and P. Waage, *J. prakt. Chem., 19,* 71 (1879); collected papers in Ostwald, *Klassiker der exakten Wissenschaften,* W. Engelmann, Leipzig, 1899, No. 104.

35. A. A. Frost, *J. Chem. Education, 18,* 272 (1941).

36. M. Manes, L. J. E. Hofer, and S. Weller, *J. Chem. Phys., 18,* 1355 (1950); see also C. A. Hollingsworth, *ibid., 27,* 1346 (1957); J. Horiuti, *Z. physik. Chem., 12,* 321 (1957).

37. H. A. C. McKay, *Nature, 142,* 997 (1938); *J. Am. Chem. Soc., 65,* 702 (1943).

38. S. Z. Roginsky, *Acta Physicochim., 14,* 1 (1941).

39. R. B. Duffield and M. Calvin, *J. Am. Chem. Soc., 68,* 557 (1946).

40. G. Friedlander and J. W. Kennedy, *Introduction to Radiochemistry,* John Wiley & Sons, New York, 1949, pp. 285–288. [Out of print.]

41. T. H. Norris, *J. Phys. & Colloid Chem., 54,* 777 (1950).

42. O. E. Myers and R. J. Prestwood in A. C. Wahl and N. A. Bonner, editors, *Radioactivity Applied to Chemistry*, John Wiley & Sons, New York, 1951.
43. G. M. Harris, *Trans. Faraday Soc.*, *47*, 716 (1951).
44. J. W. Bigeleisen, *J. Chem. Phys.*, *17*, 425 (1953); J. W. Bigeleisen and M. Wolfsberg, *ibid.*, *22*, 1264 (1954); P. E. Yankwich and A. E. Veazie, *J. Am. Chem. Soc.*, *80*, 1835 (1958).
45. M. Bodenstein, *Z. Elektrochem.*, *24*, 183 (1918); *Z. physik. Chem.*, *100*, 68 (1922).
46. M. Bodenstein, *Helv. Chim. Acta*, *18*, 743 (1935); O. K. Rice, *J. Chem. Phys.*, *4*, 53 (1936).
47. H. Gershinowitz and H. Eyring, *J. Am. Chem. Soc.*, *57*, 985 (1935).
48. F. Daniels and E. H. Johnston, *J. Am. Chem. Soc.*, *43*, 53 (1921).

9.

HOMOGENEOUS CATALYSIS

Although a great many catalytic reactions of industrial importance are heterogeneous in type in that the catalyst is present as a distinct solid phase, we shall restrict ourselves here to the subject of homogeneous catalysis; this is partly because a study of homogeneous catalysis is of more general importance in understanding the mechanism of all chemical reactions and partly because heterogeneous catalysis is an extensive, largely self-contained subject in itself. Consequently, its discussion would lead us too far afield.

The accepted definition of a catalyst is that it is a substance which changes the speed of a chemical reaction without undergoing itself any chemical change. Since a reactant or a product may also be a catalyst as well, Bell[1] suggests the definition "a catalyst is a substance which appears in the rate expression to a power higher than that to which it appears in the stoichiometric equation."

Actually many substances classified as catalysts are destroyed either as a result of the process which gives them their catalytic activity or because of subsequent combination with the products. From a practical point of view, a catalyst is a substance which changes the rate of a desired reaction, regardless of the fate of the catalyst itself. This view-point is expressed, for example, in the Friedel-Crafts reaction, where the aluminum chloride cannot be recovered unchanged because it is complexed with the products. Also aluminum chloride must be used in equivalent, not catalytic, quantities for the same reason. The concept of "catalytic quantities" depends on the assumptions of (a) no destruction of the catalyst, and (b) a fairly efficient catalyst.

The recognized fact in the Friedel-Crafts example is that the desired reaction, perhaps the acetylation of benzene, although possible is too slow by itself to be practicable. The same reaction with aluminum chloride goes quite rapidly. In the same way we speak of the base-catalyzed halogenation of ketones and similar compounds in which the catalyzing base is converted into its conjugate acid and hence destroyed as a catalyst.

$$CH_3\overset{O}{\underset{\|}{C}}CH_3 + B + Br_2 \rightarrow CH_3\overset{O}{\underset{\|}{C}}CH_2Br + BH^+ + Br^- \quad (1)$$

The catalyst is used up, but it is recognized again that we have a possible reaction which is very slow in the absence of the base and that the chief function of the base is to speed up the reaction and not to remove the resulting acid.

In the reaction similar to the Friedel-Crafts between carbon monoxide, hydrogen chloride, and benzene, however, aluminum chloride serves as more than a means of speeding the reaction:

$$C_6H_6 + CO + HCl \xrightarrow{AlCl_3} C_6H_5CHO \cdot AlCl_3 + HCl \quad (2)$$

This is because benzaldehyde is thermodynamically less stable than benzene and carbon monoxide. Hence in the absence of aluminum chloride, or a similar reagent, no appreciable reaction will occur even in an infinite time. When aluminum chloride is added, the reaction proceeds rapidly and in good yield because of the favorable equilibrium constant for the reaction

$$C_6H_5CHO + AlCl_3 \rightleftharpoons C_6H_5CHO \cdot AlCl_3 \quad (3)$$

Thus only the stability of the complex enables the reaction to go as desired.[2] Such a case cannot be considered an example of catalysis.

A somewhat less clearly defined case appears in a number of free-radical chain reactions which are initiated by the addition of small amounts of substances which readily decompose to yield free radicals. For example, benzoyl peroxide is used to polymerize styrene, and the addition of dimethyl mercury enables the pyrolysis of propane to occur at a lower temperature. The essential feature here is that the added substance must be completely destroyed in order to have any effect, dimethyl mercury, for example, breaking down into mercury atoms and methyl radicals. The methyl radicals then cause a decomposition of the propane by a chain process (Chapter 10). It is true that the rate of the desired reaction is greatly increased and that only small or "catalytic" quantities of dimethyl mercury are needed. It is quite common to speak of such added substances as catalysts though the better terms "sensitizer" and "initiator" are available.

Completely indefensible is the common practice of saying "the reaction is catalyzed by light" or "the reaction is catalyzed by heat." Such situations where the rate of a reaction is changed by an agency other than a substance should never be classed as catalytic. Also, when the rate is changed by switching to another solvent, it seems desirable to discuss the change in rate as due to a change in medium without speaking of the "catalytic effect" of the solvent. This term incidentally has a proper usage as, for example, in acid-base catalysis.

As we shall see in the discussion that follows, an important criterion of a catalyst is that it changes the mechanism of the parent reaction. Indeed without this change in mechanism, the observed change in rate could not occur. Since most catalysts increase the rate of reaction, it follows that the mechanism must change to one that is easier for the system to follow, involving in general a lower energy barrier. It is frequently said that the function of a catalyst is to lower the activation energy of a given reaction. More correctly, the catalyst changes the mechanism to one having a lower activation energy.

It is true that there are also negative catalysts which slow down the rate of a reaction. Actually, examples of this behavior do not conform to the strict definition of catalysis, since the negative catalyst is used up or permanently altered. Such substances are properly called inhibitors, and the mechanism of their action will be discussed in Chapter 10.

Homogeneous Catalysis in the Gas Phase

A familiar example of catalysis in reactions involving gases is the oxidation of sulfur dioxide to sulfur trioxide by the catalytic action of oxides of nitrogen. The actual mechanism of this reaction is quite complicated as actually carried out in the lead chamber process for sulfuric acid. However, there are several probable steps which would be quite typical of catalyzed reactions. The direct oxidation of sulfur dioxide with oxygen is a slow process

$$2SO_2 + O_2 \xrightarrow{\text{slow}} 2SO_3 \tag{4}$$

because it must occur either by a termolecular collision or by a very high-energy bimolecular or unimolecular process. But the two reactions

$$2NO + O_2 \rightarrow 2NO_2 \tag{5}$$

$$NO_2 + SO_2 \rightarrow NO + SO_3 \tag{6}$$

can both occur with reasonable speeds, since termolecular reactions with nitric oxide are well known, and since nitrogen dioxide can oxidize sulfur dioxide in a low-energy bimolecular process. Since the sum of equations

5 and 6, after doubling (6), is equal to (4), it is seen that one slow reaction is replaced by two faster ones to give the same chemical result. Nitrogen dioxide (which is in equilibrium with the tetroxide, N_2O_4) is formed as an intermediate product from which nitric oxide is regenerated again. Schematically many catalytic reactions can be represented as

$$\text{(uncatalyzed) } A + B \xrightarrow{\text{slow}} \text{products} \tag{7}$$

$$\text{(catalyzed) } A + C \xrightarrow{\text{fast}} X \tag{8}$$

$$X + B \xrightarrow{\text{fast}} \text{products} + C \tag{9}$$

where C is the catalyst and the X is the intermediate. The rate of the uncatalyzed reaction is given by

$$\text{rate} = k_1[A][B] \tag{10}$$

and for the catalyzed reaction

$$\text{rate} = k_2[A][C] \tag{11}$$

if reaction 9 is much more rapid than (8). Equation 11 predicts that the rate of the catalyzed reaction is first-order in the catalyst concentration; this is found to be true experimentally in the majority of catalyzed reactions. Actually it need not be, for, depending on the complexity of the mechanism, the rate expression may contain terms independent of the catalyst, terms involving various positive powers of the catalyst concentration, and even terms inversely proportional to the catalyst concentration.

Molecular iodine is an effective catalyst for many thermal decompositions in the gas phase. This is true, for example, in the pyrolysis of acetaldehyde, propionaldehyde, ethyl ether, ethylene oxide, methyl alcohol, formaldehyde, and diethylamine.[3] Since most pyrolytic decompositions of organic compounds occur by a free-radical mechanism,[4] the most reasonable explanation for the catalysis is that iodine molecules dissociate into iodine atoms which then attack the organic molecule. In this way a chain reaction is started. Consider the decomposition of acetaldehyde as a more specific example. The kinetics are complicated, the order of the reaction being between one and two, depending on the conditions. A free-radical chain mechanism is indicated, for example:

$$CH_3CHO \rightarrow CH_3\cdot + H\cdot + CO \tag{12}$$

$$CH_3\cdot \text{ (or H)} + CH_3CHO \rightarrow CH_4 \text{ (or } H_2) + CH_3\cdot + CO \tag{13}$$

The methyl radicals produced in (13) then carry the chain, the net result being that a mole of acetaldehyde gives essentially a mole of carbon monoxide and a mole of methane. Iodine probably enters into the reaction as follows:[3]

$$I_2 \rightleftharpoons 2I\cdot \qquad (14)$$

$$I\cdot + CH_3CHO \rightarrow HI + CH_3\cdot + CO \qquad (15)$$

$$CH_3\cdot + I_2 \rightarrow CH_3I + I\cdot \qquad (16)$$

$$CH_3\cdot + HI \rightarrow CH_4 + I\cdot \qquad (17)$$

The iodine is eventually regenerated, since the final reaction will be

$$CH_3I + HI \rightarrow CH_4 + I_2 \qquad (18)$$

Consequently the net result in the presence of iodine is the same as in its absence, acetaldehyde being decomposed into methane and carbon monoxide.

The acceleration in rate observed when even small amounts of iodine are added is several thousandfold. This is due to a lowering of the activation energy from about 50 kcal for the uncatalyzed reaction (dependent on the pressure) to 32.5 kcal for the catalyzed reaction. The lower energy reflects the greater ease of breaking an iodine-iodine bond (36 kcal) than a carbon-carbon bond (80 kcal) or a carbon-hydrogen bond (100 kcal). In a similar way the activation energy for the thermal decomposition of ethyl ether is reduced from 53 kcal to 35 kcal by adding iodine and for *i*-propyl ether from 61 kcal to 29 kcal. In these cases the rate of decomposition varies directly with the initial iodine concentration, being given by the expression

$$rate = k[CH_3CHO][I_2]_0 \qquad (19)$$

for acetaldehyde. Actually the iodine concentration varies during the reaction, but a steady state is reached during the period of experimental measurements. Other reactions catalyzed by iodine have mechanisms leading to a different dependence on the iodine concentration. The decomposition of ethylene oxide apparently is independent of the oxide concentration and second-order in iodine, whereas the decomposition of chloral is first-order in chloral and dependent on the square root of the iodine concentration (Schumacher, in reference 4).

Mitchell and Hinshelwood[5] point out that those reactions which are catalyzed by iodine are also those which are susceptible to sensitization by free-radical producers. The decomposition of acetaldehyde can be accelerated by the addition of biacetyl, a substance which readily splits into two acetyl free radicals which are unstable, giving carbon monoxide and methyl radicals.

$$CH_3—\overset{O}{\overset{\|}{C}}—\overset{O}{\overset{\|}{C}}—CH_3 \rightarrow 2CH_3—\overset{O}{\overset{\|}{C}}\cdot \qquad (20)$$

$$CH_3—\overset{O}{\overset{\|}{C}}\cdot \rightarrow CH_3\cdot + CO \qquad (21)$$

Since a chain reaction with acetaldehyde follows as in (13), small amounts of biacetyl can cause the reaction of large amounts of aldehyde. It is clear that there is not much difference in the way in which biacetyl and iodine behave except that the iodine is finally regenerated, whereas the biacetyl is permanently destroyed. In the case of the latter the greater rate is due to an increase in the frequency factor, the activation energy actually increasing somewhat.[6] Also the rate depends on the square root of the biacetyl concentration.

An example of catalysis in the gas phase not involving free radicals is the dehydration of tertiary alcohols catalyzed by hydrogen halides,[7] The reaction

$$(CH_3)_3COH \rightarrow (CH_3)_2C = CH_2 + H_2O \qquad (22)$$

has a rate constant equal to $4.8 \times 10^{11} \, e^{-65,500/RT}$ sec^{-1} at high temperatures in the vapor.[8] If hydrogen bromide is added, a great increase in rate occurs with no change in the hydrogen halide, which thus acts as a true catalyst. The rate is directly proportional to the concentration of added HBr. The rate constant becomes second-order for the catalyzed reaction and is given by $k = 9.2 \times 10^9 e^{-30,400/RT}$ M^{-1} sec^{-1}. Maccoll and Stimson[7] suggest that HBr adds to the alcohol to give a polar complex resembling an ion-pair.

For other catalyzed reactions in the gas phase, reference may be made to Schumacher[4] or to Schwab.[9] Not many true cases of catalysis are known, most examples being sensitized rather than catalyzed. There are, of course, a large number of heterogeneous reactions involving gaseous reactants and solid catalysts.[9]

Homogeneous Catalysis in Solution

The earliest observed cases of catalysis in solution and still the most frequently encountered are acid- or base-catalyzed reactions. Ester hydrolysis may be mentioned as a familiar example. The mechanism of acid-catalyzed ester hydrolysis is discussed in Chapter 12.

A more general example of acid catalysis occurs in the decarboxylation of dimethyloxaloacetic acid.[10] The dianion of this acid loses carbon dioxide, forming an enolate dianion:

$$^-O_2C-\overset{\overset{\displaystyle O}{\|}}{C}-C(CH_3)_2-CO_2^- \rightarrow {}^-O_2C-\overset{\overset{\displaystyle O^-}{|}}{C}=C(CH_3)_2 + CO_2 \quad (23)$$

Placing a proton on the carboxylate group or esterifying the carboxyl gives a monoanion which decarboxylates more rapidly:

$$RO_2C-\overset{\overset{\displaystyle O}{\|}}{C}-C(CH_3)_2-CO_2^- \rightarrow RO_2C-\overset{\overset{\displaystyle O^-}{|}}{C}=C(CH_3)_2 + CO_2 \quad (24)$$

where $R = H$ or C_2H_5. A reasonable explanation is that in losing carbon dioxide, the rest of the molecule must absorb the pair of electrons initially bonding the carboxylate group. Reducing the negative charge on the rest of the molecule must facilitate this transfer of electrons. In agreement with this explanation a number of multiply charged cations act as catalysts for the decarboxylation of the dianion of dimethyloxaloacetic acid. Presumably a complex is formed which reduces the negative charge and increases the ease of accepting a pair of electrons from the carboxyl group, as shown in (25). The catalytic efficiency of a metal ion seems to

$$O=C-C-C(CH_3)_2-CO_2^- \rightarrow O=C-C=C(CH_3)_2 + CO_2 \quad (25)$$

depend both on its positive charge and on its ability to form a stable complex of the chelate type. The metal ions may be considered to be acting as generalized acids. A number of examples of metal ion catalysis involving complex formation are known, particularly with the transition metal ions.[11]

Another example of catalysis in solution is the decomposition of hydrogen peroxide catalyzed by hydrobromic acid or bromine. Hydrogen peroxide decomposes

$$2H_2O_2 = 2H_2O + O_2 \quad (26)$$

very slowly in the absence of a catalyst. A number of solids such as platinum black and a number of ions catalyze the evolution of oxygen. In the presence of hydrobromic acid, the kinetics follow the rate equation:

$$-\frac{d[H_2O_2]}{dt} = 2k_1[H_2O_2][H^+][Br^-] \quad (27)$$

The assumed mechanism[12] is

$$H_2O_2 + H^+ + Br^- \xrightarrow[k_1]{\text{slow}} HBrO + H_2O \tag{28}$$

$$HBrO + Br^- + H^+ \underset{}{\overset{\text{fast}}{\rightleftharpoons}} Br_2 + H_2O \tag{29}$$

$$H_2O_2 + HBrO \xrightarrow[k_2]{\text{slow}} O_2 + Br^- + H^+ + H_2O \tag{30}$$

Reaction 28 probably involves oxonium complex formation with hydrogen peroxide followed by a displacement of water by bromide ion:

$$\begin{array}{c} H \\ HO{-}OH + Br^- \rightarrow H_2O + Br{-}OH \\ + \end{array} \tag{31}$$

Although there is no evidence on the details of (30), it seems likely that an exchange of positive bromine between hydroxyl ion and perhydroxyl ion occurs to give a substance which breaks down to oxygen and the ions of hydrogen bromide.

$$HO{-}Br + H_2O_2 \rightleftharpoons HO_2{-}Br + H_2O \rightarrow H^+ + O_2 + Br^- + H_2O \tag{32}$$

An objection to this mechanism is that the reaction involves a change in electron multiplicity, oxygen being paramagnetic, whereas the other substances are diamagnetic. Such reactions are usually slow (see Chapter 4). However, it is possible to change electron spin by means of collisions with other molecules. In the particular case of reaction 32 the reactants are so much less stable than the products that it might be possible for oxygen in the excited singlet state to be formed (22 kcal higher in energy than the normal triplet state) without excessive activation energy. The excited oxygen would then be converted to normal oxygen by collisions with solvent molecules.

In support of the mechanism shown in (28) to (30) Bray and his co-workers[12] were able to measure the rate at which bromine is formed initially from hydrobromic acid and hydrogen peroxide, and the rate at which bromine is destroyed in a reaction mixture to which it is added. The rate equations were found to be

$$d[Br_2]/dt = k_1[H_2O_2][H^+][Br^-] \tag{33}$$

$$-d[Br_2]/dt = k_2[H_2O_2][HBrO] \tag{34}$$

The rate constant k_1 is the same as in equations 27 and 28. The concentrations are ordinarily such that (28) is the slowest reaction. Therefore, the rate of decomposition of hydrogen peroxide is the same as the initial rate of formation of bromine except for a factor of 2, since two moles of peroxide disappear each time (28) is completed. Because of the equilibrium in (29) either bromine or hydrobromic acid can act as a catalyst.

Steady-state concentrations of bromine and hydrobromic acid are rapidly established, regardless of the starting conditions.

Although the above scheme for peroxide decomposition is similar to that for some other catalysts, or substances which react with H_2O_2,[13] another general mechanism exists in which free radicals are formed. The transition metal ions often act as catalysts.[14]

The best studied case is that in which a catalytic couple of ferrous and ferric ions exists. The reactions with the metal ions are as follows when a large excess of peroxide is present:

$$Fe^{+2} + H_2O_2 \rightarrow FeOH^{+2} + HO\cdot \tag{35}$$

$$Fe^{+3} + H_2O_2 \rightarrow Fe^{+2} + H^+ + HO_2\cdot \tag{36}$$

$$Fe^{+3} + HO_2\cdot \rightarrow Fe^{+2} + H^+ + O_2 \tag{37}$$

The overall efficiency is greatly increased, however, because of a chain decomposition of the peroxide.

$$HO\cdot + H_2O_2 \rightarrow H_2O + HO_2\cdot \tag{38}$$

$$HO_2\cdot + H_2O_2 \rightarrow O_2 + H_2O + HO\cdot \tag{39}$$

The effect of pH shows that the anions $O_2\cdot^-$ and HO_2^- are more efficient than their conjugate acids $HO_2\cdot$ and H_2O_2 in certain of the above reactions. Other metal ions with two stable valence states can also function as catalysts.

The Basis of Catalytic Action

The preceding examples show the way in which catalytic processes occur. A question which naturally arises is why any reaction which goes through several steps should ever be faster than one which goes through only one. Thus a catalytic reaction must consist of a series of steps, each more rapid than the uncatalyzed reaction. Now, if the end result of both processes is the same so that the thermodynamic tendency for the reaction is constant, how can the rate constants be different? The answer lies in the well-known fact that there is no necessary correlation between the thermodynamic tendency of a reaction to occur (i.e., the free-energy change) and the rate of the reaction. Thus coal and air are quite indifferent to each other for long periods of time in spite of the thermodynamic stability of carbon dioxide.

Nevertheless, the cases where rates and equilibria are related are sufficiently numerous so that they may be considered the rule rather than the exception. Hence catalytic reactions must either be very fast in that their rates are greater than the corresponding equilibrium constants would lead one to expect, or else the uncatalyzed reaction must be abnormally slow. Both kinds of behavior are encountered.

For example, the oxidation of thallous ion with ceric ion is a very slow process in spite of favorable values of the oxidation potential.

$$2Ce^{+4} + Tl^{+1} \rightarrow 2Ce^{+3} + Tl^{+3} \tag{40}$$

The reason for the slowness is that a three-body collision is necessary for the direct reaction to occur, no intermediate valence of $+2$ for thallium or cerium being stable. Manganous ion, however, catalyzes the oxidation by the following sequence:

$$Ce^{+4} + Mn^{+2} \rightarrow Ce^{+3} + Mn^{+3} \tag{41}$$

$$Mn^{+3} + Ce^{+4} \rightarrow Ce^{+3} + Mn^{+4} \tag{42}$$

$$Mn^{+4} + Tl^{+1} \rightarrow Mn^{+2} + Tl^{+3} \tag{43}$$

each two-body reaction occurring with normal velocity.

Acid-base catalysis is effective for the first reason mentioned above. Proton transfers are generally rapid compared to the making and breaking of most chemical bonds. Hence reactions involving proton transfers as in typical acid or base catalysis are rapid compared to similar reactions of comparable free-energy change. Consider the two reactions

$$CH_3OCH_3 + H_2O \rightleftharpoons CH_3O^- + CH_3OH_2{}^+ \tag{44}$$

$$CH_3OH + H_2O \rightleftharpoons CH_3O^- + H_3O^+ \tag{45}$$

which are not much different in free energy, (45) being favored by about 3 kcal. The equilibrium constants thus differ by a factor of about 100. The variations in the rate, however, are infinitely greater, the spontaneous hydrolysis of an ether being so slow that it is never observed, whereas the second reaction is so fast that it cannot be measured by ordinary means.[15] The difference in rates arises from the lack of steric hindrance involved in a nucleophilic displacement on hydrogen as in (45) compared to a nucleophilic displacement on carbon as in (44). Steric hindrance, that is, the repulsion of nonbonded atoms in the activated complex, is the greatest single factor in determining activation energies. Since the proton lacks the filled inner electron shells usually responsible for the repulsion and is not surrounded by other groups, it is quite free from steric effects.

Proton transfers involving oxygen-hydrogen bonds are generally rapid, but they are not instantaneous. In the ionization of water, for example,

$$2H_2O \rightleftharpoons H_3O^+ + OH^- \tag{46}$$

it can easily be calculated that the activation energy is at least as great as 13.8 kcal (the heat of the reverse reaction) and that the entropy of activation is negative. The inference is that the rate constant must be small, and indeed a direct measurement gives 5×10^{-7} M^{-1} sec^{-1} for (46) reckoned as a bimolecular reaction.[16]

This example illustrates the point that an unfavorable equilibrium constant must necessarily make a reaction slow, which on all other grounds would appear to be rapid. The reverse reaction of 46 is extremely fast,[16] having a rate constant of 1.3×10^{11} M^{-1} sec^{-1}. Here the free energy change is, of course, extremely favorable. In the same way some proton transfers of nitrogen-hydrogen bonds are indeed very fast, but many are quite slow. Thus, ammonium ion exchanges protons with solvent water quite slowly in acid solution where the only available base is water itself. However, a trace of free ammonia causes a very rapid exchange.[17] Proton transfers on carbon are well known to be slow in most cases. For example, in the groups $-NHNO_2$ and $-CH_2NO_2$, the relative rate constants[18] for proton transfer to NH_3 are 10^7 to 1.

Another type of reaction with a greater rate than expected occurs when certain large, easily polarized nucleophilic (electron-donating) reagents react with typical electrophilic (electron-seeking) reagents. Iodide ion, thiosulfate ion, thiourea, and carbanions in reaction with alkyl halides disobey the usual rule that the rate correlates with the basicity of the nucleophilic reagent, being more reactive than their basicity would warrant. For example, malonic ester anion is a weaker base than ethoxide ion, since the reaction in alcohol

$$C_2H_5O^- + CH_2(COOC_2H_5)_2 \rightleftharpoons C_2H_5OH + {}^-CH(COOC_2H_5)_2 \quad (47)$$

goes well to the right, the value of the equilibrium constant being 22. Nevertheless, the carbanion reacts some seven times faster than the ethoxide ion with ethyl bromide.[19] Even more striking is the behavior of thiourea, which has negligible basic properties but which reacts with alkyl halides several hundred times faster than respectable bases like pyridine or aniline.

The explanation seems to be that the electrons in an easily polarized system (iodine, sulfur, or negative carbon) can be distorted in the field of an electrophilic reagent so that a new bond can be formed without bringing the rest of the system into close contact. Hence steric repulsion is reduced.† For this reason iodide ion can be used to catalyze the reaction of an alkyl chloride with a slow nucleophilic reagent B.

$$B + RCl \xrightarrow{\text{slow}} BR^+ + Cl^- \quad (48)$$

$$I^- + RCl \xrightarrow{\text{fast}} RI + Cl^- \quad (49)$$

$$B + RI \xrightarrow{\text{fast}} RB^+ + I^- \quad (50)$$

By the same reasoning as above it can be predicted that easily polarized

† This includes the repulsion of non-bonded electron pairs.

groups will be more easily replaced than similar but less polarized groups. Bunnett[20] has pointed out that an additional enhancement in rate is found if both the reagent and the group to be displaced are large.

Catalysis and the Equilibrium Constant

It is recognized that the addition of a catalyst, according to the strict definition of the term, merely speeds up the attainment of equilibrium but does not change the point of equilibrium. That is, the maximum yield in a given reaction cannot be changed by adding a catalyst unless it is added in amounts sufficient to constitute a change in medium or unless it combines with one of the products. This consequence follows immediately from thermodynamic arguments, since the change in free energy which controls the equilibrium constant is dependent only on the initial and final states of the system and is independent of the path or mechanism of changing from one state to another.

Another proof, useful for kinetic purposes, of the independence of the point of equilibrium on the reaction mechanism comes from the principle of microscopic reversibility.[21] From statistical consideration it can be shown that, if a system is at equilibrium and there are a number of molecular transitions occurring between various states within the system, each set of transitions must separately be at equilibrium. That is, if there is a transition from state i to state j, then the number of molecules making the change from i to j in unit time must equal the number making the reverse change from j to i in unit time at equilibrium. Taking a chemical example in the transition $A \rightleftharpoons B$ which can go by a number of mechanisms, namely,

$$A \underset{k_2}{\overset{k_1}{\rightleftharpoons}} B \tag{51}$$

$$A + C \underset{k_4}{\overset{k_3}{\rightleftharpoons}} X \underset{k_6}{\overset{k_5}{\rightleftharpoons}} B + C \tag{52}$$

$$A + D \underset{k_8}{\overset{k_7}{\rightleftharpoons}} Y \underset{k_{10}}{\overset{k_9}{\rightleftharpoons}} B + D \quad \text{etc.} \tag{53}$$

where C and D are catalysts for the reaction and X and Y are the corresponding intermediates, then at equilibrium the number of molecules of B formed from A in unit time by each of the mechanisms shown in (51) to (53) must be equal to the number of molecules of A formed from B by the reverse of each mechanism. This sets up the condition (as can be readily verified by setting the rate of formation of each component equal to zero and using the proper activity coefficients in the rate expressions)

$$\frac{a_B}{a_A} = K_{eq} = \frac{k_1}{k_2} = \frac{k_3 k_5}{k_4 k_6} = \frac{k_7 k_9}{k_8 k_{10}} \quad \text{etc.} \tag{54}$$

Accordingly it is ensured that the equilibrium constant is unchanged for the catalyzed paths and also that certain restrictions exist for the various rate constants involved.

The great importance of the principle of microscopic reversibility in mechanism determinations is that it enables the mechanism of the reverse reaction to be known with as much accuracy as the mechanism of the forward reaction is known. For example, if it is known that for a given set of conditions A is converted to B exclusively by the catalyst C as in (52), then, under the same conditions, B must be converted to A catalyzed by C and through the same intermediate X. The possibility that the reverse reaction operates through the direct process as in (51) or, more disconcertingly, by some unsuspected new mechanism is eliminated. To illustrate, for a given reaction, if a diagram of free energy as a function of the extent of reaction is drawn showing the various intermediates and activated complexes, a diagram such as Fig. 1 will be obtained. If the energy barriers and nature of the intermediate are known for the forward reaction, the principle of microscopic reversibility enables the mechanism of the reverse reaction to be described by simply reversing the procedure starting with the products and going from right to left on the energy diagram.

More specifically, the mechanism of isomerization of nitroethane to aci-nitroethane may be cited.[22] Nitroethane is an acid with an ionization constant of 10^{-9}, but it ionizes only slowly, reacting with various bases at rates varying with the ionization constant of the base.

$$C_2H_5NO_2 + B \underset{k_2}{\overset{k_1}{\rightleftharpoons}} C_2H_4NO_2^- + BH^+ \tag{55}$$

The ions thus formed are in mobile equilibrium with aci-nitroethane which is tautomeric with nitroethane and has the proton attached to an oxygen atom. It has an acid ionization constant of 10^{-5}.

$$C_2H_4NO_2^- + BH^+ \underset{k_4}{\overset{k_3}{\rightleftharpoons}} C_2H_4NO_2H + B \tag{56}$$

Hence the conversion of the nitro to the aci form is subject to catalysis by bases and goes through the intermediate formation of the ions. It can then be deduced that the transformation of aci to nitro form also goes through the formation of the ions. Since the ionization of nitroethane is catalyzed by any base B, then the reverse of (55), the conversion of the anion to the nitro form, is catalyzed by any acid BH$^+$. This is found to be true experimentally. The relationship between the rate constants and the equilibrium constants is as follows:

$$\frac{a_{\text{aci}}}{a_{\text{nitro}}} = K_{\text{eq}} = \frac{k_1 k_3}{k_2 k_4} = \frac{K_N}{K_A} \tag{57}$$

where K_N and K_A are the ionization constants for nitroethane and aci-nitroethane, respectively.

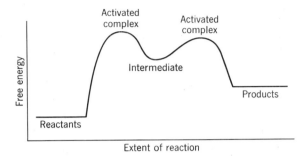

Fig. I. Free-energy profile for reaction involving an intermediate. The same intermediate and activated complexes apply to both forward and reverse reactions.

Acid-Base Catalysis

Since catalysis by acids and bases is by far the most common in homogeneous reactions, a more detailed analysis of the kinetic situations that are usually encountered will be given (for another approach see Bell,[1] who also discusses a number of the better-known examples in some detail). There are several distinct types of acid-base catalysis, almost all of which involve a proton transfer or a "prototropy" in at least one step. These will be treated in general in terms of a substrate S or SH on which the acids or bases work, S$^-$ the conjugate base of the substrate SH, B a general base, BH$^+$ its conjugate acid, HA a general acid, and A$^-$ its conjugate base. B and A$^-$ may or may not be the same as may BH$^+$ and HA. R is some other reactant not acting as a base or acid.

Case I

Prior equilibrium between substrate and hydrogen ion followed by a rate-determining reaction with another reagent:

$$S + H^+ \rightleftharpoons SH^+ \tag{58}$$

$$SH^+ + R \xrightarrow{\text{slow}} \text{products} \tag{59}$$

$$\text{rate} = k[SH^+][R] = kK_{eq}[S][H^+][R] \tag{60}$$

where K_{eq} is the equilibrium constant for (58) and k the rate constant for (59).

An important feature of this type of reaction is that the concentration of the complex SH^+ will depend only upon the hydrogen ion concentration in solution, regardless of the source of the hydrogen ion. Hence the rate also will be dependent on the hydrogen ion only and not on the concentration of any other acids in the solution. This is an example of specific hydrogen ion catalysis. The observed kinetics depend a great deal on the value of the equilibrium constant of (58) and the relative concentrations of H^+ and S. If K_{eq} is large enough, S or H^+, whichever is present in least amount, will be completely converted to SH^+. The observed kinetics will be second-order, first-order in SH^+, and first-order in R. The more usual case is for K_{eq} to be small so that S and H^+ are present in amounts initially added less that which has reacted. The observed kinetics will be third-order, first-order in H^+, S, and R. Since H^+ is frequently constant, being regenerated in step 59, the kinetics may be pseudo-second-order or pseudo-first-order, depending on the change in R. The pseudo-order constants, however, will be dependent on the hydrogen ion concentration in solution.

Examples of Case I (all with K_{eq} small) are found in the hydrolysis and alcoholysis of esters, the inversion of sucrose, the decomposition of diazoacetic ester, the hydrolysis of ethyl orthoformate, and the hydrolysis of acetals and ethers.

Case II

Prior equilibrium between substrate and hydrogen ion followed by the rate-determining proton transfer:

$$HS + H^+ \rightleftharpoons HSH^+ \qquad (61)$$

$$HSH^+ + B \xrightarrow{\text{slow}} BH^+ + SH \qquad (62)$$

$$\text{rate} = k[HSH^+][B] = kK_{eq}[H^+][HS][B] = kK_{eq}K_A[BH^+][HS] \quad (63)$$

The rate of the reaction will be proportional to the concentrations of hydrogen ion and of base, but their product in turn is proportional to the concentration of the conjugate acid BH^+. Thus the kinetics will satisfy the conditions for general acid catalysis. The rate will increase not with the hydrogen ion concentration but with the concentration of acid BH^+. Also, if a number of acids are present,

$$\text{rate} = K_{eq}[HS] \sum_i k_i K_{Ai}[BH_i^+] \qquad (64)$$

All acids will contribute to the rate to an extent determined by their concentrations, acid ionization constants, and the specific rate constants of their conjugate bases.

Examples of Case II are probably the addition reactions of the carbonyl group of aldehydes and ketones with such reagents as hydrazine, hydroxylamine, and semicarbazide. These reactions are complicated by at least one other step (the addition of the reagent to the carbonyl group) between (61) and (62). Some keto-enol transformations probably also follow the mechanism of Case II, as do the acid-catalyzed mutarotation of glucose, the depolymerization of dihydroxyacetone, and the dehydration of acetaldehyde hydrate.[23]

Case III

A prior equilibrium involving hydrogen bonding of the substrate with an acid followed by a rate-determining proton transfer:

$$HS + HA \rightleftharpoons HS \cdot HA \tag{65}$$

$$HS \cdot HA + B \xrightarrow{\text{slow}} \text{products} \tag{66}$$

$$\text{rate} = k[HS \cdot HA][B] = kK_{eq}[HS][HA][B] \tag{67}$$

Alternatively this system could operate by complex formation between B and HS followed by reaction with HA, or HA and B can attack the substrate simultaneously in a termolecular step.[24] The termolecular hypothesis seems less likely, unless either HS or B is a solvent molecule, on the basis that termolecular solute collisions are uncommon. The possibilities will be kinetically indistinguishable if K_{eq} is small. Both an acid and a base are needed for this mechanism, and, if a number of acids and bases are present, there will be a term in the rate equation for all possible pairs.

$$\text{rate} = [HS] \sum_{ij} k_{ij} K_i [HA_i][B_i] \tag{68}$$

Since a hydroxylic solvent can play the role of either an acid or base, the kinetics will frequently reduce to

$$\text{rate} = kK_{eq}[HS][HA][H_2O] \tag{69}$$

in the presence of a single acid, and to

$$\text{rate} = kK_{eq}[HS][B][H_2O] \tag{70}$$

in the presence of a single base. However, in an inert solvent both acid and base must be added.[24,25]

Examples of Case III are certain keto-enol changes such as that of acetone, the methyleneazomethine rearrangement, and the mutarotation of tetramethylglucose and similar substances in inert solvents. In nonpolar solvents Case III becomes more probable than Case II, because ionization of the acid is not required. The kinetics in such media are, however, frequently complicated, simple orders not being observed.[26]

Case IV

A prior equilibrium involving hydrogen bonding of a substrate with an acid followed by a slow step not involving a proton transfer:

$$S + HA \rightleftharpoons S \cdot HA \tag{71}$$

$$S \cdot HA + R \xrightarrow{\text{slow}} \text{products} \tag{72}$$

$$\text{rate} = k[S \cdot HA][R] = kK_{eq}[S][HA][R] \tag{73}$$

This will be an example of general acid catalysis. The mechanism seems to apply to the hydrolysis of ethylorthoacetate and ethylorthocarbonate in which the reagent R is a water molecule.

Case V

Reaction of an anion with an acid to give either the final product or an intermediate which rapidly gives the final product:

$$S^- + HA \xrightarrow{\text{slow}} SH + A^- \tag{74}$$

$$SH \xrightarrow{\text{fast}} \text{products} \tag{75}$$

$$\text{rate} = k[S^-][HA] \tag{76}$$

This is an example of general acid catalysis. For a number of acids

$$\text{rate} = [S^-] \sum_i k_i[HA_i] \tag{77}$$

This behavior is found in the conversion of the anions of the nitroparaffins into the nitro forms. Also the decomposition of the diazoacetate ion and the azodicarbonate ion seems to involve such a process.[27]

Case VI

A prior ionization of the substrate which comes to equilibrium followed by a slow reaction of the resulting anion with another reagent:

$$HS + B \rightleftharpoons S^- + BH^+ \tag{78}$$

$$S^- + R \xrightarrow{\text{slow}} \text{products} \tag{79}$$

$$\text{rate} = k[S^-][R] = kK_{eq} \frac{[SH][B][R]}{[BH^+]} = \frac{kK_{eq}}{K_B}[SH][R][OH]^- \tag{80}$$

The rate will be dependent on the ratio $[B]/[BH^+]$ which, however, depends upon the hydroxide ion concentration in aqueous solution (or in general the solvent anion). Consequently the rate will be a linear function of the hydroxide ion concentration, and the reaction will be subject to specific hydroxide ion catalysis.

Examples of Case VI include the condensation of acetone to diacetone alcohol (discussed in Chapter 12) and many similar condensations in organic chemistry, such as the Claisen, Michael, Perkin, and Aldol condensations.

In alcohol solution where many such condensations are carried out, the ethoxide ion takes the place of the hydroxide ion in the rate expression.[28] Also obeying the kinetics of (80) are the brominations of β-disulfones and β-dinitriles,[29] which have the unusual feature that the rate of reaction of the carbanion formed in (78) with bromine is slow compared to the reverse of (78).

Case VII

A slow ionization of the substrate followed by a rapid reaction of the anion to give the products:

$$HS + B \xrightarrow{\text{slow}} S^- + BH^+ \qquad (81)$$

$$S^- + R \xrightarrow{\text{fast}} \text{products} \qquad (82)$$

$$\text{rate} = k[HS][B] \qquad (83)$$

Such a reaction will show general base catalysis. For a number of bases in solution each will contribute to the rate

$$\text{rate} = [HS] \sum_i k_i[B_i] \qquad (84)$$

The kinetics indicated for Case VII are shown in the decomposition of nitramide, the halogenation, isomerization, and deuterium exchange of many organic substances containing an acidic hydrogen and in the racemization of such substances when the acidic hydrogen is on an asymmetric atom. Also general base-catalyzed is the aldol condensation of acetaldehyde (Chapter 12) which has the feature of the condensation step being faster than the reversal of (81). A similar situation occurs in the condensation of glycerinaldehyde or dihydroxyacetone.[30]

Case VIII

A prior equilibrium involving addition of a base to the substrate followed by a slow reaction of the complex with or without another reagent:

$$HS + B \rightleftharpoons B \cdot HS \tag{85}$$

$$B \cdot HS + (R) \xrightarrow{\text{slow}} products \tag{86}$$

$$rate = k[B \cdot HS][R] = kK_{eq}[B][HS][R] \tag{87}$$

This would again be an example of general base catalysis; actually most reactions of this type show considerable specificity in the base. The common examples involve addition of one base to a carbonyl group followed by the elimination of another base as in ester hydrolysis. Different products are formed, dependent on the base, and the base is generally used up. An exception is in the hydrolysis of certain activated esters such as p-nitrophenyl acetate. Here true general base catalysis by substances such as pyridine, trimethylamine, and acetate ion occurs.[31] The catalyst is regenerated. Another case is the base catalyzed alcoholysis[32] of esters, diketones, ketoesters, and the like. Here the base catalyst is regenerated:

$$
\begin{array}{ccccc}
& O & & O^- & & O \\
& \parallel & & | & & \parallel \\
R' & -C - B + OR^- \rightleftharpoons R' & -C - B \rightarrow R' & -C - OR + B^- \\
& & & | & \\
& & & O & \\
& & & R &
\end{array}
\tag{88}
$$

$$B^- + ROH \rightarrow BH + RO^- \tag{89}$$

However, this is possible only if the catalyst base is the anion of the solvent and it corresponds to a case of specific rather than general base catalysis. A similar situation is met in the exchange of O^{18} in water with the oxygen of carbonyl compounds, a reaction which is specifically catalyzed by hydroxide ion.[33]

The Brönsted Catalysis Law

For those reactions which are subject to general acid or general base catalysis an expected relationship exists between the strength of the acid or base, as determined by its ionization constant, and its efficiency as a catalyst, determined by the observed rate constant. If we denote this constant (also called the catalytic constant) by k_A or k_B for acid and base catalysis, respectively, then the relationship is best shown by the Brönsted catalysis law,[34]

$$k_A = G_A K_A{}^{\alpha} \qquad k_B = G_B K_B{}^{\beta} \tag{90}$$

where K_A and K_B are the acid and basic ionization constants and G_A, G_B, β, and α are constants characteristic of the reaction, the solvent, and the temperature. α and β are positive and have values between zero and unity. These relationships have been shown to fit the data fairly well for a number of catalytic reactions with acids and bases of various types.[35]

To illustrate (90) consider the rates and equilibrium of the following reversible reaction:

$$HS + B \underset{k_A}{\overset{k_B}{\rightleftharpoons}} S^- + HB^+ \qquad (91)$$

where HS is a constant substrate and B is a base which may be varied. The equilibrium constant is proportional to K_B in the particular solvent used, or to K_A if the reverse reaction is considered. If a change is made from B to a similar base B' which is a stronger base by two powers of 10, the equilibrium constant is increased by the same amount. Since $K_{eq} = k_B/k_A$, only one rate constant need by affected, but generally both are. For example, k_B may be increased by a power of 10, since B' is a better proton acceptor, and k_A may be decreased by a power of 10, since B'H$^+$ is a poorer proton donor or acid. This corresponds to the case

$$k_A = G_A K_A^{0.5} \qquad k_B = G_B K_B^{0.5} \qquad (92)$$

Now as to whether other bases and acids will conform to the same equation depends chiefly on how similar they are to the first two bases. A group of substituted anilines will all fit a single equation and a group of carboxylic acid anions will all fit a similar equation with slightly different constants. Since most acid-base-catalyzed reactions involve a proton transfer somewhere, the Brönsted relationship will frequently hold even for those cases involving a more complex series of reactions.

For exactness, a statistical correction should be made.[36] If for a given catalyst p is the number of equivalent protons on the acid and q is the number of equivalent positions where a proton can be accepted in the conjugate base, (90) should be written

$$(k_a/p) = G_A((q/p)K_A)^\alpha \qquad (k_b/q) = G_B((p/q)K_B)^\beta \qquad (93)$$

The current tendency is to apply the statistical correction only when p and q refer to different atoms within the molecule. That is, p is taken as 2 for oxalic acid, $(COOH)_2$, but as unity for ammonium ion, NH_4^+. The correction is made to take into account the fact that k_A increases with p but is independent of q, whereas K_A not only increases with p but decreases as q increases.

Benson[37] has shown that the logical way to apply the statistical correction, not only in acid-base reactions, but for reactions in general, is to use

Table I

DEHYDRATION OF ACETALDEHYDE HYDRATE AT 25°

(Bell and Higginson)

Acid	k_c liters/mole-min	K_a
Thymol	0.0112	3.2×10^{-11}
Hydroquinone	0.013	4.5×10^{-11}
Phenol	0.0181	1.06×10^{-10}
Resorcinol	0.026	1.55×10^{-10}
Pyrocatechol	0.049	1.4×10^{-10}
p-Chlorophenol	0.061	6.6×10^{-10}
o-Chlorophenol	0.112	3.2×10^{-9}
m-Nitrophenol	0.160	5.3×10^{-9}
2,4-Dichlorophenol	0.225	1.8×10^{-8}
o-Nitrophenol	0.334	6.8×10^{-8}
p-Nitrophenol	0.520	6.75×10^{-8}
2,4,6-Trichlorophenol	1.53	3.9×10^{-7}
Propionic	18.0	1.35×10^{-5}
Acetic	19.2	1.76×10^{-5}
β-Phenylpropionic	23.0	2.19×10^{-5}
Trimethylacetic	23.6	9.4×10^{-6}
Pentachlorophenol	24.0	5.5×10^{-6}
Crotonic	28.7	2.3×10^{-5}
Phenylacetic	33.0	4.9×10^{-5}
Cinnamic	40.5	3.8×10^{-5}
Formic	43.5	1.77×10^{-4}
β-Chloropropionic	43.9	1.04×10^{-4}
Diphenylacetic	44.2	1.15×10^{-4}
2,6-Dinitrophenol	91.0	2.6×10^{-4}
Phenoxyacetic	92.0	7.6×10^{-4}
Bromoacetic	129	1.38×10^{-3}
Chloroacetic	146	1.51×10^{-3}
2,4-Dinitrophenol	183	8.1×10^{-5}
Cyanoacetic	218	3.56×10^{-3}
Phenylpropionic	225	5.9×10^{-3}
Dichloroacetic	773	5.0×10^{-2}

the transition-state method. A part of the partition function for each reactant and for the activated complex will be the symmetry number. The proper statistical correction for the rate constant is then the ratio $\sigma_A \sigma_B / \sigma_{\ddagger}$ where A and B are the reactants. It is proposed that all rate and equilibrium constants should be considered as made up of a chemical and a statistical part which is the ratio of symmetry numbers. Thus K_A and K_B in (93) would be divided by factors such as σ_{HA}/σ_{A^-} and σ_B/σ_{BH^+}, the proton and hydroxide ion being common for all members of the series. The method is equivalent to the usual ways of making the statistical correction in simple cases. However, an assumption about the transition state must be made in dealing with the rate constants.

Table 1 shows the data obtained by Bell and Higginson[38] on the dehydration of acetaldehyde hydrate in acetone solution, using a large number of carboxylic acids and phenols as catalysts. This reaction

$$CH_3CH(OH)_2 \rightarrow CH_3CHO + H_2O \tag{94}$$

can conveniently be followed dilatometrically by the increase in volume that occurs. It is subject to general acid catalysis as well as general base catalysis. The mechanism for the acid catalysis is probably

$$CH_3C\overset{\displaystyle OH}{\underset{\displaystyle OH}{\diagup}} \; + \; HA \;\underset{\text{fast}}{\rightleftharpoons}\; CH_3{-}\overset{\displaystyle H}{\underset{\displaystyle OH}{\overset{|}{C}}}{-}\overset{\displaystyle H}{\underset{\displaystyle +}{OH}} \; + \; A^- \tag{95}$$

$$CH_3{-}\overset{\displaystyle H}{\underset{\displaystyle OH}{\overset{|}{C}}}{-}\overset{\displaystyle H}{\underset{\displaystyle +}{OH}} \; + \; A^- \;\overset{\text{slow}}{\longrightarrow}\; CH_3{-}\overset{\displaystyle H}{\underset{\displaystyle O^-}{\overset{|}{C}}}{-}\overset{\displaystyle H}{\underset{\displaystyle +}{OH}} \; + \; HA \tag{96}$$

$$CH_3{-}\overset{\displaystyle H}{\underset{\displaystyle O^-}{\overset{|}{C}}}{-}\overset{\displaystyle H}{\underset{\displaystyle +}{OH}} \;\overset{\text{fast}}{\longrightarrow}\; CH_3{-}\overset{\displaystyle H}{C}{=}O \; + \; H_2O \tag{97}$$

(95) being an equilibrium, (96) being rate-determining, and (97) rapid. Figure 2 shows the results of plotting the Brönsted equation in the form

$$\log k_c = \alpha \log K_A + \log G_A \tag{98}$$

where K_A is measured in H_2O.

The carboxylic acids and phenols with acid strengths ranging over ten powers of 10 obey the log-log plot with a mean deviation of 0.15 logarithmic unit. Catalysts of other types show considerable deviation both in a positive and negative sense. Some of these are listed in Table 2. Acids in which the electronic structure of the anion differs considerably from the structure of the neutral acid (pseudo-acids) are less active by one to two powers of 10; acids in which the structure of the anion and acid are very similar tend to show positive deviations, being more active by one to two

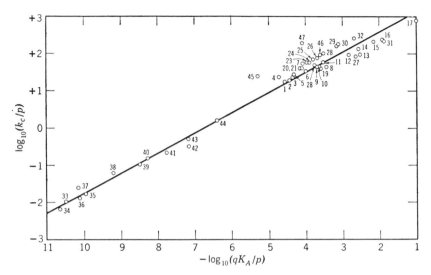

Fig. 2. Brönsted plot of catalytic constants of various acids in the dehydration of acetaldehyde hydrate. Key: (1) propionic, (2) acetic, (3) β-phenylpropionic, (4) trimethylacetic, (5) crotonic, (6) phenylacetic, (7) cinnamic, (8) formic, (9) β-chloropropionic, (10) diphenylacetic, (11) glycollic, (12) phenoxyacetic, (13) bromacetic, (14) chloracetic, (15) cyanacetic, (16) phenylpropiolic, (17) dichloracetic, (18) o-methoxybenzoic, (19) o-toluic, (20) p-hydroxybenzoic, (21) p-methoxybenzoic, (22) m-hydroxybenzoic, (23) p-toluic, (24) m-toluic, (25) benzoic, (26) p-chlorobenzoic, (27) o-chlorobenzoic, (28) m-chlorobenzoic, (29) m-nitrobenzoic, (30) p-nitrobenzoic, (31) o-nitrobenzoic, (32) salicylic, (33) thymol, (34) hydroquinone, (35) phenol, (36) resorcinol, (37) pyrocatechol, (38) p-chlorophenol, (39) o-chlorophenol, (40) m-nitrophenol, (41) 2,4-dichlorophenol, (42) o-nitrophenol, (43) p-nitrophenol, (44) 2,4,6-trichlorophenol, (45) pentachlorophenol, (46) 2,6-dinitrophenol, (47) 2,4-dinitrophenol (Bell and Higginson).

powers of 10. Carboxylic acids and phenols are acids where the anion differs moderately from the neutral acid because of the resonance:

$$R—\overset{\overset{\text{O}}{\|}}{C}—O^- \leftrightarrow R—\overset{\overset{\text{O}^-}{|}}{C}=O \quad \text{and} \quad \langle\bigcirc\rangle—O^- \leftrightarrow {}^-\langle\bigcirc\rangle=O \text{ etc.}$$

Hence they are intermediate in type, but, because of their large number and availability, they tend to establish the norm of catalytic ability.[38]

Other deviations from the Brönsted equations have been noted. A comparison of the catalytic activity of tertiary amines with primary amines of the same strength shows that the tertiary amines are more efficient catalysts.[39] This has been explained in terms of the solvation energies of

Table 2

DEHYDRATION OF ACETALDEHYDE HYDRATE
DEVIATION FROM BRÖNSTED EQUATION

(Bell and Higginson)

Catalyst	Deviation, $\log k_c$
Benzoylacetone	−1.4
Dimedone	−1.4
Nitromethane	−1.4
1-Nitropropane	−1.4
Nitroethane	−1.7
2-Nitropropane	−1.9
Benzophenone oxime	+1.2
Acetophenone oxime	+1.4
Diethyl ketoxime	+2.1
Chloral hydrate	+0.7
Water	+1.6

primary ammonium ions being greater than those of tertiary ammonium ions. Hence the measured ionization constants in water are not a true measure of the "intrinsic" proton affinity of the amine such as might be measured in the gas phase. It has been shown that the difference in hydration energies of NH_4^+ and $N(CH_3)_4^+$ is as great as 30 kcal.[40] As expected, the relative catalytic efficiencies of primary, secondary, and tertiary amines vary greatly, depending on the solvent when nonaqueous solvents are used.[41]

Linear Free-Energy Changes

The Brönsted equation is only one example of a more general phenomenon which may be called the principle of linear free-energy changes in a series of related reactions.[42] This principle may be expressed in one of several ways relating equilibrium constants and rate constants:

$$K_i = G_1 K_j^{\alpha} \qquad\qquad k_i = G_2 K_i^{\alpha} \qquad (99)$$

$$k_i = G_3 K_j^{\alpha} \qquad\qquad k_i = G_4 k_j^{\alpha}$$

where the K_i's and K_j's are a series of equilibrium constants for two related reactions involving a group of similar reactants and the k_i's and k_j's are the corresponding rate constants. The equations discussed by Hammett

$$\log k = \log k_0 + \sigma\rho \tag{100}$$

$$\log K = \log K_0 + \sigma\rho \tag{101}$$

are alternative ways of expressing the same relationships shown in (99). Here ρ is identical with α and is a measure of the susceptibility of the

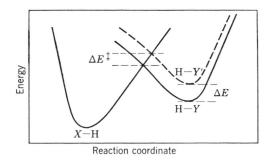

Fig. 3. Potential energy barriers in reactions: X—$H + Y \rightarrow X + H$—Y and X—$H + Y' \rightarrow X + H$—Y' (Bell, *op. cit.*).

reaction to the changes caused by *meta* and *para* substitution in the benzene ring, and σ is the log of the ratio of the ionization constant of the substituted benzoic acid to that of benzoic acid itself and is a measure of the effect of the substituent. Expressed in the more general ways of (99)–(101) the ρ's and α's may be either positive or negative and greater or less than unity.

In the Brönsted equation proper the values of α and β found experimentally range from 0.3 to 0.9. The significance of these fractional values is that, if a change from one base to another involves a certain free-energy change in the overall ionization,

$$\Delta F = RT \ln (K_B'/K_B) \tag{102}$$

Then only a fraction of this overall free-energy change will be found in the free energy of activation:

$$\Delta F^{\ddagger} = RT \ln (k_B'/k_B) = \beta \, \Delta F \tag{103}$$

That the change should be only a fraction is reasonable, since the activation step involves only a partial removal of the proton, whereas the ionization involves the complete removal. Figure 3 (taken from Bell, *Acid-Base Catalysis*, p. 159) is a diagram of the potential energy in a proton transfer reaction as a function of the distance of the proton from the two centers X and Y in the reaction

$$X—H + Y \rightarrow X\text{---}H\text{---}Y \rightarrow X + H—Y \tag{104}$$

The solid curves are obtained by a superposition of the usual Morse curves for the covalent bond energy of $X—H$ and $H—Y$ as a function of internuclear distance. They are intended to be qualitative rather than quantitative. Now, if a change is made to Y', the dotted line represents the covalent bond energy curve for $H—Y'$. If the curves for $H—Y$ and $H—Y'$ have the same shape and the same position on the distance axis, then it follows from the geometry of the system that for small changes

$$\Delta E^{\ddagger} = \beta \, \Delta E \tag{105}$$

where $\Delta E^{\ddagger}$ is the difference in the energy barriers, ΔE is the difference in the overall energy, and β is between zero and one. If the entropy factors remain the same, then (105) can be converted into rate and equilibrium constants:

$$\Delta \log k_{\mathrm{B}} = \beta \, \Delta \log K_{\mathrm{B}} \tag{106}$$

which is identical with the Brönsted relation.

Unfortunately for this simple argument, the prediction that different bases should have rate constants which differ only because the activation energies are different is not borne out by experiment. A series of acids and bases whose rate constants may obey the Brönsted equation very closely will generally have a range of activation energies and of frequency factors. This is partly because of the sometimes overlooked fact that the experimental activation energy is not identical with the height of the potential energy barrier of Fig. 3. There is, of course, a relationship such that significant changes in the energy barrier show up as changes in the activation energy. However, the activation energy, as determined by experiment, also includes all the other factors which would cause the rate of the reaction to vary with temperature (see Chapter 5).

That the arguments expressed in Fig. 3 have some validity is shown by the striking success of Hammett's $\sigma\rho$ function. It is just in the case of *meta-* and *para*-substituted benzene derivatives that the requirement that differences in rates and equilibrium be due to changes in potential energy seems best satisfied (Hammett, *Physical Organic Chemistry*, p. 121). When the entropies of activation show considerable variation, as in aliphatic compounds and *ortho*-substituted benzene derivatives, the linear free-energy relationship is most apt to break down. Factors which appear to have an adverse effect on the quantitative nature of the relationship are substitutions close to the center of reaction, substituents with different steric requirements and with differing solvation tendencies, and, in general, substituents which change the structure and geometry of the activated complex appreciably. The reactions compared must, of course, have mechanisms similar enough so that parallelisms can be expected.

In spite of these limitations the principle of linear free-energy changes is not restricted to reactions of *meta-* and *para-*substituted aromatic compounds. The existence of the Brönsted relation proves this. Other examples

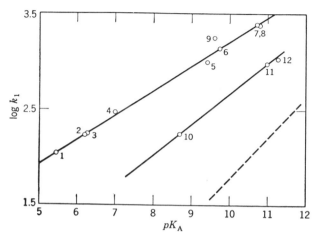

Fig. 4. First association constant k_1, of amines with silver ion, compared with the acid ionization constant of the corresponding ammonium ions. Dotted line estimated for tertiary amines. Key: (1) pyridine, (2) α-picoline, (3) γ-picoline, (4) 2,4-lutidine, (5) β-methoxyethylamine, (6) ethanolamine, (7) isobutylamine, (8) ethylamine, (9) benzylamine, (10) morpholine, (11) diethylamine, (12) piperidine (Bruehlman and Verhoek).

may be mentioned. Bruehlman and Verhoek[43] have shown that the first formation constant for a series of silver ammines shows a correlation with the basic ionization constant of the amine:

$$Ag^+ + B \rightleftharpoons AgB^+ \tag{107}$$

Thus if the log of the equilibrium constant of equation 107 is plotted against pK_A which measures the strength of the amine relative to hydrogen ion

$$H^+ + B \rightleftharpoons BH^+ \tag{108}$$

a straight line is obtained for primary amines and substituted pyridines as in Fig. 4. Secondary amines do not fall on this line but form a second straight line of their own. Tertiary amines of the trialkyl type apparently form a third line. The difference in behavior of the several classes of amines apparently relates to the different steric requirements of the silver ion and the hydrogen ion.

Figure 5 shows the result of plotting the logs of the rate constants for two similar reactions of a series of organic bromides:

$$RBr + C_5H_5N \rightarrow R\text{—}NC_5H_5^+ + Br^- \tag{109}$$

$$RBr + (NH_2)_2CS \rightarrow R\text{—}SC(NH_2)_2^+ + Br^- \tag{110}$$

Thiourea is more reactive in every case than pyridine, but the reactions are essentially of the same type as far as the rate-determining step is

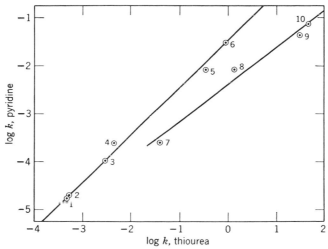

Fig. 5. Rates of reaction of organic bromides with pyridine and thiourea compared. Key: (1) β-phenoxyethyl bromide, (2) ethylene bromohydrin, (3) n-propyl bromide, (4) ethyl bromide, (5) allyl bromide, (6) benzyl bromide, (7) 2,4,6-trimethylphenacyl bromide, (8) ethyl bromacetate, (9) phenacyl bromide, (10) p-bromophenacyl bromide [Pearson, McGuire, Langer, and Williams, *J. Am. Chem. Soc.*, 74, 5130 (1952)].

concerned. In some instances the thiuronium salt which is formed first reacts further, but this is not involved in the comparison. It is seen that there is, in general, a good correlation between the rates for the bromo-ketones and bromoester, and another correlation for all other bromides. The corresponding plot for the reaction of the same bromides with iodide ion and with pyridine is very similar. The greater relative reactivity of the bromoketones and ester for the highly polarizable thiourea and iodide ion presumably means that an electrostatic factor is of importance for these reactions.[44]

Linear free-energy relationships for gas-phase reactions are also known. A linear log-log plot is formed for the rate constants for attack of $CH_3\cdot$ and $CF_3\cdot$ on a series of substrates, the reaction being hydrogen atom removal.[45]

There have been a number of fairly successful attempts to write down general equations which predict the rate of a large number of reactions from a knowledge of a small number of standard reaction rates. These are all based on the assumption of linear free-energy effects. For example, Grunwald and Winstein have correlated the rates of many solvolytic type reactions of organic compounds in various solvents by means of the equation[46]

$$\log (k/k_0) = mY \qquad (111)$$

Here k_0 is the rate constant for a given reactant in a selected standard solvent and k is the rate constant in any other solvent. The parameter m is to be determined and is characteristic of the reactant, while Y has been already evaluated and is simply $\log (k/k_0)$ for t-butyl chloride, which is taken as a standard reactant. Table 3 shows Y values for a number of solvents.

Table 3

Y VALUES FOR COMMON SOLVENTS

[A. H. Fainberg and S. Winstein, *J. Am. Chem. Soc.*,
78, 2770 (1956)]

Solvent	Y
Water	3.493
Formic acid	2.054
n-C_3F_7COOH	1.7
80% ethanol-water	0.000^a
Methanol	-1.090
Acetic acid	-1.639
90% acetone-water	-1.856
90% dioxane-water	-2.030
Ethanol	-2.033
i-Propanol	-2.73
t-Butanol	-3.26
Acetic anhydride	-3.29^b

^{a} Arbitrary standard solvent. Per cent is by volume.
^{b} Containing 2.5% acetic acid.

Other treatments[47] attempt to evaluate all reagents of a given type, say nucleophilic reagents in organic chemistry or ligands in inorganic chemistry, and to put them in quantitative order according to reactivity. Then rates of reaction, or equilibrium constants, for any reaction of a certain class can be estimated relative to a standard reaction.

An important contribution by Taft[48] shows that it is possible to discuss the reactions of aliphatic compounds and of *ortho*-substituted benzene derivatives in the same way as for *meta*- and *para*-benzene derivatives.

It is necessary to take into account separately polar, steric, and, less often, resonance effects. By picking the ratio of the rates of acid and base hydrolysis of substituted esters, steric factors are canceled and purely polar values can be obtained for a large number of substituent groups. Brown and his co-workers have greatly extended the range of predictions that can be made for the electrophilic substitution reactions of aromatic compounds.[49]

In an interesting paper Hammond[50] has shown how one can predict when a relationship between rate and overall free-energy change should exist, and how the effect of a substituent on the rate of a chemical reaction can be estimated. Two reasonable assumptions are made. One is that a correlation of rates and equilibria will be expected only if the transition state resembles the products rather than the reactants. The other is that if two states, as for example, a transition state and an unstable intermediate, occur consecutively in a reaction and have nearly the same energy, their interconversion will involve only a small degree of molecular reorganization.

In highly endothermic processes it is expected that the transition state will resemble the products most closely. Hence changes which increase the stability of the products will increase the rate of the reaction. This would be the case for reactions where the rate-determining step is the formation of reactive intermediates such as free radicals, carbonium ions, and unstable carbanions.

In highly exothermic processes it is expected that the transition state will resemble the reactants most closely. In these cases little parallelism can be hoped for between rates and equilibria. This would be the situation for the rates of reaction of highly unstable intermediates such as free radicals, etc. These entities are often non-discriminating in their reactivity towards various substrates. Prediction becomes difficult, very often the less stable of a pair of possible products being formed. Stabilizing the reactive species in some way, such as by solvation, should increase the discrimination. Varying selectivities are found, for example, in the reactions of chlorine atoms in various solvents.[51]

General and Specific Catalysis

In the Brönsted equation a low value of α signifies a low sensitivity of the catalytic constant to the strength of the catalyzing acid, whereas a large value of α indicates a high sensitivity. Since the solvated hydrogen ion is usually the strongest acid available in ordinary solvents, reactions with large values of α will show a dependence on the hydrogen ion concentration so large that catalysis by other acids will be difficult to detect. If the reaction is sufficiently fast, buffered solutions may be prepared which are basic enough so that the hydrogen ion reaction is slow, but which contain sufficient quantities of other acids to test their catalytic activities. The best procedure for detecting general acid catalysis is to measure the rate of the reaction in a series of such buffers with a constant ratio of acid to conjugate base, but of different total concentrations. If the reaction is general acid-catalyzed, the observed rate constant will increase with total concentration. If the reaction is subject to specific hydrogen ion catalysis, the rate constant will vary only slightly, since the hydrogen ion concentration is determined chiefly by the ratio which

Table 4

HYDROLYSIS OF ACETAL

[J. N. Brönsted and W. F. K. Wynne-Jones, *Trans. Faraday Soc.*, *25*, 59 (1929)]

Formic Acid	Sodium Formate	Sodium Chloride	Ionic Strength	$k_c \times 10^3$, min^{-1}
0.0296M	0.0100M	......	0.011	12.5
0.0592	0.0200	......	0.020	13.4
0.1480	0.0500	......	0.050	15.1
0.2220	0.0750	......	0.075	16.4
0.2960	0.1000	......	0.100	17.8
0.1776	0.0600	0.0400	0.100	17.6
0.0987	0.0333	0.0667	0.100	17.7
0.0222	0.0075	0.0925	0.100	18.2

remains constant. To exclude variations due to secondary salt effects on the ionization of the acid (Chapter 7), some care should be used to hold the total ionic strength constant. Although it is true that there are some mechanisms (Cases I and VI) which demand specific hydrogen or hydroxide ion catalysis, the experimental finding of specific catalysis does not prove that such a mechanism is operating, since it may be a case of a large value of α or β which makes catalysis by other acids or bases small.

A value of α or β close to zero may in some cases cause the general acid or basic catalysis of a reaction not to appear experimentally, since the solvent molecules which are present in large excess may act as acid or base. An example of this is found in the acid-catalyzed enolization of ketones where, as the acidity of the ketone increases, catalysis by acids becomes less important, finally disappearing. Thus the enolization of acetone is general acid-catalyzed, but the enolization of acetylacetone is independent of all acids, including hydrogen ion. The place of the acid in the latter case is presumably taken by water or other solvent molecules.

Table 4 shows some data on the hydrolysis of acetal in formate-ion-formic-acid buffers. The ratio of acid to ion is held constant at 2.96

(some corrections are needed because of the ionization of formic acid). The last four values show the constancy of the experimental rate constant when the ionic strength is maintained at 0.1 by the addition of sodium chloride while the total concentration of the buffer is changed. This indicates a specific hydrogen ion catalysis. The salt effect on the ionization constant of formic acid would lead to an increase in the rate constant and an apparent catalysis by formic acid molecules as the first four entries show. This is an example of the secondary salt effect.

Table 5

BROMINATION OF BROMACETYLACETONE AT 25°
CATALYSIS BY CHLORACETATE ION IN WATER

[R. P. Bell, E. Gelles and E. Möller, *Proc. Roy. Soc.*, *A198*, 308 (1949)]

Chloracetic Acid	Sodium Chloracetate	k_{obs}, min^{-1}	k_c, liters/mole-min
0.000	0.000	1.39	....
0.009	0.018	2.08	38.3
0.0275	0.055	3.46	37.6
0.058	0.116	5.96	39.4

Table 5 shows some data for the base-catalyzed bromination of mono-bromacetylacetone in aqueous solution. Various buffers of chloracetic acid and its salt show an experimental rate constant which increases linearly with the concentration of chloracetate ion, showing that the chloracetate ion serves as a catalytic base. There is also a catalysis by water acting as a base which is quite appreciable. The catalytic constant of water can be measured separately in solutions containing no chloracetate ion and is the first entry of Table 5. The catalytic constant for chlorace-tate ion is obtained by subtracting the value for water from the observed first-order rate constant and dividing by the concentration of chloracetate ion. The results are shown in the fourth column. There is no appreciable catalysis by hydroxide ion, since the buffers used are acid.

PROBLEMS

1. The hydrolysis of the hydroxylamine disulfonate ion proceeds as follows:
$$HON(SO_3)_2^{-2} + H_2O \rightarrow HONHSO_3^- + H^+ + SO_4^{-2}$$

From the following data of Naiditch and Yost determine the rate constants for the uncatalyzed reaction and for the reaction catalyzed by hydrogen ion (the former can be obtained most readily from the initial readings and the latter from the later readings). The reaction was followed by titrating the acid produced. Original concentrations: $K_2HON(SO_3)_2$, 0.0473M, no acid initially added. Temperature 25° C. [S. Naiditch and D. M. Yost, *J. Am. Chem. Soc.*, *63*, 2123 (1941).]

Time, min	Acid × 10³, moles/liter
11	1.12
24	1.34
38	1.67
58	2.03
81	2.68
100	3.39
122	4.32
144	5.32
164	6.65
189	8.93
218	12.08
256	16.17

2. The mutarotation of glucose is subject to catalysis by both acids and bases. Evaluate as well as possible the catalytic constants of water, hydrogen ion, mandelic acid, mandelate ion, pyridine, and pyridinium ion from the observed first-order rate constants given. [J. N. Brönsted and E. A. Guggenheim, *J. Am. Chem. Soc.*, *49*, 2554 (1927).]

Catalyst Conc., moles/liter		$k_{obs} × 10^3$, min^{-1}
HCl	10^{-4}	5.29
HCl	10^{-5}	5.23
$HClO_4$ 10^{-4} + 0.1 N KNO_3		5.40
$HClO_4$	0.0048	6.00
$HClO_4$	0.0247	8.92
$HClO_4$ 0.0325 + 0.2 N KNO_3		10.02

Sodium Mandelate	Mandelic Acid	
0.050	0.001	5.80
0.100	0.001	6.38
0.125	0.001	6.66
0.100	0.050	6.70
0.100	0.100	7.04

Pyridine	Pyridine Perchlorate	
0.025	0.025	7.46
0.050	0.050	9.29
0.050	0.100	9.63
0.050	0.200	9.39

K_a for mandelic acid is 4.3 × 10^{-4}. K_b for pyridine is 2.9 × 10^{-9}. Temperature, 18° C.

3. The decomposition of nitramide depends on general base catalysis, the rate-determining step being the removal of a proton.

$$NH_2NO_2 = N_2O + HO_2$$

Plot the following catalytic constants obtained for a series of negatively charged bases and obtain the constants of the Brönsted equation. Apply statistical corrections where necessary. Are there any observable deviations with structure or charge? The acid ionization constant of the conjugate acid of each base is given. All the thermodynamic and kinetic data are at 15° C. [J. N. Brönsted and K. Pedersen, *Z. physik. Chem.*, *108*, 185 (1923); E. C. Baughan and R. P. Bell, *Proc. Roy. Soc.*, *158A*, 464 (1937).]

Catalyst	K_a	k_b
Hydroxide ion	1.1×10^{-16}	1×10^6
Sec. phosphate ion	5.8×10^{-8}	86
Succinate ion	2.4×10^{-6}	1.8
Malate ion	7.8×10^{-6}	0.72
Trimethylacetate ion	9.4×10^{-6}	0.822
Propionate ion	1.3×10^{-5}	0.649
Acetate ion	1.8×10^{-5}	0.504
Tartrate ion	4.1×10^{-5}	0.165
Acid succinate ion	6.5×10^{-5}	0.320
Oxalate ion	6.8×10^{-5}	0.104
Phenylacetate ion	5.3×10^{-5}	0.232
Benzoate ion	6.5×10^{-5}	0.189
Formate ion	2.1×10^{-4}	0.082
Acid malate ion	4.0×10^{-4}	0.077
Acid tartrate ion	9.7×10^{-4}	0.036
Acid phthalate ion	1.2×10^{-3}	0.029
Salicylate ion	1.0×10^{-3}	0.021
Monochloracetate ion	1.4×10^{-3}	0.016
Prim. phosphate ion	7.6×10^{-3}	0.0079
o-Nitrobenzoate ion	7.3×10^{-3}	0.0042
Dichloracetate ion	5.0×10^{-2}	0.0007

4. In a buffered solution containing 0.05 M acetic acid and 0.05 M sodium acetate, calculate the percentage reaction due to hydrogen ion, acetic acid, and to water in acid-catalyzed reactions where the coefficient α in the Brönsted equation is equal to 0.1, 0.5, and 0.9, respectively.

REFERENCES

1. R. P. Bell, *Acid-Base Catalysis*, Oxford University Press, Oxford, 1941.
2. H. Campbell and D. D. Eley, *Nature*, *154*, 85 (1944); M. H. Dilke and D. D. Eley, *J. Chem. Soc.*, 2601 (*1949*).

234   HOMOGENEOUS CATALYSIS   Ch. 9

3. G. K. Rollefson and R. F. Faull, *J. Am. Chem. Soc.*, *59*, 625 (1937).
4. E. W. R. Steacie, *Free Radical Mechanisms*, Reinhold Publishing Corp., New York, 1946; H. J. Schumacher, *Chemische Gasreaktionen*, Steinkopff, Dresden and Leipzig, 1938.
5. J. W. Mitchell and C. N. Hinshelwood, *Proc. Roy. Soc.*, *A159*, 32 (1937).
6. A. Boyer, M. Niclause, and M. Letort, *J. Chem. Phys.*, 345 (1952).
7. A. Maccoll and V. R. Stimson, *Proc. Chem. Soc.*, 80 (1958).
8. R. F. Schutz and G. B. Kistiakowsky, *J. Am. Chem. Soc.*, *56*, 395 (1934).
9. G. M. Schwab, *Handbuch der Katalyze*, Springer, Vienna, 1940; *Catalysis*, Vols. 1–6, P. H. Emmett, editor, Reinhold Publishing Co., New York, 1954–1958.
10. R. Steinberger and F. H. Westheimer, *J. Am. Chem. Soc.*, *71*, 4158 (1949); *73*, 429 (1951).
11. (*a*) F. Basolo and R. G. Pearson, *Mechanisms of Inorganic Reactions*, John Wiley and Sons, New York, 1958, Chapter 8; (*b*) M. L. Bender, *Chem. Revs.*, *60*, 53 (1960).
12. W. C. Bray, *Chem. Revs.*, *10*, 161 (1932).
13. R. E. Connick, *J. Am. Chem. Soc.*, *69*, 1509 (1947).
14. F. Haber and J. Weiss, *Proc. Roy. Soc.*, *A147*, 332 (1934); N. Uri, *Chem. Revs.*, *50*, 375 (1952).
15. J. C. Jungers and K. F. Bonhoeffer, *Z. physik. Chem.*, *A177*, 460 (1936); Z. Luz, D. Gill and S. Meiboom, *J. Chem. Phys.*, *30*, 1540 (1959).
16. M. Eigen and L. de Maeyer, *Z. Elektrochem.*, *59*, 986 (1955).
17. R. A. Ogg, Jr., *Disc. Faraday Soc.*, *17*, 215 (1954).
18. R. P. Bell and R. G. Pearson, *J. Chem. Soc.*, 3443 (1953).
19. R. G. Pearson, *J. Am. Chem. Soc.*, *71*, 2212 (1949).
20. J. F. Bunnett, *J. Am. Chem. Soc.*, *79*, 5969 (1957).
21. R. C. Tolman, *Statistical Mechanics with Applications to Physics and Chemistry*, Chemical Catalog Co., New York, 1927.
22. R. Junell, Dissertation, Uppsala, 1935; S. H. Maron and V. K. La Mer, *J. Am. Chem. Soc.*, *61*, 2018 (1939); R. G. Pearson and R. L. Dillon, *ibid.*, *72*, 3574 (1950).
23. R. P. Bell and J. C. Clunie, *Nature*, *167*, 363 (1951).
24. T. M. Lowry and I. J. Faulkner, *J. Chem. Soc.*, *127*, 2883 (1925); C. G. Swain, *J. Am Chem. Soc.*, *72*, 4578 (1950).
25. T. M. Lowry and E. M. Richards, *J. Chem. Soc.*, *127*, 1385 (1925); C. G. Swain and J. F. Brown, Jr., *J. Am. Chem. Soc.*, *74*, 2534 (1952).
26. R. P. Bell, *Acid-Base Catalysis*, Oxford University Press, Oxford, 1941, Chapter VI.
27. C. V. King and E. D. Bolinger, *J. Am. Chem. Soc.*, *58*, 1533 (1936); C. V. King, *J. Am. Chem. Soc.*, *62*, 379 (1940).
28. E. Coombs and D. P. Evans, *J. Chem. Soc.*, 1295 (*1940*); D. S. Noyce and W. L. Reed, *J. Am. Chem. Soc.*, *81*, 624 (1959).
29. S. Widequist, *Arkiv Kemi*, *26A*, 2 (1948); L. Ramberg and E. Samen, *ibid.*, *11B*, 40 (1934).
30. K. F. Bonhoeffer and W. D. Walters, *Z. physik. Chem.*, *A181*, 441 (1938).
31. See reference 11*b* for a summary.
32. R. G. Pearson and A. C. Sandy, *J. Am. Chem. Soc.*, *73*, 931 (1951).
33. M. Cohn and H. C. Urey, *J. Am. Chem. Soc.*, *60*, 679 (1938).
34. J. N. Brönsted and K. Pedersen, *Z. physik. Chem.*, *A108*, 185 (1923).
35. R. P. Bell, *op. cit.*, Chapter V.
36. J. N. Brönsted, *Chem. Revs.*, *5*, 322 (1928).
37. S. W. Benson, *J. Am. Chem. Soc.*, *80*, 5151 (1958).
38. R. P. Bell and W. C. E. Higginson, *Proc. Roy. Soc.*, *A197*, 141 (1949).

39. R. P. Bell and A. F. Trotman-Dickenson, *J. Chem. Soc.*, 1288 (1949); A. F. Trotman-Dickenson, *ibid.*, 1293; R. G. Pearson and F. V. Williams, *J. Am. Chem. Soc.*, *76*, 258 (1954); H. K. Hall, Jr., *1810*, *79*, 5441 (1957).

40. R. G. Pearson and D. C. Vogelsong, *J. Am. Chem. Soc.*, *80*, 1038 (1958).

41. R. G. Pearson and D. C. Vogelsong, *J. Am. Chem. Soc.*, *80*, 1048 (1958).

42. L. P. Hammett, *Chem. Revs.*, *17*, 125 (1935); *Trans. Faraday Soc.*, *34*, 156 (1938); *Physical Organic Chemistry*, McGraw-Hill Book Co., New York, 1940, Chapter VII.

43. R. J. Bruehlman and F. H. Verhoek, *J. Am. Chem. Soc.*, *70*, 1401 (1948).

44. See P. D. Bartlett and E. N. Trachtenberg, *J. Am. Chem. Soc.*, *80*, 5808 (1958) for a discussion of the mechanism.

45. A. F. Trotman-Dickenson, *Chem. and Industry*, 1243 (1957).

46. E. Grunwald and S. Winstein, *J. Am. Chem. Soc.*, *70*, 846 (1948); A. H. Fainberg and S. Winstein, *ibid.*, *78*, 2770 (1956).

47. C. G. Swain and C. B. Scott, *J. Am. Chem. Soc.*, *75*, 141 (1953); J. O. Edwards, *ibid.*, *76*, 1540 (1954), *78*, 1819 (1956).

48. R. W. Taft, Jr., *Steric Effects in Organic Chemistry*, M. S. Newman, editor, John Wiley and Sons, New York, 1956, Chapter 13.

49. H. C. Brown and K. L. Nelson, *J. Am. Chem. Soc.*, *75*, 6292 (1953); H. C. Brown and L. M. Stock, *ibid.*, *81*, 3323 (1959).

50. G. S. Hammond, *J. Am. Chem. Soc.*, *77*, 334 (1955).

51. G. A. Russell, *J. Am. Chem. Soc.*, *79*, 2977 (1957).

10.

CHAIN REACTIONS

Before discussing chain reactions in general terms it will be instructive to consider in detail the classical example of the reaction of hydrogen and bromine vapor: $H_2 + Br_2 \rightleftharpoons 2HBr$. Bodenstein and Lind[1] in 1906 measured the rate in the temperature range from 200 to 300°. They found their results were in agreement with the empirical equation

$$\frac{d[HBr]}{dt} = \frac{k[H_2][Br_2]^{\frac{1}{2}}}{1 + k'[HBr]/[Br_2]} \tag{1}$$

where $k' = \frac{1}{10}$ and is temperature-independent and $k \propto e^{-40,200/RT}$ so that the activation energy is 40.2 kcal. Despite the complexity of the rate expression (1), they were able to integrate it for various initial conditions and so compare their results with the integrated form.

Thirteen years later the form of the rate expression (1) was explained by Christiansen,[2] Herzfeld[3] and Polanyi[4] in terms of the following mechanism:

$$Br_2 \xrightarrow{k_1} 2Br$$

$$Br + H_2 \xrightarrow{k_2} HBr + H$$

$$H + Br_2 \xrightarrow{k_3} HBr + Br$$

$$H + HBr \xrightarrow{k_4} H_2 + Br$$

$$2Br \xrightarrow{k_5} Br_2$$

This is a characteristic chain mechanism in that it is postulated that there are reactive intermediates or chain carriers, here H and Br atoms, at low concentration, which are involved in a cycle of reaction steps, here steps (2), (3), and (4), such that the chain carriers are reformed after each cycle and are ready to participate in another cycle. The chain carriers must be formed

236

by a chain initiation step, here the first, and may be removed by a chain breaking step, here the last or recombination of the Br atoms.

The derivation of (1) from this postulated mechanism proceeds as follows. The rate of formation of HBr can be written

$$d[\text{HBr}]/dt = k_2[\text{Br}][\text{H}_2] + k_3[\text{H}][\text{HBr}_2] - k_4[\text{H}][\text{HBr}] \qquad (2)$$

As it stands, this expression is inconvenient because it contains the not easily measured concentrations of H and Br atoms. In principle these could be eliminated by solving (2) simultaneously with other differential equations including:

$$d[\text{Br}]/dt = 2k_1[\text{Br}_2] - k_2[\text{Br}][\text{H}_2] + k_3[\text{H}][\text{Br}_2] + k_4[\text{H}][\text{HBr}] - 2k_5[\text{Br}]^2 \qquad (3)$$

$$d[\text{H}]/dt = k_2[\text{Br}][\text{H}_2] - k_3[\text{H}][\text{Br}_2] - k_4[\text{H}][\text{HBr}] \qquad (4)$$

(In (3) the rate constants k_1 and k_5 are defined in terms of Br_2 molecules.) However, the complete solution is impossible, or at least too difficult. An approximate solution is accomplished with the aid of the steady-state assumption, the validity of which has been discussed in Chapter 8. The required condition that the intermediates (chain carriers) are at low concentrations is certainly justified here. There would presumably be an induction period, but in this case it would be very small. Matsen and Franklin[3] estimate it to be about 10^{-9} sec. Therefore, let the right sides of equations 3 and 4 be zero and call these equations 3' and 4'. Subtraction of (4') from (2) gives

$$d[\text{HBr}]/dt = 2k_3[\text{H}][\text{Br}_2] \qquad (5)$$

The H atom concentration needed in (5) can be obtained by addition of (3') and (4') to get

$$2k_1[\text{Br}_2] - 2k_5[\text{Br}]^2 = 0$$

and therefore

$$[\text{Br}] = ((k_1/k_5)[\text{Br}_2])^{1/2} \qquad (6)$$

which upon substitution in (4') yields

$$\begin{aligned} [\text{H}] &= \frac{k_2[\text{H}_2]}{k_3[\text{Br}_2] + k_4[\text{HBr}]} [\text{Br}] \\ &= \frac{k_2(k_1/k_5)^{1/2}[\text{H}_2][\text{Br}_2]^{1/2}}{k_3[\text{Br}_2] + k_4[\text{HBr}]} \end{aligned} \qquad (7)$$

This then used in (5) and rearranged gives

$$\frac{d[\text{HBr}]}{dt} = \frac{2k_3k_2(k_1/k_5)^{1/2}[\text{H}_2][\text{Br}_2]^{3/2}}{k_3[\text{Br}_2] + k_4[\text{HBr}]}$$

or

$$\frac{d[\text{HBr}]}{dt} = \frac{2k_2(k_1/k_5)^{1/2}[\text{H}_2][\text{Br}_2]^{1/2}}{1 + (k_4[\text{HBr}]/k_3[\text{Br}_2])} \qquad (8)$$

This has the same form as the empirical expression (1) and will agree with it quantitatively if

$$k = 2k_2(k_1/k_5)^{1/2} \tag{9}$$

and

$$k' = k_4/k_3 = \tfrac{1}{10} \tag{10}$$

A consideration of dissociation energies shows that both steps 3 and 4 are exothermic. Since they are atom reactions, their activation energies should be small, perhaps from 1 to 3 kcal. The ratio of their rate constants could very well be temperature-independent, and any difference from unity might be due to different steric factors or entropies of activation.

From (9) it is possible to evaluate the rate constant k_2, since k_1/k_5 is the equilibrium constant, K, for the dissociation of bromine molecules. By means of thermochemical data, $K \propto e^{-45,200/RT}$, the experimental expression for k_2 results:

$$k_2 = 10^{12.3}T^{1/2}e^{-17,640/RT} \text{ liters moles}^{-1} \text{ sec}^{-1} \tag{11}$$

This has been written in this form with $T^{1/2}$ so that it may be compared readily with the simple collision formula where the number of collisions varies as $T^{1/2}$. The steric factor required to explain this rate constant is of the order of 0.1 to 0.3, which is quite satisfactory (see Chapter 4). The activation energy of 17.6 kcal is also reasonable, since from bond energies this second reaction step is endothermic to the extent of 16.3 kcal and the activation energy would have to be at least this great. The Hirschfelder rule would preduct approximately 21 kcal (see Chapter 6). It is interesting to note that the overall activation energy of the reaction, 40.2 kcal, is less than that for the first step which must be at least 45.2 kcal. In some chain reactions this difference is much more pronounced.

The proposed mechanism has now been shown to reproduce the form of the empirical rate expression, and furthermore the required activation energies and steric factors for the separate steps have been found to be reasonable. However, to be completely satisfactory, it must be shown why certain other possible reaction steps are not included, for example the following:

$$H_2 \rightarrow 2H$$
$$HBr \rightarrow H + Br$$
$$Br + HBr \rightarrow Br_2 + H$$
$$H + Br \rightarrow HBr$$
$$H + H \rightarrow H_2$$

If any of these were important the rate expression would probably be different from (1). That these are all negligibly slow as compared with other competing reactions in the mechanism can be shown as follows.

The dissociation of H_2 and HBr molecules would be slow compared with the dissociation of Br_2 because Br_2 has a much lower dissociation energy. The Br + HBr reaction would be unimportant in competition with the Br + H_2 reaction because, although the Br + H_2 reaction requires an activation energy (17.6 kcal), the Br + HBr reaction would require a much greater amount (about 42 kcal). As a chain-breaking step, recombinations of H with Br and H with H as possible competitors of the Br recombination would be determined by the relative concentrations of the atoms, the activation energies being zero. Equation 7 can be rearranged to give

$$[H]/[Br] = k_2[H_2]/(k_3[Br_2] + k_4[HBr]) \simeq k_2/k_3$$

at least in the early stages of reaction and where the starting concentrations of H_2 and Br_2 are about equal. At a typical temperature, say 300°, $k_2/k_3 \simeq e^{-17,600/RT} \simeq 10^{-6}$. Therefore, a recombination of H with Br would be only one millionth as likely as a recombination of Br with Br, and H with H only one in 10^{12} as likely. Although the H atom concentration is exceedingly low, it may be estimated that it is of the order of 10^4 times as great as its concentration in equilibrium with H_2. Whereas in this reaction there is an equilibrium of Br with Br_2, the H atom concentration is determined by the steady state involving the second, third, and fourth steps of the mechanism.

Up to now it has been supposed that the first and last steps of the mechanism are unimolecular and bimolecular, respectively. Because of the result that the steady-state assumption leads to the equilibrium relation (6) between Br and Br_2, it does not matter what mechanism is assumed for the attainment of this equilibrium. For example, the recombination of atoms may be taken as termolecular, third-order with rate = $k_5'[M][Br]^2$ where M is the third body, as in fact it has been proved directly by Rabinowitch and Lehman.[6] The first step, dissociation, must then also involve M in a bimolecular step with rate = $k_1'[M][Br_2]$. The resulting expression for the rate of HBr formation would then have $k_1'[M]$ replacing k_1 and $k_5'[M]$ replacing k_5. But only the ratio appears in (8); therefore, [M] cancels and does not make itself apparent.

Direct proof that the fifth step must be termolecular under the conditions of this chain reaction is obtained by initiating the chain photochemically rather than by thermal dissociation. The photochemical reaction has been studied by Bodenstein and Lütkemeyer[7] and by Jost and Jung.[8] By studying the reaction over a wide pressure range, Jost and Jung found that the rate could be expressed by

$$\frac{d[HBr]}{dt} = \frac{k[H_2]I_{abs}^{1/2}}{p^{1/2}[1 + (k'[HBr]/[Br_2])]} \qquad (12)$$

where I_{abs} is the intensity of light absorbed and p is the total pressure. This form may be derived by substituting for the first step of the mechanism the absorption of light quanta

$$Br_2 + h\nu \rightarrow 2Br$$

with rate equal to $2k_1'' I_{abs}$. In the steady state (6) becomes

$$2k_1'' I_{abs} = 2k_5'[M][Br]^2$$

or

$$[Br] = \left(\frac{k_1'' I_{abs}}{k_5'[M]}\right)^{\frac{1}{2}} \tag{13}$$

When this expression is substituted for (6) in the previous derivation, the result is

$$\frac{d[HBr]}{dt} = \frac{2k_2(k_1''/k_5')^{\frac{1}{2}}[H_2]I_{abs}^{\frac{1}{2}}}{[M]^{\frac{1}{2}}[1 + (k_4[HBr]/k_3[Br_2])]} \tag{14}$$

This has the same form as (12) if it is assumed that [M] is proportional to p. This is a fair assumption but not exact inasmuch as different molecules are known to have different efficiencies in causing recombination of atoms. It is the presence of $p^{\frac{1}{2}}$ in the denominator of (12) that confirms the termolecular character of the recombination process and by inference requires the thermal dissociation to be bimolecular.

Incidentally, the previous calculation that the activation energy of the second step of the thermal reaction was 17.6 kcal is in excellent agreement with the observed overall activation energy of the photochemical reaction, 17.6 kcal. This should be true because k_1'' and k_5' presumably are nearly temperature-independent.

Direct measurements of both k_1 and k_5 have been made, k_1 by shock methods and k_5 by flash photolysis (see Chapter 11). The recombination reaction is termolecular and the rate constant is given by $1.9 \times 10^8 e^{1410/RT}$ M^{-2} sec^{-1} when argon is the third body.[9] From the principle of microscopic reversibility, k_1 must then be a bimolecular process and it is found[10] to obey the equation $k = 1.39 \times 10^8 \ T^{\frac{1}{2}}(\Delta E_0/RT)^{1.97} e^{-\Delta E_0/RT}$ M^{-1} sec^{-1}. Here ΔE_0 is the heat of dissociation of bromine molecules. The rate constant is written in this way because it reproduces the experimental data over the astonishing range of 10^{27} in the rate constant, between room temperature and 2225° K. Also the equation then agrees with the theoretical equation, similar to equation 55 in Chapter 4, which can be obtained if other degrees of freedom are allowed to contribute to the activation energy besides the usual translational motion. It is assumed that the recombination reaction has zero true activation energy and hence the minimum energy for dissociation is ΔE_0. Since Br_2 has only one vibrational mode, the power of $\Delta E_0/RT$ equal to nearly 2 must mean that rotational energy can be drawn on for dissociation.[10] The frequency factor also corresponds to a p value of 0.06 if the collision diameter is $3A$.

Other chain reactions similar to that of H_2 and Br_2 are those of H_2 and Cl_2 and the reactions of Br_2 and Cl_2 with aliphatic hydrocarbons. The high reactivity of Cl atoms leads to very long chains, up to 10^6. For the same reason, the reactions of Cl_2 are difficult to study because of the great sensitivity to traces of impurities and to wall reactions. It has recently been shown that even the classical reaction of H_2 and I_2 becomes partly chain type at very high temperatures where I is sufficiently reactive.[11]

General Treatment of Chain Reactions

Chain mechanisms in general can be handled satisfactorily by the steady-state method. Instead of presenting a number of particular examples, the present aim will be to provide a general treatment that can be specialized for particular cases. The essence of the chain mechanism is the existence of unstable intermediates or chain carriers that take part in the chain propagation steps and are regenerated. The derivation of the rate expression for the overall reaction depends only on the order of each of the reaction steps with respect to the chain carriers. The functional dependence of the rate of each step on the concentrations of molecular species other than chain carriers can be indicated explicitly after deriving general formulas.

Let the mechanism be indicated as follows:

$$\text{initiation} \qquad\qquad \cdots \to m\text{R} + \cdots$$

$$\text{propagation} \qquad \text{R} + \cdots \to \text{R} + \cdots$$

$$\text{breaking} \qquad \text{R} + \text{R} + \cdots \to \cdots$$

where R represents a chain carrier, typically an atom or radical. The other molecules involved are not shown explicitly. Let the rates of the various simple steps be designated respectively by mr_i, $r_p\text{R}$, and $r_b\text{R}^2$; r_i, r_p, and r_b would be zero, first- or second-order in the various concentrations of molecules other than R. For example, r_b is the rate of chain breaking due to a recombination of radicals if the radical concentration is unity and might be $k_b\text{M}$ where k_b is a termolecular rate constant and M is the third body. In other words, the r_i, r_p, and r_b include rate constants and all concentration factors other than the radical concentrations. This mechanism would be that for a *single chain carrier with second-order chain breaking*. The steady state requires

$$mr_i + r_p\text{R} - r_p\text{R} - 2r_b\text{R}^2 = 0$$

Solving for the concentration R

$$R = (mr_i/2r_b)^{1/2}$$

This result may then be used to eliminate the unknown R in the expressions for the rates of the various steps giving for

initiation r_i

propagation $r_p(mr_i/2r_b)^{1/2}$ (15)

breaking $r_b(mr_i/2r_b) = mr_i/2$

If the product is formed only in the propagation step, the overall rate is

$$dx/dt = r_p(mr_i/2r_b)^{1/2}$$ (16)

A specific example of this kind of chain is the homogeneous thermal *ortho-para* hydrogen conversion which goes by the mechanism

$$M + H_2(o \text{ or } p) \rightarrow 2H + M$$

$$H + H_2(p) \rightarrow H_2(o) + H$$

$$M + 2H \rightarrow H_2(o \text{ or } p) + M$$

In this case

$$r_i = k_i[H_2][M]$$

$$r_p = k_p[H_2(p)]$$

$$r_b = k_b[M]$$

Substitution in (16) yields

$$dx/dt = k_p(k_i/k_b)^{1/2}[H_2]^{1/2}[H_2(p)]$$

and, therefore, a three-halves-order reaction, first-order with respect to the variable *para*-hydrogen concentration. This is a somewhat trivial case of a chain reaction, since the steady state here is equivalent to an equilibrium $H_2 \rightleftharpoons 2H$ followed by the simple reaction $H + H_2(p) \rightarrow H_2(o) + H$, this being the customary way of viewing this reaction.[12]

For *single carrier with first-order breaking* the chain-breaking step is $R + M \rightarrow \cdots$, and the steady state is

$$mr_i - r_bR = 0$$

and

$$R = mr_i/r_b$$

and the chain propagation rate becomes

$$r_pR = r_p(mr_i/r_b)$$ (17)

First-order chain breaking might occur by adsorption of chain carriers on the reaction vessel wall.

Of more importance is the case of *two chain carriers with second-order breaking*. Suppose the mechanism, involving carriers R and S, is of the form

$$\text{initiation} \qquad \cdots \to m\text{R} + n\text{S} \qquad \text{rate: } r_i$$

$$\text{propagation} \quad \begin{cases} \text{R} + \cdots \to \text{S} + \cdots & r_{p1}\text{R} \\ \text{S} + \cdots \to \text{R} + \cdots & r_{p2}\text{S} \end{cases}$$

$$\text{breaking} \qquad 2\text{R} + \cdots \to \cdots \qquad r_b\text{R}^2$$

Steady state for R and S, respectively, gives

$$mr_i - r_{p1}\text{R} + r_{p2}\text{S} - 2r_b\text{R}^2 = 0$$
$$nr_i + r_{p1}\text{R} - r_{p2}\text{S} = 0$$

Solving for R and S yields

$$\text{R} = [(m + n)r_i/2r_b]^{\frac{1}{2}} \qquad \text{S} = \frac{nr_i + r_{p1}[(m + n)r_i/2r_b]^{\frac{1}{2}}}{r_{p2}}$$

The rates of the two chain propagation steps are then

$$r_{p1}\text{R} = r_{p1}[(m + n)r_i/2r_b]^{\frac{1}{2}} \tag{18}$$
$$r_{p2}\text{S} = nr_i + r_{p1}[(m + n)r_i/2r_b]^{\frac{1}{2}} \tag{19}$$

(18) and (19) become identical if either the second radical is not produced in the initiation step ($n = 0$) or if the chains are long.

An application of these general formulas to the hydrogen-bromine reaction leads to the same result as that obtained earlier in this chapter. Here R represents the Br atom and S the H atom. $m = 2$, $n = 0$, $r_i = k_1'[\text{Br}_2][\text{M}]$, $r_{p1} = k_2[\text{H}_2]$, $r_{p2} = k_3[\text{Br}_2] + k_4[\text{HBr}]$, and $r_b = k_5'[\text{M}]$. The rate of formation of HBr is a linear combination of the two chain propagation rates (18) and (19). In particular, all of rate (18) is formation of HBr but for rate (19) only the fraction

$$\frac{k_3[\text{Br}_2] - k_4[\text{HBr}]}{k_3[\text{Br}_2] + k_4[\text{HBr}]}$$

leads to a net production of HBr. Therefore,

$$\frac{d[\text{HBr}]}{dt} = r_{p1}\text{R} + \left(\frac{k_3[\text{Br}_2] - k_4[\text{HBr}]}{k_3[\text{Br}_2] + k_4[\text{HBr}]}\right) r_{p2}\text{S}$$

$$= k_2[\text{H}_2]\left(\frac{k_1'[\text{Br}_2]}{k_5'}\right)^{\frac{1}{2}} + \left(\frac{k_3[\text{Br}_2] - k_4[\text{HBr}]}{k_3[\text{Br}_2] + k_4[\text{HBr}]}\right) k_2[\text{H}_2]\left(\frac{k_1'[\text{Br}_2]}{k_5'}\right)^{\frac{1}{2}}$$

$$= \frac{2k_2(k_1'/k_5')^{\frac{1}{2}}[\text{H}_2][\text{Br}_2]^{\frac{1}{2}}}{1 + (k_4[\text{HBr}]/k_3[\text{Br}_2])}$$

which agrees with equation 8.

The general formulas which have been obtained so far for the rate of propagation, equations 16, 17, and 18, are special cases of the proportionality

$$\text{rate of propagation} \propto r_p (r_i/r_b)^{1/w} \tag{20}$$

where w is the order of the chain-breaking process with respect to the chain carrier. Qualitatively, the rate of propagation is proportional to the rate of the propagation step (at unit chain-carrier concentration) involving the chain carrier which participates in the chain-breaking process. And it is proportional to the $1/w$-th power of the ratio of the rate of initiation to the rate of breaking, each at unit carrier concentration. The chain-breaking process is of more importance in determining the form of the expression than is the initiation process. In particular, it makes no difference whether the carrier first formed is or is not the one involved in the chain-breaking process.

Apparent Activation Energy of Chain Reactions

Suppose that each of the rate processes has a simple Arrhenius temperature dependence:

$$r_i \propto e^{-E_i/RT}$$

$$r_p \propto e^{-E_p/RT}$$

$$r_b \propto e^{-E_b/RT}$$

Substitution in (20) yields

$$\text{rate of propagation} \propto e^{-[E_p + (1/w)(E_i - E_b)]/RT}$$

Therefore, the apparent activation energy of the chain propagation is

$$E_a = E_p + (1/w)(E_i - E_b) \tag{21}$$

It might have been expected that the apparent activation energy would be at least as great as that of the initiation process E_i. However, (21) shows that E may be smaller either because of a possible non-zero value for E_b or, more likely, because of a second-order chain-breaking process where $w = 2$. This fact has already been illustrated in the hydrogen-bromine reaction discussion. In the *ortho-para* hydrogen conversion mentioned above, $E = 58.7$ kcal, although E_i is presumably about 103 kcal. With $E_b = 0$ and $w = 2$, E_p would then be by (21) about 7 kcal, which is in agreement with the directly measured value of Geib and Harteck.[13]

It is sometimes thought that an unusually high frequency factor for a reaction is evidence that the reaction involves a chain. Conversely, the efficiency of a chain reaction is sometimes ascribed to a high frequency factor for the overall rate constant. This is definitely not the case for steady-state conditions.[14] An examination of overall reaction rates, which would be equal to the rate of propagation for long chains, shows that the frequency factors would be normal. The efficiency of chain mechanisms is due to the lowering of the activation energy, as in equation 21 when w is greater than unity.

Chain Length

The concept of chain length is of great importance but somewhat indefinite. It may be defined theoretically as the number of successful chain-propagation steps resulting from a single original chain carrier. Or it may be considered the number of cycles before the chain is interrupted. This number of cycles would be one half of that calculated on the basis of the number of steps if two chain carriers were involved. The chain length might also be taken as the number of molecules of the desired product produced from each initial chain carrier. This definition would yield a lower value for the chain length of the hydrogen-bromine reaction than would the first definition because one of the propagation steps involves a reverse reaction.

Experimental definitions of chain length are often used. Perhaps the best example is that, for a photochemical reaction the chain length is the quantum yield. For a thermal reaction there may be used the ratio of the overall rate of reaction to the rate of the first step, the first step being set equal to the rate of the completely inhibited reaction, that is, with chains cut short by the rapid removal of chain carriers by nitric oxide or other inhibitors. Or, if the reaction is initiated by the decomposition of some other substance, the independently determined rate of decomposition of the initiator may be used. This would include substances such as the aliphatic azo compounds, the peroxides, and hydroperoxides.

For the general chain mechanisms presented above let the chain length v be defined as the rate of a propagation reaction divided by the rate of formation of chain carriers. There results for *single carrier with second-order breaking*

$$v = r_p/(2nr_i r_b)^{1/2} \tag{22}$$

for *first-order breaking*

$$v = r_p/r_b \tag{23}$$

for *two carriers with second-order breaking*

$$v = r_{p1}/[2(m + n)r_i r_b]^{1/2} \tag{24}$$

or more generally

$$v \propto (r_p/r_i)(r_i/r_b)^{1/w} \tag{25}$$

where w is the order of the chain-breaking process. To have a long chain, r_p should be large and r_b small. In addition, if the chain breaking is second-order, a small rate of initiation is favorable for long chains inasmuch as the carriers in different chains are not so likely to interact and recombine, owing to their low concentration.

Chain Transfer Reactions

A process which can be of great importance in determining the nature of a chain reaction, particularly the chain length, is the chain-transfer reaction. This is the name applied to the process whereby a chain carrier reacts with a molecule to give a new species which is not the normal chain carrier.[15] An example would be the reaction of a free radical with the solvent for a chain reaction in solution.

$$R \cdot + SH \rightarrow RH + S \cdot \tag{26}$$

It can be seen that a new free radical is formed so that the chain is not necessarily broken. However, the original chain is stopped and a new one started. The reactivity, or lack of reactivity, of $S \cdot$ may play a part in the observed kinetics.

The most important case of chain transfer occurs in polymerization reactions.[16] The formation of an addition polymer is a chain process carried by free radicals, anions, or cations. There is an initiation act which produces an active radical or ion. This species adds to the double bond of an olefinic compound to produce a new active radical or ion which can add again. Hence a chain reaction occurs in which the chain is the actually growing polymer molecule. The chain is broken by some termination reaction such as the recombination or disproportionation of two free radicals, or neutralization of the ionic charge.

Take as a specific example the radical polymerization of a styrene initiated by the thermal decomposition of ABN (azo-isobutyronitrile):

$$\underset{\overset{|}{CN}}{(CH_3)_2C}-N{=}N-\underset{\overset{|}{CN}}{C(CH_3)_2} \xrightarrow{k_i} 2\underset{\overset{|}{CN}}{(CH_3)_2C} \cdot + N_2 \tag{27}$$

Calling this radical $R \cdot$

$$R \cdot + CH_2{=}CHAr \xrightarrow{k_p} R'{-}CH_2{-}CHAr \cdot \tag{28}$$

$$R'{-}CH_2{-}CHAr \cdot + CH_2{=}CHAr \xrightarrow{k_p} R'{-}CH_2{-}CHAr$$
$$-CH_2{-}CHAr \cdot \tag{29}$$

$$2R'{-}(CH_2CHAr)_nCH_2CHAr \cdot \xrightarrow{k_t} \text{recombination or}$$
$$\text{disproportionation} \tag{30}$$

It is found experimentally that the overall rate of polymerization, or disappearance of the monomer, is first-order in the monomer concentration and half-order in the initiator. This result can be deduced theoretically from the above mechanism if it is assumed that the rate constant for propagation, k_p, is independent of the size of the growing polymer chain. The usual steady-state theory for the concentration of all free radicals then leads to

$$\text{rate} = (k_i/k_t)^{1/2} k_p [\text{ABN}]^{1/2} [\text{monomer}] \tag{31}$$

which is the same as equation 16.

In the absence of chain transfer, the molecular weight, or degree of polymerization of the polymer, is directly proportional to the chain length. If disproportionation of the two polymer radicals is the means of termination, that is,

$$2\text{R---CH}_2\text{---CHA}r\cdot \rightarrow \text{R---CH}{=}\text{CHA}r + \text{R---CH}_2\text{---CH}_2\text{A}r \tag{32}$$

then the chain length is the degree of polymerization. From (22) this becomes

$$\nu = \frac{k_p[\text{monomer}]}{(k_i k_t[\text{ABN}])^{1/2}} \tag{33}$$

This indicates that chains started late in the reaction have a different length from those initiated earlier when the monomer and catalyst concentrations are high. Actually there will be a statistical distribution of chain lengths since (33) only gives the average for any instantaneous set of concentrations.[17]

If a transfer reaction occurs with the solvent, for example,

$$\text{R---CH}_2\text{---CHA}r\cdot + \text{CCl}_4 \xrightarrow{k_s} \text{R---CH}_2\text{---CHClA}r + \text{Cl}_3\text{C}\cdot \tag{34}$$

the average chain length becomes

$$\nu = \frac{k_p[\text{monomer}]}{(k_i k_t[\text{ABN}])^{1/2} + k_s[\text{S}]} \tag{35}$$

Equation 35 follows from the definition of chain length as rate of propagation divided by rate of formation of chain carriers, since a new chain is started by the trichloromethyl radical.

Chain transfer in which a hydrogen atom is abstracted from a polymer chain is also of great importance in producing cross linking and branches in polymers.

Typical values of k_p are of the order 10^3 M^{-1} sec^{-1} at 80° C[18]. For k_t a typical figure is 10^7 M^{-1} sec^{-1}. Since k_i for ABN is 1.5×10^{-4} sec^{-1} at this temperature, it can be calculated that the degree of polymerization is about 1000 for 10^{-3} M initiator and one molar concentration of monomer. Activation energies for propagation are of the order of 5 to 10 kcal and about zero for termination.[18] The reaction of two large radicals is considerably slower than for two methyl or ethyl radicals in agreement with the transition-state theory.

Inhibition

The slowing down of a reaction upon addition of a constituent to the reaction mixture is called inhibition. The constituent causing the effect is an inhibitor. Occasionally such an effect is referred to as negative catalysis, but, since the inhibitor usually undergoes chemical change, inhibition is the preferred term. Inhibition always occurs in reversible reactions where, upon approaching an equilibrium, the addition of a product of the reaction decreases the net rate of reaction. But the more spectacular cases are those in which a mere trace of inhibitor can cause a marked decrease in rate. Such an effect can be explained in at least two ways: the inhibitor may combine with a catalyst and prevent it from operating. This is probably true in many enzyme reactions. The second method of inhibition is that of chain breaking by the inhibitor. This process would be expected to be first-order in the chain carrier and first-order in the inhibitor corresponding to a bimolecular process.

$$R + \text{inh} \rightarrow \cdots$$
$$\text{with the total rate of breaking} = r_b R$$
$$= k_b R[\text{inh}] + k_b' R$$

where the second term must be included because, when the inhibitor is absent, there is certainly some other chain-breaking process. Then

$$r_b = k_b[\text{inh}] + k_b'$$

and substitution into (20) with $w = 1$ results in

$$\text{rate of propagation} \propto \frac{r_p r_i}{k_b[\text{inh}] + k_b'}$$

$$= \frac{k}{k'[\text{inh}] + 1} \qquad (36)$$

or in terms of chain length

$$\nu = \frac{k''}{k'[\text{inh}] + 1} \qquad (37)$$

where the k's may be functions of various reactant concentrations but independent of the inhibitor concentration. Relations 26 and 27 have been verified for the inhibitory action of various compounds on several chain reactions such as the autoxidation of sodium sulfite in aqueous solution, the autoxidation of benzaldehyde, and the decomposition of hydrogen peroxide.[19] The theory is further verified by the experimental detection of oxidation products of the inhibitor presumably produced in the chain-breaking process.

The exact mode of action of inhibitors is still uncertain in most cases.[20] Probably several mechanisms operate, such as combination of the chain carrier with the inhibitor to form an unreactive free radical, or hydrogen abstraction from the inhibitor to produce the same effect. The latter process seems plausible for the cases in which easily oxidized substances such as phenols and amines are used to inhibit autoxidation.

This reaction, autoxidation, is one of the most important known to chemistry and has been the subject of a very large number of studies.[21] It refers to the slow reaction with molecular oxygen of most organic and some inorganic substances at moderate temperature conditions (under 150° C) and usually in the liquid state. It is sometimes an undesirable reaction to be inhibited and sometimes the desired reaction in large-scale commercial processes.

Usually the process can be represented by the following reaction mechanism.

$$\text{initiator} \xrightarrow{k_i} 2R' \tag{38}$$

$$R' + O_2 \longrightarrow R'O_2 \cdot \tag{39}$$

$$R'O_2 \cdot + RH \longrightarrow R'O_2H + R \cdot \tag{40}$$

$$R \cdot + O_2 \xrightarrow{\text{fast}} RO_2 \cdot \tag{41}$$

$$RO_2 \cdot + RH \xrightarrow{k_p} RO_2H + R \cdot \tag{42}$$

$$2RO_2 \cdot \xrightarrow{k_t} \text{inactive products} \tag{43}$$

Thus a chain reaction occurs with a hydroperoxide as the first product. This may be oxidized further, or may decompose thermally to start new chains. If the assumption is made that reaction 41 is fast but (42) is slow, because of the lesser reactivity of $RO_2 \cdot$ compared to $R \cdot$, then the above mechanism leads to the rate equation (for long chains)

$$\text{rate} = (k_i/k_t)^{\frac{1}{2}}k_p[\text{initiator}]^{\frac{1}{2}}[RH] \tag{44}$$

The termination step involves two $RO_2 \cdot$ rather than $R \cdot$ because of the higher steady-state concentration of the former.

Experimentally the rate of oxygen consumption is found to obey equation 44 at oxygen pressures above 50 mm or so. At lower oxygen pressures, a dependence on oxygen is found. This is expected since, if less oxygen is dissolved in the liquid phase, it will no longer be true that (41) is fast compared with (42). The initiation can be produced by the thermal decomposition of substances naturally present or added to the reaction mixture. Photoinitiation is also possible. The square root dependence on the initiator concentration, or on the light intensity, is generally found, at least under well-defined experimental conditions.

Metal ion catalysis (see Chapter 9) is often used to increase the rate of initiation if due to the decomposition of peroxides.[22] These will usually

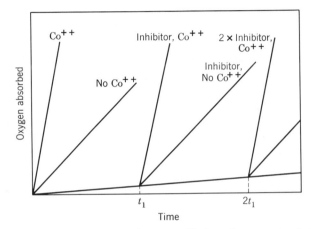

Fig. I.   Effect of catalyst and inhibitor on oxidation of cumene in glacial acetic acid (H. S. Blanchard).

be formed as a result of reaction in any case. The mechanism is illustrated by a study of the autoxidation of cumene initiated by ABN.[23] The effect of cobaltous acetate catalysis and of inhibition by 2,6-di-t-butyl-p-cresol is shown in Fig. 1. The inhibitor produces an induction period of very low rate, with the length of the induction period proportional to the amount of inhibitor. Inhibition ceases when the inhibitor is all consumed at the end of the induction period. Cobaltous ion increases the rate greatly but, at low concentrations, has no effect on the length of the induction period. This must mean that the inhibitor is used up only by the radicals produced from ABN and that the catalyst metal ion does not produce acceleration by increasing the rate of decomposition of ABN.

The reactivity of a carbon-hydrogen bond to the atom transfer reaction 42 is greatly affected by the presence of a double bond or aromatic ring, or by allyl substituents on the carbon atom. This is because the resulting free radical R· is greatly stabilized. Addition of RO_2· to the double bond can also occur.

The slowing down of some decomposition reactions by nitric oxide is another example of inhibition. Szwarc and his co-workers[24] have completely inhibited the chains in certain organic vapor decompositions by the use of toluene as a carrier and inhibitor. This is important as a means of getting precise information about the chain-initiation reactions and hence of bond dissociation energies, a subject which will be briefly discussed next.

Bond Dissociation Energies

The above term refers to the energy required to rupture one bond in a molecule, usually in the gaseous state. Atoms or radicals are formed. An example would be

$$H_2O \rightarrow H\cdot + HO\cdot \qquad (45)$$

This process requires 118 kcal per mole and represents a fairly large value indicating a stable bond. It is clear that chemical kinetics greatly depends upon the strength of bonds being broken or made in the course of the reaction. A knowledge of bond energies is of considerable use in estimating the probability of certain reactions occurring and even of estimating their rate.

It was shown in Chapter 6 that there is no rigorous a priori method of calculating activation energies, except that they must always be as great as the endothermicity of the reaction. There is one case, however, in which the bond dissociation energy is probably equal to the activation energy. This is for the case of the elementary act of decomposition into unstable atoms or radicals. Reaction 45 mechanistically would not be unimolecular but rather bimolecular, because of the small number of degrees of freedom in the water molecule. However, for a more complicated molecule the corresponding reaction would more probably be unimolecular.

$$C_6H_5CH_2Br \rightarrow C_6H_5CH_2\cdot + Br\cdot \qquad (46)$$

In either case the experimental evidence suggests that the reverse reaction has zero activation energy, though usually not a zero temperature dependence. This means that the activation energy (minimum energy required for decomposition, according to theory) is the bond energy itself.

A large amount of data on bond energies for dissociation into various atoms and radicals is available.[25] From thermodynamic arguments, the bond energies can be calculated from heats of formation of the original compound and the various atoms and radicals, and it is often these values that are tabulated. The experimental results come from several main sources, of which one is kinetic in that the activation energy for reaction

46 or 45 is taken as the bond energy. To obtain this activation energy it is necessary to inhibit any chains produced by the products of the desired reaction since chain reactions will have a complex activation energy.

Another source of bond energy data is spectroscopic, in which extrapolation of the vibrational spectrum to a continuum gives a value for the heat of dissociation. This method only works well for diatomic molecules, but is valuable for giving heats of formation of atoms.

Electron impact methods have recently been used to get heats of formation of many radicals.[26] The mass spectrograph provides information about the energetics of reactions such as

$$R—H \rightarrow R^+ + H + e \qquad (47)$$

$$R\cdot \rightarrow R^+ + e \qquad (48)$$

The latter result is obtained under conditions where the radical $R\cdot$ is known to be present in the ionization chamber of the mass spectrograph. From (47) and (48) it is possible to get the heat of formation of the radical if the heat of formation of R—H is known.

Another kind of bond energy which is sometimes useful is the coordinate bond energy.[27] This refers to the so-called "heterolytic" split of a molecule into groups as contrasted to the "homolytic" split of equations 45 and 46.

$$Ni(CO)_4 \rightarrow Ni(CO)_3 + CO \qquad (49)$$

$$CuCl_2 \rightarrow CuCl^+ + Cl^- \qquad (50)$$

Data for such reactions in the gas phase are often indirectly available. For ionic reactions the coordinate bond energies are more pertinent than the usual bond energies. However, ionic reactions are rare in the gas phase, and in solution a reaction such as (50) is greatly complicated by the intervention of the solvent.

In conclusion it must be pointed out that often tables of bond energies are used in which average values for a C—H or a C—C bond are listed. Such average values can be used as a rough guide, but it must be borne in mind that the dissociation energy of a bond depends not only on the atoms forming the bond but also on the groups attached to these atoms. Thus the carbon-carbon single bond energy varies from 85 kcal for ethane or diamond to 10 kcal for hexaphenylethane.

Rice-Herzfeld Mechanisms of Organic Molecule Decomposition

The thermal decomposition of gaseous hydrocarbons, ketones, aldehydes, alcohols, ethers, etc., were formerly thought to be unimolecular reactions inasmuch as the kinetics were generally first-order. However, Rice and co-workers,[28] using the Paneth mirror technique, detected free radicals in these decomposition reactions. Rice then proposed that these were chain reactions, Rice and Herzfeld[29] discussing the possible kinetic mechanisms. In particular it was shown that, despite the complexity of a chain mechanism, it was still possible to have simple first-order kinetics.

They presented general mechanisms as follows:

(1) $$M_1 \rightarrow R_1 + M_2$$

(2) $$R_1 + M_1 \rightarrow R_1H + R_2$$

(3) $$R_2 \rightarrow R_1 + M_3$$

(4) $$R_1 + R_2 \rightarrow M_4$$

(5) $$2R_1 \rightarrow M_5$$

(6) $$2R_2 \rightarrow M_6$$

where only one of the three chain-breaking steps (4), (5), and (6) would be expected to be important in a given reaction. If (5) or (6) is the chain-breaking step, the mechanism is a particular case of the "two-carrier with second-order breaking" mechanism discussed above. The rate can be expressed by equation 18. Substitution for r_{p1} of $k_2[M_1]$ if (5) is the chain-breaking step, or of k_3 if (6) is the chain-breaking step (R and S referring to R_1 and R_2 in the first case and R_2 and R_1 in the second case) leads to the rate of reaction. Using step (5)

$$-d[M_1]/dt = k_1[M_1] + k_2(k_1/2k_5)^{1/2}[M_1]^{3/2} \qquad (51)$$

Using step (6)

$$-d[M_1]/dt = 2k_1[M_1] + k_3(k_1/2k_6)^{1/2}[M_1]^{1/2} \qquad (52)$$

In long chains the first terms on the right become negligible and the rates are three-halves- and one-half-order, respectively. In order to explain first-order kinetics, Rice and Herzfeld assume that two unlike radicals react to break the chain, that is, process 4. This mechanism can be put in more general terms, similar to our other general formulas, as follows, for *two carriers with second-order breaking involving different carriers*:

initiation	$\cdots \rightarrow mR + nS$	rate: r_i
propagation	$R + \cdots \rightarrow S + \cdots$	$r_{p1}R$
	$S + \cdots \rightarrow R + \cdots$	$r_{p2}S$
breaking	$R + S + \cdots \rightarrow \cdots$	r_bRS

Application of the steady-state approximation gives for the first propagation step

$$\text{rate} = r_{p1}R = \left(\frac{m-n}{4}\right)r_i + \sqrt{\left(\frac{(m-n)}{4}r_i\right)^2 + \frac{(m+n)r_i r_{p1} r_{p2}}{2r_b}} \tag{53}$$

and a similar expression for the second propagation step. This simplifies if the chains are long, in which case the rate of propagation, and the rate of reaction, becomes

$$r_{p1}R \simeq \sqrt{(m+n)r_i r_{p1} r_{p2}/2r_b} \tag{54}$$

It is interesting to note that (54) is the same as (18) if r_{p1} is replaced by $(r_{p1}r_{p2})^{1/2}$. Also relation 20 still applies if r_p is understood to be the geometric mean of r_{p1} and r_{p2}.

In the Rice-Herzfeld mechanism, using their breaking step $R_1 + R_2 \rightarrow M_4$, substitution in (54) gives for the reaction rate in the long-chain approximation

$$-d[M_1]/dt \simeq (k_1 k_2 k_3/2k_4)^{1/2}[M_1] \tag{55}$$

which is the desired first-order result. However, this is not the only way to get a first-order rate. If Rice and Herzfeld's step (1) were bimolecular and the chain breaking were (6), first order would also result. Various other possibilities have been discussed by Goldfinger, Letort, and Niclause.[30]

As an example of the Rice-Herzfeld mechanism the classical example of the pyrolysis of acetaldehyde may be discussed.[31] The stoichiometry of this reaction is closely given by the simple equation

$$CH_3CHO \rightarrow CH_4 + CO \tag{56}$$

Traces of ethane and hydrogen are also formed. The reaction shows the characteristics of a chain reaction in that propylene inhibits the rate. However, nitric oxide, oxygen, and iodine accelerate the reaction as do many sensitizers such as azomethane and di-*t*-butyl peroxide. Metal mirror tests show the presence of free radicals as does *ortho-para* hydrogen conversion.

By using isotope tracers, such as a mixture of CH_3CHO and CD_3CDO, it is possible to show that the reaction is in part intramolecular, because of an excess of pure CH_4 and CD_4 formed, and in part a free-radical process.[32] The latter appears to be 75 to 85 per cent of the total reaction in the absence of accelerators.

Photolysis of acetaldehyde at temperatures above $300° C$ gives high quantum yields of carbon monoxide, up to 300.[33] The rate law for the formation of CO is given by

$$\text{rate} = k_1 I + k_2 I^{1/2}[CH_3CHO] \tag{57}$$

where I is the intensity of absorbed light. The first term corresponds to the direct, intramolecular process and the second to a chain reaction. The constant k_2 (not a rate constant only) has a temperature coefficient corresponding to an apparent activation energy of 10 kcal.

If the changes in pressure are plotted as a function of time in the usual way it is found that the order of the reaction is close to 2 (about 1.90). However, Letort[34] showed that the initial rate of the reaction varied with the initial pressure of aldehyde according to the equation

$$\log R_0 = 1.50 \log P_0 + \text{constant} \tag{58}$$

The reaction is initially then of $\frac{3}{2}$ order. The initial rate constant is given by $k = 2.3 \times 10^{12} e^{-46,000/RT}$ cm$^{3/2}$ mole$^{-1/2}$ sec^{-1}.

The Rice-Herzfeld mechanism to fit the data would involve the following steps.

$$CH_3CHO \xrightarrow{k_1} CH_3 \cdot + CHO \cdot \tag{59}$$

$$CHO \cdot \xrightarrow{k_2} H \cdot + CO \tag{60}$$

$$H \cdot + CH_3CHO \xrightarrow{k_3} H_2 + CH_3CO \cdot \tag{61}$$

$$CH_3 \cdot + CH_3CHO \xrightarrow{k_4} CH_4 + CH_3CO \cdot \tag{62}$$

$$CH_3CO \cdot \xrightarrow{k_5} CH_3 \cdot + CO \tag{63}$$

$$2CH_3 \cdot \xrightarrow{k_6} C_2H_6 \tag{64}$$

Reactions 62 and 63 are then the chain-carrying steps, with (59) being the initiating reaction and (64) the terminating. Applying the steady-state treatment to the radicals $CH_3 \cdot$ and $CH_3CO \cdot$ leads to the result

$$[CH_3 \cdot] = (k_1/k_6)^{1/2}[CH_3CHO]^{1/2} \tag{65}$$

and $\qquad [CH_3CO \cdot]/[CH_3 \cdot] = k_4[CH_3CHO]/k_5 \tag{66}$

For long chains the overall rate must be equal to the rate of reaction 62, or 63.

$$\text{rate} = k_4(k_1/k_6)^{1/2}[CH_3CHO]^{3/2} \tag{67}$$

The result is in agreement with the three-halves-order found experimentally.

It is now necessary to see if the energetics of the individual steps are in agreement with the overall activation energy and if the rates are of the proper magnitude to be reasonable. The activation energy is

$$E_a = E_4 + \tfrac{1}{2}(E_1 + E_6) \tag{68}$$

The energy E_1 can be calculated if heats of formation of the methyl and formyl radicals are known, as well as that of acetaldehyde. The heats for the radicals are known approximately for formyl and quite well for methyl.[35] The difference in heats of formation gives ΔH for reaction 59 as about 74 kcal, which may then be taken as E_1. This is a reasonable value when other carbon-carbon bond energies are considered. The unsaturated carbonyl group will lower the energy of dissociation compared to ethane. The carbon-hydrogen bond energy will be stronger in CH_3CHO than the carbon-carbon bond energy.

The energy E_6 is experimentally equal to zero for methyl radicals (Chapter 6). The value of E_4 cannot be estimated very accurately. However, reaction 62 is exothermic by about 15 kcal from heats of formation. Hence the activation energy may be guessed as $11.5 - 0.25$ (15) according to Semenov's rule. This gives 8 kcal for E_4. A better estimate comes from considering the photochemical results. By the Rice-Herzfeld mechanism the initiation reaction is now replaced by

$$CH_3CHO \xrightarrow{h\nu} CH_3\cdot + HCO\cdot \tag{69}$$

which is essentially temperature-independent. Hence the activation energy for the photolysis of acetaldehyde is simply $E_4 + \tfrac{1}{2}E_6$. This gives E_4 a value of 10 kcal.

Combining the information

$$E_a = 10 + \tfrac{1}{2}(74 + 0) = 47 \text{ kcal} \tag{70}$$

which is in excellent agreement with the experimental value of 46 kcal. The expected frequency factor will be

$$A_4(A_1/A_6)^{\frac{1}{2}} = 10^{12}(5 \times 10^{13}/5 \times 10^{13})^{\frac{1}{2}} = 10^{12} \tag{71}$$

where the units are moles/cm^3 and seconds. The experimental value of A_6 is used, and rough estimates for the unimolecular step A_1 and the bimolecular step A_4 using transition state theory (Chapter 5).

It remains to show that (64) was a good choice for the termination reaction rather than a process involving the acetyl radical. The rate and activation energy of reaction 63 have been estimated by several workers from competition experiments.[36] The results agree in giving an activation energy of 13.5 kcal for the decomposition of $CH_3CO\cdot$ into $CH_3\cdot$ and CO. The frequency factor is about 10^{10} sec^{-1}. Putting these figures back into equation 66 together with $k_4 = 10^{12} e^{-10,000/RT}$ cm^3 mole^{-1} sec^{-1}, enables us to calculate the ratio of acetyl radical to methyl radical for a given temperature and concentration of acetaldehyde. A typical experimental temperature is in the range of 800° A and at a pressure of one atmosphere, $[CH_3CHO] \simeq 10^{-5}$ moles per cm^3. This gives a figure of 10^{-4} for $[CH_3CO\cdot]/[CH_3\cdot]$ and justifies the selection of reaction 63 as the termination step.

In spite of the good agreement shown between theory and experiment, it must be admitted that it is only the initial rate which agrees with the above predictions. Actually, the greater part of the reaction is more nearly second-order, as was pointed out. Also the isotope labeling experiments, using substantial amounts of CH_3CHO and CD_3CDO mixtures, produce substantial amounts of CH_2D_2 which cannot be explained by the Rice-Herzfeld mechanism.[32]

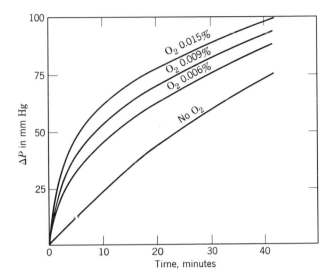

Fig. 2.   Sensitization by oxygen of acetaldehyde decomposition at 477° C and 226 mm initial pressure (M. Letort).

It was mentioned that oxygen has an accelerating effect on the rate of decomposition of acetaldehyde, as does nitric oxide. Figure 2 shows the considerable initial increase in rate due to small amounts of oxygen.[37] Ten parts per million by volume of O_2 will double the rate of decomposition, or up to 4000 molecules of aldehyde can be decomposed by 1 molecule of oxygen. This is the length of a chain induced by the oxygen, probably by a reaction of hydrogen abstraction.

$$CH_3CHO + O_2 \rightarrow CH_3CO\cdot + HO_2\cdot \qquad (72)$$

Under other circumstances oxygen can be an inhibitor of chain reactions. For example, in the reaction between H_2 and Cl_2, oxygen is a strong inhibitor as it is in many polymerization reactions. In these cases it is the formation of radicals such as $HO_2\cdot$ and $RO_2\cdot$, which are relatively unreactive compared to the radicals $H\cdot$ and $R\cdot$, which accounts for the inhibition. At the high temperature of aldehyde pyrolysis and in other high temperature reactions oxygen can start chains much faster than it stops them. This is also true in combustion, flame, and explosion phenomena. Some of the special kinetic characteristics of the latter kinds of reaction will now be considered.

Branching Chain Reactions

Suppose that a chain carrier R undergoes a reaction which yields two chain carriers instead of the customary one. Considering only the simplest mechanism with a single kind of carrier and first-order breaking:

initiation	$\cdots \rightarrow m\text{R} + \cdots$	rate: r_i
propagation	$\text{R} + \cdots \rightarrow \text{R} + \cdots$	$r_p\text{R}$
branching	$\text{R} + \cdots \rightarrow 2\text{R} + \cdots$	$r_s\text{R}$
breaking	$\text{R} + \cdots \rightarrow \cdots$	$r_b\text{R}$

For the steady state

$$mr_i + r_s\text{R} - r_b\text{R} = 0$$

Then

$$\text{R} = mr_i/(r_b - r_s)$$

and the rate of propagation is

$$r_p\text{R} = mr_i r_p/(r_b - r_s) \tag{73}$$

Comparing this formula with (17), the only difference is that the rate of branching per unit concentration of R is subtracted from the rate of breaking. For higher-order chain breaking or chains involving more than one carrier the corresponding expressions are somewhat more complicated. Nevertheless, it is a useful qualitative idea that branching amounts to a negative breaking.

If the functional dependence of r_b and r_s on temperature, pressure, and composition is such that $r_b - r_s$ approaches zero, the rate should become indefinitely large. This is the characteristic of an *explosion limit*. The explanation of observed explosion limits can then be attempted with the aid of the idea of branching chain reactions.[38] An older explanation of explosion limits, or ignition temperatures, considers that explosion will occur whenever the reaction rate becomes so fast that the heat evolved in the reaction cannot be dissipated sufficiently rapidly. The temperature, and therefore the rate, would then increase indefinitely. This is referred to as a *thermal* explosion.

In the low-pressure explosion region as found in the reaction of oxygen with hydrogen, carbon monoxide, phosphine, phosphorus, etc., the branching-chain theory has been successfully applied, the low-pressure limit being due to chain breaking at the wall of the container and the upper explosion limit being due to the balance between chain branching and chain breaking in the gas phase. Explosions at higher pressures may be thermal.

The nuclear-fission chain reaction,[39] in which neutrons are chain carriers, has an analogy to branching-chain combustion processes. Owing to the escape of neutrons from the reactor, analogous to loss of atom or radical chain carriers at the reaction vessel wall, there is a critical size below which no reaction takes place. Correspondingly, low-pressure explosion limits in combustion reactions are dependent on the size of the reaction vessel.

PROBLEMS

1. Consider whether a chain mechanism involving hydrogen and iodine atoms analogous to that in the hydrogen and bromine reaction could compete with the direct bimolecular reaction of hydrogen and iodine. Use the Hirschfelder rules (Chapter 6) to estimate activation energies.

2. Consider the chain mechanism for autoxidation of a hydrocarbon RH. (a) Using the steady-state approximation derive an expression for $-d[O_2]/dt$ under conditions of very low oxygen pressure. Use the initiation reaction

$$2ROOH \xrightarrow{k_i} R\cdot + RO_2\cdot + H_2O_2$$

and the termination reactions

$$2R\cdot \text{ or } R\cdot + RO_2\cdot \rightarrow \text{inactive products}$$

(b) What relationships between the various activation energies would be necessary for long chains?

REFERENCES

1. M. Bodenstein and S. C. Lind, *Z. physik. Chem.*, *57*, 168 (1907).
2. J. A. Christiansen, *Kgl. Danske Videnskab. Selskab.*, *Mat.-fys. Medd.*, *1*, 14 (1919).
3. K. F. Herzfeld, *Ann. Physik*, *59*, 635 (1919).
4. M. Polanyi, *Z. Elektrochem.*, *26*, 50 (1920).
5. F. A. Matsen and J. L. Franklin, *J. Am. Chem. Soc.*, *72*, 3337 (1950).
6. E. Rabinowitch and H. L. Lehman, *Trans. Faraday Soc.*, *31*, 689 (1937).
7. M. Bodenstein and H. Lütkemeyer, *Z. physik. Chem.*, *114*, 208 (1925).
8. W. Jost and G. Jung, *ibid.*, *B3*, 83 (1929); see discussion in G. K. Rollefson and M. Burton, *Photochemistry and the Mechanism of Chemical Reactions*, Prentice-Hall, New York (1939), Chapter 11.

9. W. G. Givens, Jr., and J. E. Willard, *J. Am. Chem. Soc.*, *81*, 4773 (1959).

10. D. F. Hornig and H. B. Palmer, *J. Chem. Phys.*, *26*, 98 (1957).

11. J. H. Sullivan, *J. Chem. Phys.*, *30*, 1292 (1959).

12. A. Farkas, *Z. physik. Chem.*, *B10*, 419 (1930).

13. K. H. Geib and P. Harteck, *Z. physik. Chem.*, Bodenstein Festband, 849 (1931).

14. C. N. Hinshelwood, *Nature*, *180*, 1233 (1957).

15. P. J. Flory, *J. Am. Chem. Soc.*, *59*, 241 (1937).

16. For general discussions see (*a*) C. Walling, *Free Radicals in Solution*, John Wiley and Sons, New York, 1957; (*b*) F. S. Dainton, *Chain Reactions*, Methuen and Co., London, 1956; (*c*) P. J. Flory, *Principles of Polymer Chemistry*, Cornell University Press, Ithaca, 1953; (*d*) A. V. Tobolsky, *J. Am. Chem. Soc.*, *80*, 5927 (1958).

17. M. Dole, *Introductory Principles of Statistical Thermodynamics*, Prentice-Hall, New York, 1954, Chapter 3.

18. G. M. Burnett and H. W. Melville, *Chem. Revs.*, *54*, 225 (1954).

19. H. L. J. Bäckstrom, *J. Am. Chem. Soc.*, *49*, 1460 (1927); H. N. Alyea and H. L. J. Bäckstrom, *ibid.*, *51*, 90 (1929); K. Jen and H. N. Alyea, *ibid.*, *55*, 575 (1933).

20. G. S. Hammond, C. E. Boozer, C. E. Hamilton, and J. N. Sen, *J. Am. Chem. Soc.*, *77*, 3238 (1955); J. L. Bolland and P. ten Have, *Trans. Faraday Soc.*, *43*, 201 (1947).

21. For recent reviews (*a*) see reference 16*a*, Chapter 9; and (*b*) G. A. Russell, *J. Chem. Ed.*, *36*, 111 (1959).

22. A. Robertson and W. A. Waters, *Trans. Faraday Soc.*, *42*, 201 (1946).

23. H. S. Blanchard, *J. Am. Chem. Soc.*, *82*, 2014 (1960). See also reference 21(*b*).

24. M. Szwarc, *Chem. Revs.*, *47*, 75 (1950).

25. (*a*) T. L. Cottrell, *Strengths of Chemical Bonds*, Butterworths Scientific Publications, London, 1954; (*b*) M. Szwarc and D. Williams, *Proc. Roy. Soc.*, *A219*, 353 (1953) and earlier papers by Szwarc; (*c*) N. Semenov, *Some Problems in Chemical Kinetics and Reactivity*, Vol. I, Pergamon Press, New York, 1958; (*d*) A. G. Gaydon, *Dissociation Energies and Spectra of Diatomic Molecules*, Chapman and Hall, London, 1952.

26. D. P. Stevenson, *Disc. Faraday Soc.*, *10*, 35 (1951); J. L. Franklin and E. H. Field, *Electron Impact Phenomena*, Academic Press, New York, 1957.

27. F. Basolo and R. G. Pearson, *Mechanisms of Inorganic Reactions*, John Wiley and Sons, New York, 1958, Chapter 2.

28. F. O. Rice and K. K. Rice, *The Aliphatic Free Radicals*, Johns Hopkins Press, Baltimore, 1935.

29. F. O. Rice and K. F. Herzfeld, *J. Am. Chem. Soc.*, *56*, 284 (1934).

30. P. Goldfinger, M. Letort and M. Niclause, in *Contrib. étude structure mol.*, Vol. commem. Victor Henri, Desoer, Liége, 1947–1948, pp. 283–296.

31. For recent reviews see M. Letort, *Chimie et Ind.*, *76*, 430 (1956); and W. D. Walters, in Friess, A. L., and Weissberger, A., *Technique of Organic Chemistry*, Vol. 8, Interscience Publishers, New York, 1953, p. 291.

32. P. D. Zemany and M. Burton, *J. Phys. and Coll. Chem.*, *55*, 949 (1951); L. A. Wall and W. J. Moore, *ibid.*, 965; *J. Am. Chem. Soc.*, 73, 2840 (1951).

33. J. A. Leermakers, *J. Am. Chem. Soc.*, *56*, 1537 (1934); R. E. Dodd, *Trans. Faraday Soc.*, *47*, 56 (1951).

34. M. Letort, *J. Chim. Phys.*, *34*, 265 (1937); *Compte Rendu*, *199*, 351 (1934).

35. References 25*b* and 25*c*.

36. D. H. Volman and W. M. Graven, *J. Am. Chem. Soc.*, *75*, 3111 (1953); J. G. Calvert and J. T. Gruver, *ibid.*, *80*, 1313 (1958).

37. M. Letort, *J. Chim. Phys.*, *34*, 428 (1937).

38. N. Semenov, *Chemical Kinetics and Chain Reactions*, Oxford University Press, London, 1935; B. Lewis and G. von Elbe, *Combustion, Flames and Explosions of Gases*, Academic Press, New York, 1951; N. Semenov, *Some Problems in Chemical Kinetics and Reactivity*, Vol 2, Princeton University Press, Princeton, 1959. Translated by M. Boudart.

39. H. Soodak and E. C. Campbell, *Elementary Pile Theory*, John Wiley and Sons, New York, 1950. [Out of print.]

11.

THE STUDY OF RAPID REACTIONS

One of the important recent developments in chemical kinetics has been a great interest in the study of very rapid reactions. Chemical events with half-lives as short as 10^{-7} seconds have been studied with reasonable accuracy. It is obvious that rather special methods must be used to investigate systems as labile as this. In fact, a large part of the interest in rapid reactions results from the availability of electronic methods of measuring times in the millisecond, microsecond, and even millimicrosecond ranges.

Even though short times can be easily measured, there is still a difficult problem of detecting a change in the system owing to reaction in this time interval. Furthermore, the problem of mixing together the reagents to initiate reaction is a severe limitation in that a finite time of about one millisecond is required for homogeneous mixing, even in the best circumstances. For shorter times it is necessary to start with the reactants already uniformly distributed and in a state of equilibrium. The equilibrium may then be disturbed and subsequent events followed, or alternatively the consequences of the dynamic nature of a chemical equilibrium may be utilized. A favorite technique is to invoke some other time-dependent phenomenon, with a characteristic time of the same order as that of the half-life of the chemical reaction. The interplay of these two times then produces observable results which depend on their relative values.

The procedure followed in this chapter will be first to discuss flow methods which represent the best way of following fast reactions in which the reagents cannot be previously mixed. Then a general discussion of the theory of rapid reactions will be given. Finally a brief summary of some of the special experimental methods used will be presented.

Flow Methods

Open systems sometimes offer special advantages in kinetic studies. Such a system is characterized by a continuous flow of reactants through a reactor space and is widely used for industrial processes because of the ease of handling large quantities of reactants in a limited reactor space and because of the possibility of continuous operation. This is particularly advantageous for reactions involving solid catalysts, where emptying and filling operations may change the activity of the catalyst.

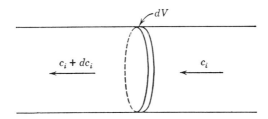

Fig. I.   Cylindrical volume element in a flow system.

For the measurement of reaction rates, mixtures of reactants of known initial composition flow through a region of reaction of known volume at a known rate. Analysis of the mixture is made after passage through the reaction space or at fixed points in the space. It is assumed that there is complete mixing of the reactants before entering the zone of reaction. Specially designed chambers just before the reaction space may be used to facilitate mixing.

If after the flow enters the reaction space, partial mixing between different parts of the flow stream occurs, such as would be caused by diffusion and convection, the results are difficult to interpret.† There are two limiting cases which can generally be approached, however, corresponding to no longitudinal mixing and to complete longitudinal mixing. These can be treated mathematically in a straightforward way.

The basic equations for flow systems with no mixing can be derived as follows: Consider a tubular reactor space of constant cross-sectional area A as shown in Fig. 1 with a steady flow u of reaction mixture expressed as volume per unit time. Select a small cylindrical volume unit dV such that the concentration of a component i entering the unit is c_i and the concentration leaving the unit is $c_i + dc_i$. Within the volume unit the component is

† Reference 1 discusses the equations involved.

changing in concentration owing to chemical reaction with a rate equal to r_i. This rate is of the form of the usual chemical rate equation and is a function of the rate constants of all reactions involving the component i and of the various concentrations in the volume unit. The change in the number of moles of the component i with time in the volume unit is given by

$$dn_i/dt = r_i\,dV - u\,dc_i \qquad (1)$$

After a time the concentration will generally become constant for each component within the element. That is, a steady state will be set up throughout the reactor volume so that the composition remains uniform with time at any point. The composition from point to point will be different, however. In this steady state the condition will be that

$$r_i\,dV = u\,dc_i \qquad (2)$$

Equation 2 may be integrated to give

$$\frac{V}{u} = \int_{c_0}^{c} \frac{dc_i}{r_i} \qquad (3)$$

where c_0 is the concentration of the component entering the reaction space and V is the total volume of the reactor up to the point where the concentration is c. For a tubular reactor this volume is Al where l is the distance from the entrance of the reactor to the point in question.

To use equation 3, it is necessary to have a definite expression for the rate r. Consider the case of a first-order reaction:

$$A \xrightarrow{k_1} B \qquad (4)$$

where the concentration of component A is equal to c.

$$r = dc/dt = -k_1 c \qquad (5)$$

Putting this value of r in (3) and integrating gives

$$k_1 = (u/V)\ln(c_0/c) \qquad (6)$$

The resemblance of this equation to that for a first-order reaction in a closed system

$$k_1 = (1/t)\ln(c_0/c) \qquad (7)$$

is obvious. The result can be generalized in that the integrated equation for a reaction of any order in a flow system with no mixing is the same as for a closed system except that the time variable is replaced by V/u. Since this can also be written as lA/u, it may be said that a distance coordinate *l* replaces the time coordinate as a variable. Special cases have been discussed by Harris.[2]

The case of complete mixing as in a stirred flow reactor has been discussed thoroughly by Denbigh.[3] Here the composition becomes uniform throughout the entire volume of the reactor as a result of efficient stirring. The volume element dV may be replaced by V, the total volume, in equation 1, and dc_i may be replaced by $(c - c_0)$. Dividing through by V, equation 1 becomes

$$dc/dt = r - (u/V)(c - c_0) \qquad (8)$$

which gives the equation for the approach to the steady state. At the steady state dc/dt is zero and the composition in the reactor becomes constant and remains so, independent of time. The basic equation then becomes

$$r = (u/V)(c - c_0) \qquad (9)$$

A similar equation can, of course, be derived for each component in the system if the proper expression for r is used. It is to be noted that an analysis for c and c_0 in a system where u and V are known gives a numerical value for r. This is the value of the rate for the particular constant conditions in the reactor. By changing the initial concentrations or the flow u, the rate can be found for other conditions. This enables an explicit determination of the form of the rate equation. No integration is necessary.

Consider again the first-order reaction of equation 4. In a stirred flow reactor if B is initially absent, equation 9 becomes

$$k_1(a - x) = (u/V)x \qquad (10)$$

so that a knowledge of a, the initial concentration, and x, the concentration of the product in the steady state, enables the rate constant k_1 to be evaluated. Changing the initial concentration and the flow would simply result in a change in x so that $ux/(a - x)$ remains constant.

The extension of (9) to more complex rate equations can readily be seen, and indeed an important advantage of the stirred flow reactor is that complicated rate expressions can be handled since integration is avoided.[4] However, the method depends upon knowing the concentrations of all the reactants, intermediates, and products for a complete solution. In a simple case analysis for one component might suffice, but as the complexity increases the number of separate analyses that must be made also increases.

Another important advantage of the stirred flow method in solutions is that the reaction can be carried out under constant conditions of solvent composition, ionic strength, catalyst concentration, etc. The method has a potential advantage in that a transient intermediate may be built up to an optimum concentration and then maintained for a long enough period to be detected and measured. The disadvantages are that the method is relatively slow. Large volumes of solution must go to waste while the steady state is being established, and each experiment gives essentially only one point in establishing the rate equation. Hammett and his co-workers[5] have investigated the method from an experimental point of view and give valuable details on apparatus and operation. The method has been applied to the simple hydrolysis of an ester with base and to the complex bromination of acetone with hypobromite with encouraging results. Johnson and Edwards[6] discuss the use of a number of stirred flow reactors arranged in series.

The more conventional flow systems without mixing also have the advantage of maintaining unstable reactants and intermediates at steady concentrations so that they may be detected and measured. However, the most important advantage of unstirred flow methods is that they may be applied to very rapid reactions. The time coordinate, it will be recalled, was replaced by the quantity V/u. If the volume of the reactor is reduced to a small value and a high rate of flow is used, the equivalent reaction time can be reduced to values as small as 0.001 second. Thus reactions with half-lives of less than 0.01 second may be measured if a suitable means of analysis can be found.

Among the methods of analysis which have been used is colorimetry,[7] in which a colored reaction product is detected or an indicator color change is observed during the course of the reaction. Conductance and electromotive force measurements can also be used.[8] The temperature rise in an exothermic reaction can be utilized,[9] or the reaction mixture issuing from the reactor space may be quenched in a suitable manner and standard methods of analysis applied.[10] Since different lengths in a tubular reactor represent different values of V, a series of measurements at different points may be made to obtain in one experiment a number of concentration-equivalent-time data.

Chance[11] has developed an ingenious variation of the unstirred flow method in that the reaction time is continuously varied by changing u and holding V constant. In this accelerated flow method the output from a photoelectric colorimeter is fed to a cathode-ray oscilloscope, which then sweeps out a complete time-concentration record which may be photographed. The method was developed for very rapid enzyme reactions and requires only small quantities of reactants.

An even simpler procedure is to use the stopped flow method.[12] Here a sample is injected into the flow system in the usual way, but flow is suddenly stopped. By oscilloscopic methods the concentration can again be studied as a function of time after the flow ceases. The method is very economical of reagents and takes advantage of the rapid mixing offered by the flow method. By using special chambers with multiple entry ports arranged tangentially, it is possible to mix two reagents efficiently in a time of less than one millisecond. All fast flow methods are based on the pioneering work of Hartridge and Roughton.[13]

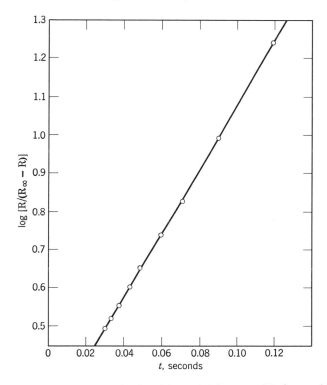

Fig. 2. Decomposition of carbonic acid at 22° C [Pearson, Meeker, and Basolo].

As an example, Fig. 2 shows some data obtained in a flow study[14] of the moderately rapid reaction

$$H_2CO_3 \rightarrow H_2O + CO_2 \tag{11}$$

The experimental procedure is to mix aqueous solutions of sodium bicarbonate and hydrochloric acid (about $0.005M$) in equivalent amounts. Measurements of the electrical conductivity are then made as a function of time as calculated from the flow rate. The reaction follows a first-order course from which a rate constant of 20.9 sec^{-1} can be calculated. The resistance increases with time as the partly ionized carbonic acid is converted to un-ionized carbon dioxide.

Steady-State Methods

The flow method is an example of a steady-state method in that, at each point in the observation tube, a constant set of concentrations is maintained which can be measured with reasonable leisure. Similar steady states for reactive species are often encountered in kinetic studies. The measurement of steady-state concentrations enables rate constants for rapid reactions to be calculated.[15]

Suppose a reactant is introduced into a reacting system with a constant rate of r_1 moles/liter-second and reacts chemically with a rate of r_2 moles/liter-second. If r_2 increases with the concentration of the reagent, as is usually true, then a steady state will be produced. The equation governing this will be

$$dc/dt = 0 = r_1 - r_2 \qquad (12)$$

Take the case of a simple first-order reaction again where $k_2 c = r_2$. The steady state will be given by

$$c_s = r_1/k \qquad (13)$$

so that if r_1 is known and c_s is measured then k can be calculated. The reagent may be introduced by electrolytic generation, by flow, by photolysis of a substrate, or by pyrolysis. The latter two procedures are useful for free radicals. Generally r_1 can be determined by suitable calibration procedures and can be of the order of $10^{-4} - 10^{-6}$ M/sec. If c_s can be measured down to low concentrations (10^{-5} M or below) then k values corresponding to half-lives of less than one second can be measured. In the case of electrolytic generation, or flowing in of the reagent, mixing is the determining factor. In the case of photolysis or pyrolysis, where mixing is already complete, the fastest reaction that can be measured is determined by the lowest value of c_s that can be measured. By electron spin resonance, free radicals can be detected down to 10^{-6} M or lower in favorable cases and thus a way for studying their rapid reactions is open.[16]

Encounter-Controlled Reactions

An interesting question now arises as to how rapid a rapid reaction may be. This question is answered for a unimolecular reaction by stating that the unimolecular rate constant could be as great as 10^{12}–10^{14} sec^{-1} in the limit. The case of a bimolecular reaction is more important. The obvious answer here is that the maximum rate of a bimolecular process is equal to the rate of encounter of the reacting molecules. That is, reaction occurs at the first collision of the molecules. Such reactions may be called encounter-controlled and the problem now is to calculate the maximum rate of encounter for pairs of molecules. Four separate situations may be visualized in which encounters are either in the gas phase or in solution, and either between non-interacting molecules or between molecules which attract or repel each other strongly.

For non-interacting molecules in the gas phase the encounter frequency is simply the kinetic molecular collision number. The second-order rate constant is then given by

$$k = \left(\frac{8\pi RT}{N\mu}\right)^{1/2} \frac{\sigma_{AB}^2 N}{1000} \text{ M}^{-1} \text{ sec}^{-1} \tag{14}$$

(Compare equation 28 in Chapter 4.)

If the gaseous molecules interact with each other at a distance, then this will influence their frequency of collision. If the interaction potential is known, the collision number can be calculated in simple cases.[17] The problem involves calculating the possible orbits of a pair of particles with the given potential. Some of these orbits lead to collision of the two particles and others do not. A critical impact parameter b_{AB} exists which will lead to collision.

The parameter b_{AB} is analogous to σ_{AB} in equation 14 but is not a constant since it depends on the relative velocity, $\bar{r}$, of the two particles. The total number of collisions per unit volume per second is given by

$$Z_{AB}' = \int \int \pi b_{AB}^2 \bar{r} \, dn_A \, dn_B \tag{15}$$

Here dn_A is the number of molecules of type A having a velocity within a specified range. It must be expressed as a distribution function, such as equation 2 of Chapter 4 if the velocities are Maxwellian. If b_{AB} were constant and the velocity distributions Maxwellian, (15) would give simply equation 12 of Chapter 4.

In the case of an ion and a nonpolar neutral molecule, the potential energy is given by

$$U = -\frac{q^2\alpha}{2d^4} \tag{16}$$

where d is the distance apart, α is the polarizability of the molecule and q is the charge on the ion. This leads to a critical impact parameter (defined as the distance between the initial part of the orbit of one particle and a parallel line passing through the center of the other particle)

$$b_{AB} = \left(\frac{4q^2\alpha}{\bar{r}^2\mu}\right)^{1/4} \tag{17}$$

where μ is the reduced mass.[18] The collision number from the integration of (15) becomes

$$Z_{AB}' = \left(\frac{4\pi^2 q^2 \alpha}{\mu}\right)^{\frac{1}{2}} n_A n_B \qquad (18)$$

whence the second-order rate constant can be calculated as

$$k = \left(\frac{4\pi^2 q^2 \alpha}{\mu}\right)^{\frac{1}{2}} \frac{N}{1000} \qquad (19)$$

For a typical case, this constant is several times larger than the corresponding figure for non-interacting systems given by (14).

The above results are valid in the gaseous phase. For condensed media the situation is considerably different. As mentioned in Chapter 7, collisions in solution tend to be repeated many times after the first encounter between two molecules because of the solvent cage effect. What is desired is not the total number of collisions, but the number of first collisions or encounters. This is the quantity which fixes the maximum rate of a bimolecular reaction. For non-interacting particles these encounters are determined by the random, Brownian movements of the particles through the viscous medium, and hence are diffusion-controlled. The basic equations for diffusion are Fick's two laws.

$$\phi = -D\nabla n \qquad (20)$$

$$\frac{\partial n}{\partial t} = \nabla \cdot \phi = -D\nabla^2 n \qquad (21)$$

Here ϕ, the flux, is the number of particles moving across a boundary of unit area per second and D is the diffusion coefficient.

For the case of spherical symmetry and the boundary conditions $n = n_0$ at $t = 0$ for all values of d, and $n = n_0$ at $d = \infty$ and $n = 0$ for $d = \sigma_{AB}$ for t greater than zero, equations 20 and 21 can be solved.[19]

$$n = n_0 \left[1 - \frac{\sigma_{AB}}{d} \, \text{erfc} \left(\frac{d - \sigma_{AB}}{2(Dt)^{\frac{1}{2}}}\right)\right] \qquad (22)$$

$$\phi = \frac{Dn_0}{\sigma_{AB}} \left(1 + \frac{\sigma_{AB}}{(\pi Dt)^{\frac{1}{2}}}\right) \qquad (23)$$

In equation 22 erfc (x) stands for the complement of the error function of x which is tabulated for various values of x.[20] In equation 23 the second term usually becomes small in comparison to the first term in a very short time. This corresponds to $(\partial n/\partial t)$ being equal to zero, a stationary state. At any instant of time ϕ is the flux of B molecules into a single A molecule and n_0 is the instantaneous bulk concentration of B molecules. The diffusion coefficient $D = D_A + D_B$ because of the simultaneous movement of both particles.

The total rate is clearly $4\pi\sigma_{AB}{}^2\phi n_A$ and accordingly the second-order rate constant becomes, using only the first term of (23),

$$k = \frac{4\pi\sigma_{AB}DN}{1000} \tag{24}$$

a result first obtained by Smoluchowski.[21] Putting in reasonable values of $\sigma_{AB} = 5 \times 10^{-8}$ cm and $D = 10^{-5}$ cm^2 sec^{-1} gives a rate constant of 4×10^9 M^{-1} sec^{-1}. The difference between this and the total collision frequency of about 10^{11} represents the number of repeated collisions between two molecules after the first encounter. It may be noted that an approximate equation for the diffusion coefficient of a spherical molecule of radius σ is

$$D = \frac{RT}{6\pi\eta\sigma N} \tag{25}$$

where η is the viscosity of the medium. Hence a viscous medium will slow down the rate of encounter-controlled reactions. Furthermore, such reactions will have an activation energy because of the temperature dependence of the viscosity.

If forces exist between the reacting molecules, then (20) and (21) must be modified to include the motion due to these forces. The flux can also be written as $\phi = n\bar{r}$, where $\bar{r}$ is the relative velocity. There will be a contribution to this relative velocity due to the force. When acceleration is negligible, the viscous force equals the force of attraction (or repulsion). This leads to

$$\phi = \frac{nDN}{RT} \nabla U - D\nabla n \tag{26}$$

The solution of equation 21 when 26 is inserted is extremely difficult in general. However, one important case can be solved[22] for the stationary state, that is, when $(\partial n/\partial t) = 0$. This is the case of two ions where the potential energy is expressed by Coulomb's equation

$$U = \frac{q_A q_B}{\epsilon d} \tag{27}$$

where ϵ is used for the dielectric constant of the medium. The solution for the bimolecular rate constant becomes

$$k = \frac{4\pi q_A q_B ND(N/1000)}{\epsilon RT(e^{q_A q_B N/\epsilon RT\sigma_{AB}} - 1)} \tag{28}$$

The limiting values of (28) are of interest. When the electrostatic energy $q_A q_B N/\epsilon \sigma_{AB}$ is small compared to RT, the thermal energy, (28) reduces simply to (24). When the electrostatic energy is large compared to the thermal energy and negative (oppositely charged ions), (28) becomes

$$k = \frac{4\pi q_A q_B N^2 D}{1000\epsilon RT} = \frac{4\pi \Lambda N}{1000\epsilon} \tag{29}$$

The molar conductance Λ of the electrolyte which the ions A and B may be considered to form has been substituted in the right-hand side of (29). This is the equation first derived by Langevin[23] and often used to estimate the maximum rate of reaction of oppositely charged ions. The usual units of Λ must be divided by 9×10^{11} which is the factor converting from practical to e.s.u. For H^+ and OH^- in water (29) leads to a second-order rate constant of 5×10^{10} M^{-1} sec^{-1}, somewhat less than the observed value (p. 210).

If the electrostatic energy is large and positive (similarly charged ions), then equation 28 predicts a small value for k. For example, if A and B are univalent, σ_{AB} is 3.5 A and the solvent is water, the value of the rate constant is calculated to be about one-third of that for a pair of neutral molecules with the same value of D. For highly charged ions or solvents of low dielectric constant, the slowing down due to electrostatic repulsion can become very much larger.

METHODS FOR STUDYING RAPID REACTIONS

In this final section an attempt will be made briefly to describe the general principles underlying some of the methods used to study rates of very fast chemical, or in some cases physical, processes. The treatment is necessarily incomplete and reference to some of the original papers is necessary for any details.[24]

Quenching of Fluorescence

An electronically excited molecule may emit a quantum of energy by fluorescence or may transfer its excitation energy by collision with other molecules. The latter process is called quenching. The competition between these two possible events has often been used as a method for testing theories of encounter-controlled reactions. Presumably quenching does not require an activation energy, other than that of diffusion, and the chance of reaction occurring on the first collision is high. On the other hand, it is well known that different molecules have different quenching efficiencies, more complex molecules being more efficient.

The overall system may be represented as follows, where F is the fluorescing molecule and Q is the quenching molecule:

$$F + h\nu \longrightarrow F^* \tag{30}$$

$$F^* \xrightarrow{k_f} F + h\nu' \tag{31}$$

$$Q + F^* \xrightarrow{k_q} F + Q^* \tag{32}$$

The fluorescent yield, or ratio of intensities of light emitted to light absorbed, is given by

$$\frac{I_f}{I_a} = \frac{k_f(F^*)}{k_f(F^*) + k_q(F^*)(Q)} \tag{33}$$

Rearranging we obtain the Stern-Volmer equation.[25]

$$\frac{I_a}{I_f} = 1 + \frac{k_q(Q)}{k_f} \tag{34}$$

A plot of the reciprocal of the fluorescent yield against the concentration of the quencher gives the ratio of rate constants. Often k_f can be evaluated from an analysis of the absorption spectrum, in which case k_q can readily be found. Table 1 shows some results on the quenching of the fluorescence of β-napthylamine by carbon tetrachloride in the gas phase and in solution obtained by Rollefson and Curme.[26]

Table I

QUENCHING OF β-NAPHTHYLAMINE BY CCl_4

[G. K. Rollefson and H. G. Curme]

Medium	k_q, $M^{-1} sec^{-1}$
Gas	5.9×10^{10}
Isoöctane	$2.0 \times 10^{11} e^{-1600/RT}$
Cyclohexane	$4.5 \times 10^{11} e^{-2470/RT}$

These results are calculated on the basis of a value of $6 \times 10^7 sec^{-1}$ for k_f in the gas phase. This value in turn is based upon quenching experiments using oxygen (an efficient quencher) with the assumption that every collision leads to deactivation. This assumption is probably good only to a factor of 2 or 3. It is of interest, however, to see that the gas-phase value agrees with the collision theory and is temperature-independent. The solution rate constants are smaller and have apparent activation energies similar to, but not identical with, the activation energies for diffusion.

An interesting variation of the above appears when the excited molecule is capable of chemical reaction without deactivation.[27] This is most common for acid-base reactions. For example, β-napthol is a weak acid and may be converted to its anion when in the excited state.

$$ROH^* + B \rightleftharpoons RO^{-*} + BH^+ \tag{35}$$

Since ROH^* and RO^{-*} have different fluorescent spectra, the extent to which (35) has occurred may be detected experimentally as a function of various concentrations of B and BH^+. The results depend on the interplay of the rate constants for fluorescence and of the proton transfers of (35).

Table 2

ACID-BASE REACTIONS OF ROH, β-NAPTHOL, 25° C

[A. Weller]

Reaction	K_{eq}	k, M^{-1} sec^{-1}
$H_3O^+ + RO^{-*}$	650	4.8×10^{10}
$HCOOH + RO^{-*}$	0.11	2.8×10^8
$CH_3COOH + RO^{-*}$	0.011	3.3×10^7
$CH_3COO^- + ROH^*$	88	2.9×10^9
$HCOO^- + ROH^*$	8.8	2.4×10^9
$H_2PO_4^- + ROH^*$	0.20	6.0×10^8

Table 2 shows some rate constants obtained by Weller using the above approach.[28] It must be remembered that these constants are for excited molecules and would not be the same for normal molecules. The equilibrium constants for (35) are also different for the excited molecule which is a much stronger acid than the normal molecule.

Polarography and Other Diffusion-Coupled Methods

A coupling of the rate of a chemical reaction with another time-dependent process is a favorite procedure, as the previous examples indicate. Ordinary diffusion is a rate process that may be used to study rapid reactions. For example, in polarography the observed current is usually governed by the rate of diffusion of the reducible substance to the cathode. Imagine a chemical reaction occurring in which a reducible material present in small amounts at equilibrium is formed from a nonreducible substance present in large amounts. The formation of formaldehyde from its hydrate is such a case.

$$CH_2(OH)_2 \underset{k_2}{\overset{k_1}{\rightleftharpoons}} CH_2O + H_2O \tag{36}$$

Only the free formaldehyde is reducible and at equilibrium only 0.04 per cent of all the aldehyde is unhydrated. Terms corresponding to the chemical reaction are inserted into the diffusion equation (21). For first-order or pseudo-first-order reactions the equations can be solved.[29]

An increased current called the kinetic current is found, compared to the current which would be due to the equilibrium amount of reducible formaldehyde. In other words, diffusion of the hydrate plus the dissociation (36) produces the reducible material at the electrode faster than diffusion of free formaldehyde alone. The kinetic current depends inversly on a parameter $\mu = (D/k_2)^{1/2}$ called the thickness of the reaction layer. It is necessary that this parameter be at least 20 A or so for the diffusion equations to be meaningful. Since D is about 10^{-5} this means that k_2 can be no larger than 10^9 sec^{-1}. On the other hand, if k_2 is rather small, the diffusion current will be so small as to be unmeasurable.

The reaction pattern produced when one reagent is allowed to diffuse into an atmosphere of another reagent has been used to study rapid gas-phase reactions. This is the method of "dilute flames" developed by Polanyi and his co-workers.[30] The amount of reaction as a function of distance may be measured from the spread of solid deposit in the reaction of, say, sodium vapor with a halogen or alkyl halide. Or the resonance absorption and fluorescence of the sodium might be used to measure its depletion.

An interesting variation is the measurement of the temperature pattern produced during the course of a highly exothermic reaction.[31] Suppose a finite amount of reagent A diffuses from a small opening into a large atmosphere of reagent B. A steady state is reached governed by the equation

$$D_A \nabla^2(A) - k(B)(A) = 0 \qquad (37)$$

If B is in excess the solution of (37) is

$$(A) = \frac{b}{4\pi D_A d} e^{-\left(\frac{k(B)}{D_A}\right)^{1/2} d} \qquad (38)$$

where b is the total number of moles of A introduced and all assumed to be reacted. By means of a movable thermocouple it is possible to measure the temperature increase at each point in the system. The temperature change follows the same exponential form given by (38). The method has been used for the reactions of amines with BF_3 where rate constants of the order of 10^9 to $10^{10} \text{ M}^{-1} \text{ sec}^{-1}$ have been found in the gas phase.[32]

The Rotating Sector Method

In photochemical reactions a periodic interruption of the beam of light can lead to interesting kinetic consequences. This is the rotating sector method for studying the lifetimes of photochemically produced reactive species.[33] As indicated earlier, a steady state is usually produced when constant illumination is used. Such a steady state is not produced instantaneously however. Take the case of a chain reaction in which a free radical is produced photochemically and disappears by a bimolecular reaction. The approach to the steady state is given by

$$\frac{d(\text{R·})}{dt} = kI - 2k_t(\text{R·})^2 \tag{39}$$

where I is the intensity of absorbed light. The steady-state solution is for $d(\text{R·})/dt = 0$.

$$(\text{R·})_s = (kI/2k_t)^{1/2} \tag{40}$$

The time-dependent solution of (39) is

$$(\text{R·}) = (\text{R·})_s \tanh(t/\tau) \tag{41}$$

where τ is $(2kIk_t)^{-1/2}$ which also happens to be the mean chain life.

The interpretation of (41) is that a time several times greater than τ is needed to establish the steady state. If the light source is interrupted with a period much greater than τ, then for each cycle the steady state will be reached early in the cycle. The overall rate of reaction will be given by the usual treatment for chain reactions with a light intensity equal to I, but only operating half the time. If the flashing period is much less than τ, the full steady state corresponding to I will never be reached. Instead, a concentration $(\text{R·})_s$ corresponding to a uniform intensity of $I/2$ will be developed. In other words

$$(\text{R·})_s = (kI/4k_t)^{1/2} \tag{42}$$

which is greater by 40 per cent than that given by one-half of the value in equation 40. For values of the flashing period of the order of τ, a transition between the two limiting values is observed. From this behavior, the value of τ can be found. From τ it is then possible to calculate termination rate constants for free radicals.

Magnetic Resonance Methods

Nuclear magnetic resonance (NMR) and electron paramagnetic resonance (EPR) offer ways of studying very rapid exchange reactions in an equilibrium system.[34] It is necessary that at least two different environments be accessible to the magnetically active nucleus or electron. These environments must be such that quite different spectra would be observed for a system in each environment in the absence of exchange or with slow exchange. Rapid exchange leads to an averaging of the spectra in a way which is predictable by the solution of the Bloch equations for magnetic resonance as modified by the inclusion of the exchange phenomenon.[35]

A simple example is provided by the system hydrogen peroxide-water in which two major environments exist for the proton.[36] Figure 3 shows how the NMR spectrum of such a system, approximately equimolar,

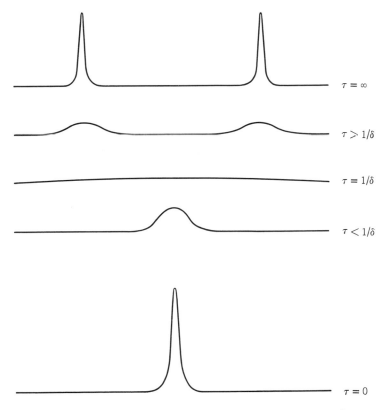

Fig. 3. The effect of chemical exchange on NMR line-width. Resonance frequency increasing from left to right.

would appear for various circumstances. In this case the important variables are τ, the mean lifetime of a proton in each environment, and δ, the frequency difference in cycles per second between the two signals in the absence of exchange. In general, values of τ of the order of $1/\delta$ can be found. For τ much larger than $1/\delta$, two separate signals are observed, and for τ much less than $1/\delta$ one averaged signal is observed. Since δ for protons is generally of the order of 100 cycles per second, τ is of the order of 0.01 seconds for easy measurement. Acid-base catalysis or change in temperature can be used to bring about changes in τ. Table 3 shows some proton transfer rate constants for amines obtained by NMR methods.

The rate constant k_1 is for the direct transfer of a proton from the ammonium ion to the amine. The rate constant k_2 is for the interesting process in which a water molecule mediates between the two reactants.

$$R_3NH^+ + \overset{\overset{\displaystyle H}{|}}{O}\!\!-\!\!H + NR_3 \overset{k_2}{\longrightarrow} R_3N + H\!\!-\!\!\overset{\overset{\displaystyle H}{|}}{O} + HNR_3^+ \qquad (43)$$

Such an indirect proton transfer is supported not only by the NMR spectra but also by other evidence.[37] It is quite reasonable since the water

Table 3

RATES FOR AMINE-AMMONIUM ION PROTON EXCHANGE AT 22° C

$B + BH^+ \rightleftharpoons BH^+ + B$

System	k_1, $M^{-1}\,sec^{-1}$	k_2, $M^{-1}\,sec^{-1}$
$NH_3 + NH_4^+$	10.6×10^8	0.9×10^8
$CH_3NH_2 + CH_3NH_3^+$	2.5×10^8	3.4×10^8
$(CH_3)_2NH + (CH_3)_2NH_2^+$	0.4×10^8	5.6×10^8
$(CH_3)_3N + (CH_3)_3NH^+$	0.0×10^8	3.1×10^8

Data from A. Loewenstein and S. Meiboom, *J. Chem. Phys.*, 27, 1067 (1957); and S. Meiboom, A. Loewenstein, and S. Alexander, *ibid.*, 29, 969 (1958).

molecule would be hydrogen bonded to both reactants both before and after the exchange. Furthermore, from Table 3 it can be seen how the sterically hindered amines take more and more advantage of this path compared to sterically free ammonia.

If one of the two environments for the nucleus gives a much broader line than the other, in the absence of exchange, then the important variables are the width of the broad line in cycles per second and τ, the exchange lifetime.[38] Such a broad line would be obtained from a paramagnetic environment. Also if the nucleus whose NMR signal is being investigated has a quadrupole moment, then an electrically unsymmetrical environment would produce a broad line.[39] A rather small amount of the broad line environment will mix in to a large amount of the narrow line environment to give an overall broadening. It has been possible to measure the mean lifetimes of various anions and molecules bound to cations by this line-broadening method. Exchange times down to a microsecond or less may be determined.

The width of a resonance signal is related to the lifetime of a given magnetic state. The reciprocal of the half-width at half-height is called T_2, the transverse relaxation time. In many cases, including the H_2O—H_2O_2 system where two lines are still distinguishable, the line width is simply related to the exchange lifetime.

$$\frac{1}{T_2} = \frac{1}{T_2'} + \frac{1}{\tau} \tag{44}$$

In this equation T_2' is the line width for the case of no exchange. The line broadening is an example of the Heisenberg uncertainty principle which states that $(\Delta E)(\Delta t) \simeq h$. Hence a shortening of the lifetime, Δt, means an increase in ΔE, or in this case, an increase in the width of the line. This argument also holds for electron spin resonance.

A paramagnetic ion produces a very short relaxation time in nuclei closely associated with it. This implies a broad line. The overall effect depends on both the rate of relaxation and on the rate of exchange. In general, whichever rate is the slower will be the rate that can be measured.[38a]

Ion-Molecule Reactions in the Mass Spectrometer

Gas-phase reactions involving ions are rare in ordinary chemical processes. Ionizing radiation, however, initiates many ionic reactions and these are of increasing interest. A convenient way to study gas-phase ionic reactions is in the mass spectrometer.[40] Positive ions are formed in the ionization chamber and expelled into an accelerating electric field. Here they can react with neutral molecules to produce secondary ions. Usually such secondary ions are considered a nuisance to be avoided. They may also be studied to give rate data on the reactions producing them.

The experimental quantity observed is the ratio of mass spectral currents for the primary and secondary ions. Also the pressure of the neutral reactant, the length of the path of the ions through the neutral molecules, and the accelerating voltage must be known. From the theory of ion-molecule reactions in the gas phase, as outlined on pp. 269–70, it is possible to calculate what the ratio of ion-currents should be for a given case.[17] An ion-induced dipole interaction is assumed and an equation corresponding to (15) is integrated. However, in this case the velocity distribution in one dimension is not Maxwellian but depends on the accelerating voltage and is a function of distance. If it is assumed that reaction occurs at every collision and that all the secondary ions are collected, the ion-current ratio can be calculated.

Conversely, from the measured ion-current ratio a second-order rate constant for the ion-molecule reaction can be calculated to compare with the theoretical one given by equation 19. Table 4 shows some rate constants obtained by mass spectral studies. Values for k both from the experimental results and as calculated from the charges, masses, and polarizabilities of the molecules are given. The agreement is seen to be good.

Table 4

RATE CONSTANTS FOR ION-MOLECULE REACTIONS

Reaction	k exp, $M^{-1} sec^{-1}$	k theor, $M^{-1} sec^{-1}$
$H_2^+ + H_2 = H_3^+ + H$	1.7×10^{12}	1.2×10^{12}
$He^+ + H_2 = HeH^+ + H$	0.37×10^{12}	1.1×10^{12}
$H_2^+ + O_2 = HO_2^+ + H$	4.6×10^{12}	1.3×10^{12}
$HCl^+ + HCl = H_2Cl^+ + Cl$	0.27×10^{12}	0.56×10^{12}

k theor is for a Maxwellian distribution of velocities.

Data from H. Gutbier, *Zeit. Naturforsch.*, *12A*, 499 (1957); and D. P. Stevenson and D. O. Schissler, *J. Chem. Phys.*, *23*, 1353 (1955); *29*, 282 (1958).

In many cases involving more complicated molecules the theory of ion-dipole attractions cannot be used because the ion-current does not depend on the accelerating field in the proper way.

Relaxation Methods

If a labile system is perturbed from equilibrium by a small amount, it will approach equilibrium in a first-order process characterized by a relaxation time τ which is the time needed for the system to traverse a fraction $1/e$ of its path to equilibrium. This concept was first used by Maxwell to describe physical phenomena. Lately it has been widely applied to chemical systems.[41]

Consider the ionization in water of a weak acid, HA, and let a be the total concentration of acid and x the concentration of ions. Let $\Delta x =$

$$HA + H_2O \underset{k_2}{\overset{k_1}{\rightleftharpoons}} H_3O^+ + A^- \tag{45}$$

$x - x_e$, where x_e refers to equilibrium. Then we have the equations, when Δx is small,

$$\frac{dx}{dt} = k_1(a - x) - k_2 x^2 \tag{46}$$

$$k_1(a - x_e) = k_2 x_e^2$$

$$\frac{d\Delta x}{dt} = -(k_1 + 2k_2 x_e)\Delta x = -\frac{\Delta x}{\tau} \tag{47}$$

The relaxation time is seen to be a function of the forward and reverse rate constants and the concentration of ions.

The perturbation from equilibrium may be the result of a sudden change in temperature, pressure, or strong electric field. This must be accomplished in a time much less than τ, that is to say, in 10^{-6} to 10^{-7} seconds. By oscilloscopic means the approach to the new equilibrium can be measured over times of a few microseconds to a few milliseconds. Spectroscopic or conductance measurements may be used.

It is often convenient to use periodic disturbances such as an ultrasonic wave or an alternating electric field. In the case of ultrasonics it is the rapidly changing pressure which produces the change in the chemical equilibrium. Since this is periodic, an alternating response is established in the chemical system which is partly in phase and partly out of phase with the perturbation. The out-of-phase component produces an absorption of energy in the system from the sound wave or the electric field. This absorption can be measured as a function of the frequency of the perturbation and the relaxation time of the system found. Maximum absorption usually occurs when the frequency is equal to $1/\tau$.

A complex system will have several relaxation times which can be separated to lead to values of several rate constants if a mechanism for the reaction of the system is assumed.[42] The name relaxation spectroscopy has been applied to this kind of study by Eigen who has made measurements on a number of systems in aqueous solution. Bimolecular rate constants up to 10^{11} M^{-1} sec^{-1} can be measured. For example, the rate of ionization of acetic acid at $25°$ C is given by $k_1 = 8 \times 10^5$ sec^{-1} and $k_2 = 4.5 \times 10^{10}$ M^{-1} sec^{-1}, as in equation 45.

The absorption of ultrasonic waves has been widely used to get the rates of very rapid intramolecular processes, such as the interconversion of rotational isomers of organic molecules.[43] Also the rates of formation and dissociation of hydrogen bonded species may be studied by ultrasound.[44] The classical chemical example is the dissociation of nitrogen tetroxide

$$N_2O_4 \underset{k_2}{\overset{k_1}{\rightleftharpoons}} 2NO_2 \tag{48}$$

where it was early recognized by Einstein that the anomalous velocity and excess sound absorption at certain frequencies were due to the reaction 48. An approximate value for the rate was obtained by ultrasonic techniques, but the reaction is better studied by the next method to be described.

Shock Methods

A shock method is similar to the single-impulse relaxation method in that a perturbation is suddenly applied to a pre-mixed system in stable or metastable equilibrium. However, here the perturbation is violent so that departure from equilibrium becomes great. Usually the applied disturbance is in the form of a shock wave which passes through the system with supersonic velocity. In the case of liquid or gaseous systems such a shock wave could be caused by the detonation of an explosive charge.[45]

However, shock methods are most common for gases. A device called a shock tube is conveniently used. This consists of a tube containing an inert gas at high pressure and a reactive gas at low pressure separated from each other by a diaphragm. When the diaphragm is ruptured, the expansion of the high pressure gas produces a shock wave which passes through the low pressure gas, adiabatically compressing it. Very high temperatures, up to 2000° C can be produced. This change in temperature can be calculated from the thermodynamic properties of the gases and from a measurement of the shock velocity.

The change in temperature is produced in a very short time, of the order of that for a few molecular collisions. Subsequently chemical reactions occur in the high temperature gas. These can be followed by photoelectric light absorption methods. Reaction times of several hundred microseconds are available for study. The method has been used to study the rates of dissociation of simple molecules, reactions simple to interpret.[46] The more complicated reactions of complex molecules and of mixtures are also of interest.

Flash Photolysis

Another type of extreme disturbance in a pre-mixed system is produced by a sudden irradiation with a very intense burst of light energy. This method is called flash photolysis.[47] Up to 10^5 joules of energy may be absorbed in a few hundred microseconds, or more typically several hundred joules in a few microseconds. The light is produced by the firing of a flash lamp at high voltages and currents.

The light energy absorbed causes both electronic excitation and chemical reaction. Unlike ordinary photolysis, the excited or reactive species are produced in rather high concentrations. This makes it possible to identify and study them by measuring their characteristic absorption spectra. This is usually done by means of a weaker flash following the initial flash and by using photographic or photoelectric methods.

The kinetics of the excited molecules or free radicals can then be followed in the time following the initial flash. First-order rate constants as large as 10^5 sec^{-1} and second-order rate constants as large as 10^{11} M^{-1} sec^{-1} can be measured. The method works equally well in the gas phase and in solution.

PROBLEMS

1. Show that equations 22 and 23 are solutions of (20) and (21).

2. Calculate the rate constant for the recombination of hydrogen ions and acetate ions using equation 29. Compare with the experimental value. Estimate the forward rate constant for the reaction

$$Co(NH_3)_5OH^{2+} + H_3O^+ \rightleftharpoons Co(NH_3)_5H_2O^{3+} + H_2O$$

The acid ionization constant of the aquo complex is equal to 2×10^{-5} at 15° C.

REFERENCES

1. I. Förster and K. H. Geib, *Ann. Physik*, *20*, 250 (1934); H. M. Hurlburt, *Ind. Eng. Chem.*, *36*, 1012 (1944).
2. G. M. Harris, *J. Phys. & Colloid Chem.*, *51*, 505 (1947).
3. K. G. Denbigh, *Trans. Faraday Soc.*, *40*, 352 (1944); *44*, 479 (1948); B. Stead, F. M. Page, and K. G. Denbigh, *Disc. Faraday Soc.*, *2*, 263 (1947).
4. L. P. Hammett and H. H. Young, *J. Am. Chem. Soc.*, *72*, 280 (1950).
5. J. Saldick and L. P. Hammett, *J. Am. Chem. Soc.*, *72*, 283 (1950); M. J. Rand and L. P. Hammett, *J. Am. Chem. Soc.*, *72*, 287 (1950).
6. J. D. Johnson and L. J. Edwards, *Trans. Faraday Soc.*, *45*, 286 (1949).
7. H. Hartridge and F. J. W. Roughton, *Proc. Roy. Soc.*, *A104*, 376 (1923); F. J. W. Roughton and G. A. Millikan, *Proc. Roy. Soc.*, *A155*, 258 (1936); G. A. Millikan, *Proc. Roy. Soc.*, *A155*, 277 (1936).
8. R. N. J. Saal, *Rec. trav. chim.*, *47*, 73 (1928).
9. V. K. La Mer and C. L. Read, *J. Am. Chem. Soc.*, *52*, 3098 (1930).
10. P. D. Bartlett, F. E. Condon, and A. Schneider, *J. Am. Chem. Soc.*, *66*, 1531 (1948).
11. B. Chance, *J. Franklin Inst.*, *229*, 455, 737 (1940); *J. Biol. Chem.*, *179*, 1249 (1949); *180*, 865 (1949).
12. B. Chance, *Rev. Sci. Instr.*, *22*, 619 (1951); Q. H. Gibson, *Disc. Faraday Soc.*, *17*, 137 (1954).
13. For detailed discussions see *Techniques of Organic Chemistry*, Vol. 8, S. L. Friess and A. Weissberger, eds., Interscience Publishers, New York, 1953, pp. 669–738.
14. R. G. Pearson, R. E. Meeker, and F. Basolo, *J. Am. Chem. Soc.*, *78*, 709 (1956).
15. R. G. Pearson and L. H. Piette, *J. Am. Chem. Soc.*, *76*, 3087 (1954); P. S. Farrington and D. T. Sawyer, *ibid.*, *78*, 5536 (1956).
16. L. H. Piette and W. C. Landgraf, *J. Chem. Phys.*, *32*, 1107 (1960); R. W. Fessenden and R. H. Schuler, *ibid.*, *33*, 935 (1960).
17. (*a*) P. Langevin, *Ann. chim. phys.*, *5*, 245 (1905); (*b*) G. Gioumousis and D. P. Stevenson, *J. Chem. Phys.*, *29*, 294 (1958).
18. Reference 17 and H. Eyring, J. O. Hirschfelder, and H. S. Taylor, *J. Chem. Phys.*, *4*, 479 (1936).
19. E. L. Lederer, *Kolloid-Z.*, *44*, 108 (1928); *46*, 169 (1928).
20. See P. Delahay and G. L. Stiehl, for example, *J. Am. Chem. Soc.*, *74*, 3500 (1952).
21. M. V. Smoluchowski, *Physik. Z.*, *17*, 557, 583 (1916); *Zeits. physikalische Chem.*, *92*, 129 (1917).

22. P. Debye, *Trans. Electrochem. Soc.*, *82*, 265 (1942); M. Eigen, *Zeits. f. physikalische Chem.*, *1*, 176 (1954); V. K. LaMer and J. Q. Umberger, *J. Am. Chem. Soc.*, *67*, 1099 (1945); see A. Weller, *Zeits. f. physikalische Chem.*, *13*, 335 (1957) for extension to the non-stationary case.

23. P. Langevin, *Ann. chim. phys.*, *28*, 433 (1903); L. Onsager, *J. Chem. Phys.*, *2*, 599 (1934).

24. General discussions are to be found in *Disc. Faraday Soc.*, *17* (1954) and *Zeits. f. Elektrochem.*, *64* (1960).

25. O. Stern and M. Volmer, *Physik. Z.*, *20*, 183 (1919).

26. G. K. Rollefson and H. G. Curme, *J. Am. Chem. Soc.*, *74*, 3766 (1952).

27. T. Förster, *Naturwiss.*, *36*, 186 (1949); *Zeits f. Elektrochem.*, *54*, 42, 531 (1950); A. Weller, *ibid.*, *64*, 55 (1960).

28. A. Weller, *Zeits. f. physikalische Chem.*, *17*, 224 (1958).

29. For references to work on kinetics by polarographic means see R. Brdička, *Zeits. f. Elektrochem.*, *64*, 16 (1960); J. Koryta, *ibid.*, *64*, 23 (1960); P. Delahay, *New Instrumental Methods in Electrochemistry*, Interscience Publishers, New York, 1954, Chapter 5.

30. M. Polanyi, *Atomic Reactions*, Williams and Norgate, London, 1932.

31. D. Garvin, V. P. Guinn, and G. B. Kistiakowsky, *Disc. Faraday Soc.*, *17*, 32 (1954).

32. G. B. Kistiakowsky and R. Williams, *J. Chem. Phys.*, *23*, 334 (1955); J. Daen and R. A. Marcus, *ibid.*, *26*, 162 (1957).

33. A. Berthoud and H. Bellenot, *Helv. Chim. Acta*, *7*, 307 (1923); G. M. Burnett and H. W. Melville, *Chem. Revs.*, *54*, 225 (1954).

34. H. S. Gutowsky, D. W. McCall, and C. P. Schlicter, *J. Chem. Phys.*, *21*, 279 (1953); H. S. Gutowsky and A. Saika, *ibid.*, 1698 (1953); R. A. Ogg, Jr., *ibid.*, *22*, 560 (1954); *Disc. Faraday Soc.*, *17*, 215 (1954).

35. H. M. McConnell, *J. Chem. Phys.*, *28*, 430 (1958).

36. M. Anbar, A. Loewenstein, and S. Meiboom, *J. Am. Chem. Soc.*, *80*, 2630 (1958).

37. M. Eigen, *Disc. Faraday Soc.*, *17*, 194 (1954); C. G. Swain and M. M. Labes, *J. Am. Chem. Soc.*, *79*, 1084 (1957); C. G. Swain, J. T. McKnight, and V. P. Kreiter, *ibid.*, 1088 (1957).

38. (*a*) R. G. Pearson, J. W. Palmer, M. M. Anderson, and A. L. Allred, *Zeits. f. Elektrochem.*, *64*, 110 (1960); (*b*) H. M. McConnell and S. B. Berger, *J. Chem. Phys.*, *27*, 230 (1957).

39. O. E. Myers, *J. Chem. Phys.*, *28*, 1027 (1958); H. G. Hertz, *Zeits. f. Elektrochem.*, *64*, 53 (1960).

40. D. P. Stevenson and D. O. Schissler, *J. Chem. Phys.*, *23*, 1353 (1955); *24*, 926 (1956); F. H. Field, J. L. Franklin, and F. W. Lampe, *J. Am. Chem. Soc.*, *79*, 2419, 2665 (1957).

41. M. Eigen, *Disc. Faraday Soc.*, *17*, 194 (1954); *Zeits. f. Elektrochem.*, *64*, 115 (1960); L. DeMaeyer, *ibid.*, *64*, 65 (1960); R. G. Pearson, *Disc. Faraday Soc.*, *17*, 187 (1954).

42. R. A. Alberty and G. C. Hammes, *Zeits. f. Elektrochem.*, *64*, 124 (1960).

43. J. Lamb, *Zeits. f. Elektrochem.*, *64*, 135 (1960).

44. W. Maier, *Zeits. f. Elektrochem.*, *64*, 132 (1960).

45. R. Schall, *Z. angew. Physik*, *2*, 252 (1950); H. G. David and S. D. Hamann, *Trans. Faraday Soc.*, *55*, 72 (1959).

46. D. Britton, N. Davidson, and G. Schott, *Disc. Faraday Soc.*, *17*, 58 (1954); S. H. Bauer and M. R. Gustavson, *ibid.*, 69 (1954).

47. R. G. W. Norrish and G. Porter, *Disc. Faraday Soc.*, *17*, 40 (1954); G. Porter, *Zeits. f. Elektrochem.*, *64*, 59 (1960).

12.

SOME REACTIONS WHOSE MECHANISMS HAVE BEEN INVESTIGATED BY KINETIC AND OTHER METHODS

In this chapter we shall discuss some reactions whose mechanisms have been investigated in detail. It will be noted that a variety of methods have been used in these investigations. The examples chosen will show that the kinetic method, including not only determinations of reaction rates and orders but also the changes in these rates and orders with changing conditions, is the best approach to reaction mechanisms. However, they will also show that kinetics is incomplete by itself and must be liberally supplemented with other studies. Frequently there are several mechanisms which are kinetically indistinguishable; frequently the mechanisms deduced from kinetics alone are too vague as to the exact processes taking place.

The classic example of a reaction whose mechanism has been exhaustively studied is the work of Ingold and Hughes on the reactions of alkyl halides with nucleophilic reagents. This work is now familiar enough and has been reviewed adequately enough so that we shall not discuss it in detail.[1] However, some review is desirable, both because of the admirable way in which this example interweaves kinetic and other evidence to obtain a mechanistic solution, and because Hughes and Ingold introduced a terminology useful for discussing mechanisms in general. For example, a nucleophilic reagent is one which will react with a positive center; it is a base in the general sense since it will contain an unshared pair of electrons. An electrophilic reagent is one which will react with a nucleophilic reagent; it has a positive center which will accept a pair of electrons. Generalized acids are electrophilic reagents, but the reverse is not necessarily true.

The reaction of an alkyl halide with a base

$$RX + B \rightarrow RB^+ + X^- \tag{1}$$

is a nucleophilic substitution reaction, since one nucleophilic reagent, B, displaces another, X^-. There are two kinetically distinguishable ways in which this can happen. Either the rate of the reaction is dependent on the concentration of B, or it is not. Since the rate always depends on the alkyl halide, the overall result is either a second-order reaction or a first-order one. Paralleling these kinetic observations are two mechanisms, the S_N2 and S_N1 processes of Hughes and Ingold. The S_N2 mechanism (substitution, nucleophilic, bimolecular) involves a direct attack on the alkyl halide by B. This occurs at the face of the carbon tetrahedron opposite to the halogen atom which is to be displaced. The stereochemical result is inversion of configuration. The S_N1 mechanism involves a two-step process; a slow, rate-determining ionization of the alkyl halide

$$RX \rightarrow R^+ + X^- \qquad \text{slow} \tag{2}$$

and a rapid, follow-up reaction of the electrophilic positive ion R^+ (a carbonium ion) with the nucleophilic reagent

$$R^+ + B \rightarrow RB^+ \qquad \text{fast} \tag{3}$$

The first step is considered a unimolecular process since only one molecule changes its covalence. As in any ionization there will be a participation by solvent molecules or other species in stabilizing the ions as they form. In this sense reaction 2 has sometimes been called polymolecular.

The stereochemical results of the S_N1 mechanism depend on the nature of R^+ since there may be factors which tend to hold a particular configuration (plural asymmetry centers or neighboring groups). However, if R is a simple alkyl radical, it is to be expected that the carbonium ion will form a planar, symmetrical configuration. This can then react with B to give either stereoisomer so that, if RX were optically active at the start, a racemic product would be formed. Racemization may not be complete in spite of the symmetrical configuration of the carbonium ion, since the environment may not be the same on the two sides of the ion (halide ion may still be present on the front side). Thus reaction can occur to give preferentially more of one stereoisomer.

Though an identification of the S_N1 and S_N2 mechanisms, respectively, with the observed first- and second-order reactions is natural, it may not always be correct. Thus it may be that the solvent acts as the nucleophilic reagent in a typical S_N2 process, as in the alcoholysis of an alkyl halide. Since the solvent is in great excess, its concentration will not change and the kinetics will be pseudo-first-order. Or, in an S_N1 reaction, if the rate of the reverse of reaction 2 is great enough because of the presence of a constant excess of halide ion, for example, the kinetics will turn out to be second-order.

Accordingly, the following experimental procedures have been used to establish proof of mechanism:[2] (1) kinetic methods—to determine orders; (2) structural changes in RX—to determine effect on rates and orders; (3) changes in the reagent—to determine effect on rates; (4) changes in solvent—to determine effect on rates, orders, and products; (5) salt additions to solvent—to determine effect on rates, orders, and products; (6) stereochemistry—to determine amounts of retention, racemization, or inversion. To this list may be added another technique: (7) use of radioactive isotopes—to show relation between rate of exchange and rate of reaction.[3] The results of applying this battery of methods clearly establishes the existence of the S_N1 and S_N2 processes in general, and often permits an assignment of one mechanism or the other to any given example.

The danger of trying to select a mechanism on the basis of kinetic evidence alone is shown by the behavior of alkyl bisulfate ions towards alkaline hydrolysis:

$$RSO_4^- + OH^- \rightarrow ROH + SO_4^{-2} \tag{4}$$

When R is ethyl or methyl this reaction is second-order, depending on the hydroxide ion concentration.[4] When R is isopropyl the reaction is very nearly first-order, and when R is secondary butyl the hydrolysis is first-order, independent of the pH, for low concentrations of hydroxide ion.[5] This suggests that for methyl and ethyl bisulfate the S_N2 mechanism is operating but for the secondary bisulfates a switch over to the S_N1 mechanism has occurred. Such a result would be in accord with the accepted theory that a secondary carbonium ion would be more stable than a primary carbonium ion. In direct opposition to this conclusion is the observation that, when optically active secondary butyl bisulfate ion is used, the product obtained is optically pure secondary butyl alcohol with an inversion of configuration.[6] This now suggests that we are dealing with an S_N2 mechanism in which a water molecule is the nucleophilic reagent instead of hydroxide ion. In this case the greater nucleophilic activity of the stronger base seems to be overcome by the coulombic repulsion between two negatively charged ions which leads to a low-frequency factor for their reaction (see Chapter 7). This does not happen to the same extent for the primary bisulfate ions, however, the difference being the greater bond strength which requires the stronger nucleophilic reagent for rupture. At higher concentrations of hydroxide ion the expected second-order reaction appears with the secondary bisulfates.[7]

REFERENCES

1. C. K. Ingold, *Structure and Mechanism in Organic Chemistry*, Cornell University Press, Ithaca, 1953, Chapter 7; J. Hine, *Physical Organic Chemistry*, McGraw-Hill Book Co., New York, 1956, Chapter 5.
2. L. C. Bateman, E. D. Hughes, and C. K. Ingold, *J. Chem. Soc.*, 979 (1940).
3. W. Koskoski, H. Thomas, and R. D. Fowler, *J. Am. Chem. Soc.*, *63*, 2451 (1941).
4. G. H. Green and J. Kenyon, *J. Chem. Soc.*, 1389 (1950).
5. R. L. Burwell, Jr., *J. Am. Chem. Soc.*, *74*, 1462 (1952).
6. R. L. Burwell, Jr., and H. E. Holmquist, *J. Am. Chem. Soc.*, *70*, 878 (1948).
7. G. M. Calhoun and R. L. Burwell, Jr., *J. Am. Chem. Soc.*, *77*, 6441 (1955).

A. Hydrolysis of Ethylene Chlorohydrin and Ethylene Oxide

In this section we shall review the extensive work that has been done on the mechanism of the reversible reaction†

$$CH_2\text{—}CH_2 + OH^- \underset{k_2}{\overset{k_1}{\rightleftharpoons}} CH_2\text{—}CH_2 + Cl^- + H_2O \tag{1}$$
$$\underset{OH \quad Cl}{} \qquad\qquad \underset{O}{}$$

The formation of ethylene oxide from ethylene chlorohydrin is a reversible one and both rate constants, k_1 and k_2, have been measured in water at a series of temperatures, as has the equilibrium constant.[2] Since reaction 1 goes 98 per cent to completion to the right at 20°, the reverse reaction is usually carried out under acid conditions, as shown in (2).

$$CH_2\text{—}CH_2 + H^+ + Cl^- \underset{k_4}{\overset{k_3}{\rightleftharpoons}} CH_2\text{—}CH_2 \tag{2}$$
$$\underset{O}{} \qquad\qquad \underset{OH \quad Cl}{}$$

We might guess that the function of the acid is to neutralize hydroxide ion and so drive reaction 1 to completion to the left by a mass action effect. Actually, the mechanism is changed as well, that in acid solution being substantially different from that in neutral or basic media. Only the rate constant k_3 has been measured,[3] attempts to measure k_4 being hindered by

† For a discussion of this reaction and a number of similar ones involving ethylene oxides see reference 1.

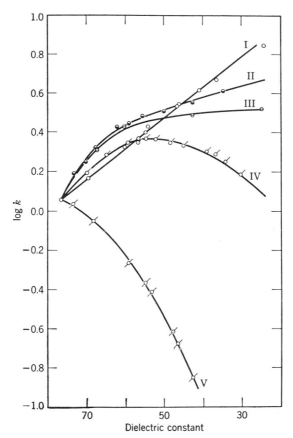

Fig. I. Rate of reaction of ethylene chlorohydrin with base in mixtures of solvents. (I) water-dioxane, (II) water-*i*-propanol, (III) water-*t*-butanol, (IV) water-ethanol, (V) water-methanol (Stevens, McCabe, and Warner).

the occurrence of other reactions. From thermodynamic and kinetic arguments, however, we may write

$$K_1 = \frac{[C_2H_4O][Cl^-]}{[C_2H_4OHCl][OH^-]} = \frac{k_1}{k_2} \tag{3}$$

$$K_2 = \frac{[C_2H_4OHCl]}{[C_2H_4O][H^+][Cl^-]} = \frac{k_3}{k_4} = \frac{1}{K_1 K_w} \tag{4}$$

where K_w is the ion product of water, so that the four rate constants are not independent.

We shall return later to reaction 2 in acid and describe first the experimental results pertinent to (1). The forward reaction has been studied by a number of workers.[4] The reaction is first-order in ethylene chlorohydrin and first-order in hydroxide ion when sodium or barium hydroxide is used as a base. When a buffer of carbonate-bicarbonate ion is used, the rate appears to depend only on the hydroxide ion concentration and not on the carbonate concentration. This is shown more clearly in the work of Bergkvist[5] on styrene β-bromohydrin and β-chlorohydrin, and the results may be assumed applicable to the simpler ethylene derivative. Hence we have an example of a specific hydroxide ion reaction rather than general base catalysis. Warner[4] and his co-workers have studied in detail the effect of changing solvent composition and ionic strength. There is no ionic strength effect, unless very concentrated salt solutions are used, as would be expected for a reaction between an ion and a neutral molecule. Changing the solvent from water to water-dioxane mixtures increases the rate somewhat as would be expected for such a reaction (see Chapter 7). Adding alcohol, however, introduces another factor since the rate first increases and then decreases with increasing alcohol content, as shown in Fig. 1. The decrease is greatest for methyl alcohol, less for ethyl, still less for isopropyl, and least for t-butyl alcohol. The nature of the products, however, is unchanged in water-alcohol mixtures.

Ethylene chlorohydrin is less reactive than ethylene bromohydrin or iodohydrin and more reactive than the fluorohydrin.[6] Also alkyl substitution on either carbon atom increases the rate of the reaction so that both α- and β-propylene chlorohydrin are more reactive than ethylene chlorohydrin.

The stereochemistry of the reaction has been worked out in great detail for substituted ethylene oxides. Thus the *erythro* derivatives of 3-bromo-2-butanol, chloromalic acid, and stilbene bromohydrin give the *trans* oxides (Winstein and Henderson[1]). The analogous *threo* compounds yield the *cis* oxides. When the bromobutanol is optically active, an active

$$\text{(5)}$$

$$\text{(6)}$$

trans oxide can be obtained from the *erythro* compound. The *cis* oxide being internally compensated is inactive even when prepared from an active *threo* halohydrin.

The nature of the ring closure is also indicated by the differences in behavior of the *cis* and *trans* halohydrins of cyclohexane. The *trans* chlorohydrin reacts 150 times faster with alkali than the *cis* chlorohydrin, the activation energy for the latter being 4 kcal greater.[7] Furthermore, only the *trans* compound gives an oxide; the *cis* chlorohydrin in this and similar cases produces carbonyl compounds with alkali.[8] In other words, for oxide formation to occur the hydroxyl group must be *trans* to the halogen atom, and the oxygen becomes attached to the carbon atom originally holding the halogen with an inversion of configuration.

All these facts seem consistent with a detailed mechanism proposed by Winstein and Lucas.[9] The assumed mechanism is a two-step process involving a prior equilibrium between hydroxide ion and the hydroxyl group of the halohydrin, followed by a rate-determining displacement of halide ion by negatively charged oxygen, as shown in (7) and (8). The

$$
\begin{array}{c}
\text{OH} \\
| \\
\text{CH}_2\text{---CH}_2 + \text{OH}^- \\
| \\
\text{Cl}
\end{array}
\rightleftharpoons
\begin{array}{c}
{}^-\text{O} \\
| \\
\text{CH}_2\text{---CH}_2 + \text{H}_2\text{O} \\
| \\
\text{Cl}
\end{array}
\qquad \text{fast} \qquad (7)
$$

$$
\begin{array}{c}
{}^-\text{O} \\
| \\
\text{CH}_2\text{---CH}_2 \\
| \\
\text{Cl}
\end{array}
\xrightarrow{k_5}
\begin{array}{c}
\text{O} \\
\diagup \quad \diagdown \\
\text{CH}_2\text{---CH}_2 + \text{Cl}^-
\end{array}
\qquad \text{slow} \qquad (8)
$$

displacement is an intramolecular example of the usual bimolecular reaction leading to inversion of configuration at the carbon holding the halogen. The rate equations are

$$
d[\text{Cl}^-]/dt = k_5[\text{CH}_2\text{ClCH}_2\text{O}^-] = k_5K_3[\text{CH}_2\text{ClCH}_2\text{OH}][\text{OH}^-] \qquad (9)
$$

where K_3 is the equilibrium constant for (7). The observed second-order rate constant k_1 is equal to k_5K_3 and depends upon the acidity of the halohydrin, K_3 being greater for a more acid hydroxyl, and upon the reactivity of the halide towards displacement. It may be noted that the rate constant k_5 depends upon the basicity of the oxide ion so that k_5 and K_3 may change with substituents in an opposing manner. The effect of alkyl substituents on the carbon atom holding the halogen is opposite to that found for bimolecular displacements which are usually slowed down by alkyl groups. Evidently the steric factors operating in a bimolecular displacement are of lesser importance in this intramolecular example, since the displacing group is already in a position close enough for reaction.

When the solvent is a mixture of water and another hydroxylic solvent, SH, account must be taken of the side reaction

$$SH + OH^- \rightleftharpoons S^- + H_2O \tag{10}$$

as well as of changes in the activity coefficients as compared to dilute aqueous solutions. This leads to the expression (McCabe and Warner.[6])

$$k_{exp} = \frac{k_1 f_{CH_2ClCH_2OH} f_{OH^-}}{[(a_{SH}K_{SH}/a_{H_2O}K_w) + 1]f^{\ddagger}} \tag{11}$$

where $a_{H_2O}K_w = a_{H^+}a_{OH^-}$ and $a_{SH}K_{SH} = a_{H^+}a_{S^-}$. The ratio of activity coefficients would be expected to increase as the polarity of the solvent decreases because of the greater spreading of charge in the activated complex. This is observed in mixtures of water and dioxane where K_{SH} is zero. However, for the alcohols, K_{SH} can be large enough to cause a net decrease in the value of k_{exp} compared to k_1 in pure water. The results shown in Fig. 1 are explicable on this basis if methanol and perhaps ethanol are stronger acids than water. It has been shown that methanol is a stronger acid than water both in water and in the solvent i-propanol.[10] However, in these solvents ethanol seems to be somewhat weaker than water. The relative acid strengths are definitely a function of the medium.[10b] Ethanol always seems weaker as an acid than methanol. The mechanism proposed in (7) and (8) accounts for the fact that oxide continues to be the product of the reaction even in alcoholic solvents.

We may now consider the reverse reaction between ethylene oxide and chloride ion. The existence of this reaction is easily demonstrated since a mixture of the two reactants will develop an appreciable alkalinity. If the mechanism of (7) and (8) is correct, the mechanism of the reverse reaction is known since from the principle of microscopic reversibility all that must be done is to reverse the sequence of events shown there. Hence the opening of the oxide ring proceeds through a rate-determining displacement of oxygen by chloride ion, followed by a rapid picking up of a proton from a water molecule. The displacement process must again go with inversion of configuration.

All of this is in agreement with experiment. The reverse reaction in neutral or alkaline solution is first-order in chloride ion and first-order in ethylene oxide (Porret;[2] Brönsted, Kilpatrick, and Kilpatrick[3]). Stereochemically the ring opening is clearly shown to be *trans* for substituted oxides. Cyclohexene oxide gives *trans* chlorohydrin and *trans* bromohydrin with hydrochloric and hydrobromic acids. *trans*-Epoxy-succinic acid gives the *erythro* halohydrin with acid and the *cis* epoxy acid gives the *threo* halohydrin. The *trans* oxide of stilbene gives the *erythro* and the *cis* oxide gives the *threo* halohydrin (Winstein and Henderson[1]). In each case the oxide is opened with acid to give the original halohydrin from which the oxide can be made. This is required by the reversal of the mechanism of formation. Actually, the results when acids are used are not strictly comparable since the mechanism is changed. However, the stereochemical results are inferred to be the same in neutral or basic solution as in acid. This is indicated by the behavior of oxides when they react with nucleophilic reagents other than halide ion under alkaline conditions. Again the addition takes place in a *trans* manner.[11] However, some differences in the stereochemistry in acid and basic media will be mentioned in a later section.

If the rate of disappearance of ethylene oxide in an acid aqueous solution containing chloride ion is studied, the kinetics are found to be complex. Four terms can be identified in the rate equation (Brönsted et al.[3]):

$$\frac{\text{rate}}{[C_2H_4O]} = k_0 + k_2[Cl^-] + k_0'[H^+] + k_2'[H^+][Cl^-] \qquad (12)$$

The first term is assumed to be due to a reaction of a water molecule opening the oxide ring. The second term is a similar reaction involving chloride ion and is the reverse reaction of equation 1. The next two terms are assumed to be due to reactions involving the oxonium complex of the oxide. If nucleophilic reagents other than water and chloride ion are present, analogous terms for the reaction of each of them will appear in the rate equation. There seem to be then two distinct and general reactions, shown in (13) and (14), where B is a nucleophilic reagent. As would be

$$\begin{array}{c} O \\ \diagup \; \diagdown \\ CH_2 \!\!-\!\!\!-\!\!CH_2 + B \rightarrow \overset{+}{B}\!\!-\!\!CH_2\!\!-\!\!CH_2\!\!-\!\!O^- \end{array} \qquad (13)$$

$$\begin{array}{c} H \\ O^+ \\ \diagup \; \diagdown \\ CH_2 \!\!-\!\!\!-\!\!CH_2 + B \rightarrow B^+\!\!-\!\!CH_2\!\!-\!\!CH_2OH \end{array} \qquad (14)$$

expected, the rate constant for the reaction of the oxonium complex is greater than that for the neutral oxide by a factor of 500 to 10,000. This is because a weaker base is being displaced from its attachment to carbon, ROH compared to RO$^-$. As the nucleophilic character of B increases both rate constants increase. Nucleophilic character rather than base strength is the important factor. Thus iodide ion, thiourea, and thiosulfate ion are very reactive, whereas hydroxide ion is only moderately reactive. The reaction of the hydroxide ion is detectable only in strong alkali.[12] However, hydroxide ion is not as unreactive as was previously thought, its rate constant being 100 times larger than that for chloride ion, for ethylene oxide itself, and for substituted ethylene oxides.[12,13] The uncatalyzed reaction with chloride ion also can be studied only by having a neutral or basic solution with a high concentration of chloride ion.

Replacement of one of the hydrogens of ethylene oxide by CH_3, CH_2OH, or CH_2Cl in each case increases the rate of oxide ring opening by chloride ion. In the case of an unsymmetrical oxide, ring opening can occur in one of two ways, as shown in (15). Under neutral or basic

$$\begin{array}{ccc} R' & O & R'' \\ \diagdown & \diagup \diagdown & \diagup \\ & C\text{----}C & \\ \diagup & & \diagdown \\ R & & R''' \end{array} \rightarrow \begin{array}{c} R' \\ \diagdown \\ R\text{---}C\text{---}C\text{---}R'' \\ \diagup \\ HO \end{array} \begin{array}{c} B \\ \diagup \\ \diagdown \\ R''' \end{array} \text{ or } \begin{array}{c} B \\ \diagdown \\ R'\text{---}C\text{---}C\text{---}R''' \\ \diagup \\ R \end{array} \begin{array}{c} R'' \\ \diagup \\ \diagdown \\ OH \end{array} \quad (15)$$

conditions the direction of ring opening is almost always such that the least-substituted carbon atom is attacked by the nucleophilic reagent. This appears to be a steric hindrance effect and is in accord with the mechanism's being a bimolecular displacement reaction. Such reactions are generally slowed down by saturated substituents on the carbon atom involved.[14] The effect cannot be a purely electrostatic one since methyl and hydroxy-methyl groups, for example, should have different polarities.

When B is the hydroxide ion it is not possible to tell by ordinary methods whether attack occurred at the least substituted carbon atom. However, Long and Pritchard[15] used H_2O^{18} for the hydrolysis of propylene oxide and butylene oxide under basic conditions. By mass spectrometric studies of the products it was possible to show that reaction was mainly at the primary carbon as expected.

The basic hydrolysis of ethylene chlorohydrin has also been studied[16] in D_2O. Relative rates of reactions in H_2O and D_2O can often throw light on the mechanisms of the reactions.[17] The analysis of the results has been well developed for the case of acid-base reactions. For example, in the case of reactions subject to specific hydrogen ion catalysis (Case I, p. 213), it is found that rates are generally faster in D_2O than in H_2O by factors ranging from 25 to 100 per cent.[17] In terms of a two-step mechanism

$$S + D^+ \rightleftharpoons SD^+ \qquad (16)$$

$$SD^+ + R \xrightarrow{\text{slow}} \text{products} \qquad (17)$$

the explanation is that the equilibrium constant for (16) is more favorable for formation of the reactive species SD^+ than the corresponding reaction to form SH^+ in light water. This is equivalent to saying that the ratio of acid ionization constants, K_{SH^+}/K_{SD^+}, is greater than unity. Here K_{SH^+} is the equilibrium constant of the acid measured in H_2O and K_{SD^+}, the same constant (for the reverse reaction of (16)) measured in D_2O.

Experimentally it is found that such ratios are in fact greater than unity.[18] The ratios generally increase as the strength of the acid, SH^+, decreases. Most of the data available are limited to acids of strengths fairly easy to measure and less is known about the isotope effect for very strong acids such as SH^+ would represent if it were the conjugate acid of an ester. However, the ratios are still found to be greater than unity for some acids stronger than H_3O^+.[18b]

In the case of a specific hydroxide ion catalyzed reaction (Case VI, p. 216) the rate law becomes the same as for the ethylene-chlorohydrin reaction.

$$\text{rate} = k_H K_{eq}[SH][OH^-] \qquad (18)$$

K_{eq} is the same as K_3 in equation 9 and may also be written as K_{SH}/K_{H_2O} in aqueous solutions, where K_{SH} is the acid ionization constant of the reacting substrate SH and K_{H_2O} is written for the ion product of light water. In heavy water three factors may be changed, the rate constant k_D, and the equilibrium constants K_{SD} and K_{D_2O}. The isotope effect which is observed depends on the changes in all three properties.

$$\frac{(k_{obs})D_2O}{(k_{obs})H_2O} = \frac{k_D K_{SD} K_{H_2O}}{k_H K_{SH} K_{D_2O}} \qquad (19)$$

The ratio of ion products K_{H_2O}/K_{D_2O} has been measured[19] and has the unusually large value of 6.5. Again, if k_D and k_H are very close, it is likely that the isotope effect in the ionization constant of K_{SH} will be less than 6.5 and the rate will be greater in heavy water than in light water. This is what is observed in the case of ethylene chlorohydrin,[16] the ratio being 1.54. In this system it is also possible to measure K_{SH} and K_{DH}. The chlorohydrin has an ionization constant of 4.9×10^{-15} in H_2O and 1.0×10^{-15} in D_2O.[16a] The ratio is 4.9 which also enables the ratio of k_D/k_H to be calculated as 1.15.

It may be noted that, since the isotope effect on K_{SH}/K_{SD} is a function of the strength of K_{SH}, it is not always necessary that the reaction rate be faster in D_2O than in H_2O even if the near equality of k_D and k_H holds. If SH is a sufficiently weak acid (pK_a of the order 18 or greater) it may well be that a greater isotope effect than the 6.5 for water will be found. No clear example of this is known as yet, but indirect evidence exists.[20]

If the rate-determining step in an acid-base reaction involves a proton transfer, it will be found to be slower in D_2O than in H_2O, if the proton transfer becomes a deuteron transfer. This will be discussed in more detail in a later section. If the proton is still the atom reacting, even though the solvent is D_2O, the rate may be faster in D_2O, for example,

$$CH_3NO_2 + OD^- \rightarrow CH_2NO_2^- + HOD \qquad (20)$$

is some 40 per cent faster in D_2O than in H_2O.[21]

For the reaction between ethylene chlorohydrin and hydroxide ion, the studies of Porret[2] on the forward rate, the reverse rate, and the equilibrium, are of interest. In water at 20° a solution originally one molar in sodium chloride and ethylene oxide reaches the following equilibrium concentrations:

$$C_2H_4OHCl + OH^- \rightleftharpoons C_2H_4O + Cl^- + H_2O \qquad (21)$$
$$\text{0.0175}M \quad \text{0.0175}M \quad \text{0.983}M \quad \text{0.983}M$$

This corresponds to a value of K_1 of $(0.983)^2/(0.0175)^2 = 3.15 \times 10^3$. The rate constant k_1 was found to be 5.5×10^{-3} liter/mole-sec, and k_2 was determined experimentally as 1.8×10^{-6}. Thus the kinetic equilibrium constant equal to k_1/k_2 is 3.06×10^3, in excellent agreement with the value found by analysis.

All constants were determined over a range of temperature from 0 to 50°. The dependence of the rate constants on temperature could be expressed by the equations $k_1 = 10^{14.5}e^{-22,600/RT}$ and $k_2 = 10^{8.2}e^{-18,900/RT}$ (McCabe and Warner[6] give a somewhat more reliable value of the activation energy for the forward reaction, their expression being $k_1 = 10^{15.0}e^{-23,300/RT}$.) From the kinetic data we can find ΔH^0 for the overall reaction as 4.0 ± 0.4 kcal by using the relationship $\Delta H^0 = E_{\text{forward}} - E_{\text{reverse}}$. Also ΔS^0 may be calculated from the difference in the entropies of activation as 30 ± 1 E.U.

An attempt to get ΔH^0 and ΔS^0 from a plot of log K versus $1/T$ is not very successful because the plot is not linear but curved. From the slope values of ΔH^0 between three and four kilocalories can be found. The nonlinearity is not due to the complexity of the mechanism as Porret assumed but is a consequence of the low value of ΔH^0. This makes the variation of the equilibrium constant, with temperature small and difficult to measure. Thermodynamically the small value of ΔH^0 is due to two compensating effects. The strained nature of the carbon-oxygen bonds in ethylene oxide makes that compound energetically unstable. On the other hand, the carbon-oxygen bond is normally more stable than the carbon-chlorine bond, particularly in the type of replacement involved here, viz.:

$$RO^- + R'Cl \rightarrow ROR' + Cl^- \qquad (22)$$

The positive value of ΔS^0 is due chiefly to the increase in the number of particles, its value being roughly equal to the entropy of the mole of water produced in the reaction.

Kinetically the positive entropy change is due to the difference in the probability factors of the forward and reverse reactions, 10^{15} compared to 10^{10}. It must be remembered that in the mechanism of equations 7 and 8, the measured rate constant k_1 is complex, including an equilibrium constant. However, for equilibria of the type involved

$$OH^- + HA \rightleftharpoons H_2O + A^- \qquad (23)$$

where HA is a weak acid of the approximate strength of water, ΔH^0 and ΔS^0 may be expected to be small. Therefore, the large value (10^{15}) of the pre-exponential term in k_1 is probably due to kinetic effects. Both the forward and reverse reactions involve a nucleophilic displacement on carbon. The reverse reaction is bimolecular, whereas the forward reaction is unimolecular, equation 8. It will be remembered that the frequency factors for bimolecular reactions are ordinarily about 10^{11} (in seconds), whereas for unimolecular reactions the factor is usually about 10^{13}. It may be concluded that for a reaction of a given type, a unimolecular process is favored over a bimolecular process (at unit concentration) by a probability factor of 100 or so. The activation energy for the unimolecular process may, of course, be prohibitively high in a given case, so that the bimolecular process is favored. In the present instance the unimolecular (intramolecular) displacement of chloride ion by alkoxide ion is much faster than the bimolecular replacement of chloride ion by external hydroxide ion. It is of interest to quote the data of Grant and Hinshelwood[22] on the alkaline hydrolysis of ethyl chloride in ethanol. Here, for a bimolecular displacement of chloride ion by alkoxide, the rate constant is given by $k = 10^{11}e^{-23,000/RT}$ liters mole-sec.

Several other examples of how unimolecular reactions are favored over bimolecular reactions may be mentioned. Freundlich and Salomon[23] and Salomon[24] have studied reactions of the following type:

$$X(CH_2)_nNH_2 \rightarrow \overset{+}{\underline{(CH_2)_nNH_2}} + X^- \qquad (24)$$

where a cyclic ammonium ion is formed from a haloamine. For $n = 2$, 3, 4, or 5 the values of the frequency factors are quite large, being 10^{12} to 10^{15} sec^{-1} in water, the smaller values of n having the larger frequency factors. In other solvents the frequency factors are also conspicuously larger than for a similar bimolecular reaction such as

$$C_2H_5NH_2 + C_2H_5Br \rightarrow (C_2H_5)_2NH^+ + Br^- \qquad (25)$$

run in the same solvent. Reactions such as (25) are generally characterized by very low frequency factors (see Chapter 7). In the lactonization of hydroxybutyric acid catalyzed by hydrogen ion,

$$HO-CH_2-CH_2-CH_2-COOH \overset{H^+}{\longrightarrow} \underline{CH_2-CH_2-CH_2-C}{=}O \qquad (26)$$
$$\underline{O}$$

the frequency factor, 10^9, in water[25] is about one thousandfold larger than that observed for the acid-catalyzed esterification of separate alcohol and acid molecules.[23] The greater ease of an intramolecular displacement compared to an intermolecular displacement has been commented on by Winstein and Grunwald.[27] This is, in fact, the basis of the well-known "neighboring group effect" in which solvolytic reactions at a saturated carbon atom can be greatly accelerated by neighboring atoms containing unshared pairs of electrons, or even by groups such as methyl if migration can occur. The names anchimeric[28] and synartetic[29] assistance have been coined for such phenomena.

The basic hydrolysis of ethylene chlorohydrin is the prototype for such neighboring group effects. Another important group of examples is to be found in the intramolecular catalysis of hydrolytic reactions of esters, amides, and related compounds.[30] A case of this class of reactions is the hydrolysis of phthalamic acid for which evidence for phthalic anhydride has been demonstrated as an intermediate by Bender, Chow, and Chloupek.[31]

$$\text{(27)}$$

$$\text{(28)}$$

The symmetrical anhydride intermediate has been proven by an ingenious double isotope labeling technique in which C^{13} is placed in the amide group and H_2O^{18} is used as a solvent. The presence in the product acid of equal amounts of adjacent and nonadjacent C^{13}—O^{18} pairs is explained by the mechanism.

In a similar case the internal carboxylate group of p-nitrophenyl glutarate catalyzes the hydrolysis of the p-nitrophenyl group.[32] The efficiency is shown by the fact that it would require something like 600 M acetate ion externally to produce the same acceleration in rate for p-nitrophenyl acetate. It is often found that the intramolecular process is favored because of large differences in the entropy of activation. The advantage of an intramolecular process can always be increased over an intermolecular one by going to greater dilution of the reactants.

Actually, of course, an intramolecular reaction will not always be favored over an analogous bimolecular process. The determining factors are whether the two reactive centers in the molecule can be brought close enough together without a large expenditure of energy, and whether they will frequently come close enough together. It might be anticipated that if a cyclic intermediate can be formed, other things being equal, the probability of the intermediate being formed will vary inversely with the number of atoms in the cycle. Thus a three-membered ring will be more probable than a four-membered ring, which in turn is more probable than a five-membered ring. However, from bond-angle consideration it is

Table I

ALKALINE HYDROLYSIS OF BROMOACIDS IN $0.25M$ SODIUM BICARBONATE
(Rates of lactone formation at 25°)
(Lane and Heine)

Acid	$\Delta H^{\ddagger}$, kcal	$\Delta S^{\ddagger}$, E.U.	k, min^{-1}
α-Bromopropionic	29.7	11	2.5×10^{-5}
β-Bromopropionic	28.7	13	2.1×10^{-4}
α-Bromocaproic	29.2	10	3.3×10^{-5}
β-Bromocaproic	27.1	12	2.1×10^{-3}
γ-Bromovaleric	22.2	11	0.33

known that three- and four-membered rings are energetically less stable than rings having a greater number of atoms. Hence a higher activation energy would be required to form the smaller rings. The observed rates would then depend on a combination of these two factors.[33] A maximum in the rate is generally found for five-membered ring intermediates. The results of Lane and Heine[34] are shown in Table 1 for the alkaline hydrolysis of some aliphatic bromoacids. The rates of formation of lactone were determined for three- four- and five-membered rings

$$Br(CH_2)_nCOO^- \rightarrow \begin{array}{c} (CH_2)_nC{=}O + Br^- \\ \hline \quad O \quad \end{array} \qquad (29)$$

together with the heats and entropies of activation. The high activation energies for the α and β haloacids are related to the strain of the three- and four-membered rings. The γ acid which gives a five-membered ring has a much lower activation energy since the ring is strainless. At the same time there is a decided decrease in the probability factor. The net result is that the rate of formation of the five-membered ring is greatest. If these studies had been continued with longer chain acids, there is no doubt that the rate would start to fall off again and approach very low values as the number of atoms in the chain increased. This is more clearly shown in the work of Salomon[33] on the haloamines. The falling off in rate is not because of a high activation energy (since strainless many-membered rings can be formed), but because of a decrease in the probability factor. The increased probability of unimolecular reactions over bimolecular reactions is valid only for small-ring intermediates (3 to 6 members). In these systems the high probability may be related to a high effective concentration of one reactive group in the neighborhood of the second reactive group.

We shall now turn to a consideration of reaction 2, the opening of ethylene oxide under acid conditions, as shown in (30). Much less work

$$CH_2\!\!-\!\!CH_2 + H^+ + Cl^- \underset{k_4}{\overset{k_3}{\rightleftharpoons}} CH_2OHCH_2Cl \qquad (30)$$
$$\diagdown O \diagup$$

of a quantitative nature has been done on this reaction; however, several points of interest may be discussed.

The value of the equilibrium constant K_2 is calculable from the ion product of water and K_1 at $20°$ as 3.3×10^{10}, so that the reaction goes to completion at reasonable $[H^+]$. The value of ΔH_2^0 and ΔS_2^0 may also be found from the data on K_1 and K_w as -18 kcal and -4 E.U., respectively. The change in entropy is not large in spite of the fact that three particles become one in the reaction. This is because of the considerable entropy decrease associated with the solvation of the ions, each of which effectively "freezes" a number of water molecules around it. The negative heat-content change is due to the breaking of the energy-rich three-membered ring, other energy effects approximately canceling.

Kinetically only k_3 has been studied and that only at a single temperature, $20°$ C. Its value is 2.1×10^{-2} liters^2/mole^2-sec (Brönsted et al.,[3] k_3 is the same as k_2' in equation 12). From the equilibrium constant it may then be calculated that k_4 is 6.4×10^{-13} sec^{-1}. Cowan, McCabe, and Warner[35] have studied the neutral and acid hydrolysis of ethylene chlorohydrin and the other ethylene halohydrins. The product is not oxide but ethylene glycol in every case. The reactions were first-order and independent of the hydrogen ion concentration. A possible mechanism might be hydrolysis to the oxide

$$CH_2OHCH_2Cl \underset{k_3}{\overset{k_4}{\rightleftharpoons}} CH_2\!\!-\!\!CH_2 + H^+ + Cl^- \qquad (31)$$
$$\diagdown O \diagup$$

followed by ring opening of the oxide by water and hydrogen ion

$$H^+ + CH_2{-}CH_2 + H_2O \xrightarrow{k_6} CH_2OHCH_2OH + H^+ \qquad (32)$$
$$\diagdown O \diagup$$

The rate of the second step has been measured by Brönsted and the Kilpatricks. The value of k_6, which is the same as k_0' in equation 12 is 5.3×10^{-3} at 20°. In solutions dilute in chloride ion, reaction 32 will be faster than the reversal of (31). Hence, in the hydrolysis of the chlorohydrin by this mechanism, formation of the oxide will be the rate-determining step and the measured first-order rate constant will be equal to k_4. Data on the hydrolysis of the chlorohydrin was obtained only at high temperatures, the rate constant at 65° being 1.36×10^{-7} sec^{-1}. From the activation energy, 26.2 kcal, a good estimate of the rate constant at 20° can be made as 6×10^{-10}. Since this is a thousand times greater than the value of k_4 calculated from the equilibrium constant, it may be safely stated that the oxide is not an important intermediate in this hydrolysis. Cowan, McCabe, and Warner correctly believed that they were studying the direct reaction of water with the chlorohydrin,

$$H_2O + CH_2OHCH_2Cl \rightarrow CH_2OHCH_2OH + H^+ + Cl^- \qquad (26)$$

though their evidence was based on a different argument from the one given above.

For at least one other halohydrin there is stereochemical evidence also that the neutral or slightly acid hydrolysis proceeds by a direct displacement of halogen by a water molecule. Kuhn and Ebel[36] in a study of the two diastereoisomeric chloromalic acids found that I, which when treated with sodium hydroxide gave pure *trans* oxide, was hydrolyzed by water alone to racemic tartaric acid, as shown in (33). The other isomer II gave

(33)

pure *cis* oxide with alkali and *meso* tartaric acid in water, as shown in (34).

$$
\begin{array}{c}
\text{COOH} \\
|\\
\text{HC}-\text{OH} \quad meso \\
|\\
\text{COOH} \qquad\qquad\qquad\qquad \text{HC}-\text{OH} \\
|\qquad\qquad\; \text{H}_2\text{O} \qquad\nearrow\qquad |\\
\text{HC}-\text{Cl} \qquad\qquad 100\% \qquad \text{COOH} \\
|\qquad\qquad\qquad\qquad\qquad\qquad\qquad\qquad\qquad\qquad\qquad\qquad (34)\\
\text{HO}-\text{CH} \qquad\qquad\qquad\qquad\qquad \text{H}\quad\text{COOH}\\
|\qquad\qquad\qquad \text{NaOH} \qquad\searrow\quad |\\
\text{COOH} \quad 100\% \qquad\qquad\qquad\qquad \text{C}\\
\text{II} \qquad\qquad\qquad\qquad\qquad\qquad\qquad\qquad\quad \diagup\;\diagdown\\
\qquad\qquad\qquad\qquad\qquad\qquad\qquad\qquad\qquad \text{O}\quad cis\\
\qquad\qquad\qquad\qquad\qquad\qquad\qquad\qquad \diagup\\
\qquad\qquad\qquad\qquad\qquad\qquad\qquad\qquad \text{C}\\
\qquad\qquad\qquad\qquad\qquad\qquad\qquad \diagup\;\diagdown\\
\qquad\qquad\qquad\qquad\qquad\qquad\text{H}\qquad\text{COOH}
\end{array}
$$

Since the *cis* oxide can be hydrolyzed by water to give pure racemic tartaric acid and the *trans* oxide on hydrolysis gives predominantly *meso* acid, it is evident that the oxide cannot be a major intermediate in the neutral hydrolysis of these chlorohydrins. Since the formation of glycol *via* the oxide involves two inversions of configuration, a glycol of the same relative configuration as the starting chlorohydrin would be obtained. The observed results are consistent with one inversion of configuration in the hydrolysis of the chloromalic acids, so that the normal backside displacement of chloride ion by a water molecule must have occurred.

For more reactive halides it may be possible that oxides are formed as intermediates.[37] This would be most likely for halohydrins with substituents on the carbon atom bearing the halogen so that the normal bimolecular displacement is hindered. Also alkyl substituents on the other carbon atom seem to have a stabilizing influence on the ethylene oxide ring. Alkyl substituents on either carbon will increase the rate of epoxide formation. Thus an increase in rate of five- to twentyfold for each methyl group has been observed in going from ethylene to tetramethyl-ethylene chlorohydrin.[38]

Small alkyl groups in general increase the stability of cyclic compounds over corresponding open-chain systems.[39] The increased stability may show up in equilibrium constants, rates of ring closing, or rates of ring opening. A number of factors may be important in various cases. Two important general ones seem to operate.[40] One is that in alkyl-substituted open-chain compounds, there is an increased barrier to internal rotations compared to the unsubstituted case. In forming the ring the entropy of these internal rotations is lost. Less entropy will then be lost by the alkyl compound since it had less to lose in the first place. The second factor is that there is a relative decrease in the number of repulsive interactions of groups on neighboring carbon atoms in going from the open-chain to the cyclic compound for alkyl-substituted systems.

The whole subject of the effect of adding acid or base to the equilibrium between ethylene oxide and ethylene chlorohydrin may now be summarized. Thermodynamically ethylene chlorohydrin is more stable than ethylene oxide and aqueous hydrochloric acid. Adding base to neutralize the acid will, however, drive the reaction to completion, forming the oxide. Kinetically the effect of base is to change the mechanism of opening and closing the oxide ring which results in a more favorable ratio of rate constants. In the presence of acid the opening of the oxide ring is a third-order process involving the steps shown in (35) and (36).

$$CH_2\!\!-\!\!CH_2 + H^+ \rightleftharpoons CH_2\!\!-\!\!CH_2 \quad\quad \text{fast} \quad\quad (35)$$
$$\underset{O}{\diagdown\diagup} \quad\quad\quad \underset{\underset{H^+}{O}}{\diagdown\diagup}$$

$$CH_2\!\!-\!\!CH_2 + Cl^- \rightarrow CH_2OHCH_2Cl \quad\quad \text{slow} \quad\quad (36)$$
$$\underset{\underset{H^+}{O}}{\diagdown\diagup}$$

This sequence is faster by 10^4 than the mechanism of direct attack by chloride ion on the neutral oxide. The closing of the oxide ring in the absence of base would be by a reversal of (36) and (35). This, however, is a slow process involving such a high activation energy that other reactions occur (hydrolysis to glycol). In alkali the ring closure is accelerated by a factor of 10^{10} by making the displacing agent the negatively charged alkoxide ion, as shown in (37). This increase in rate is due to a lowering

$$CH_2\!\!-\!\!CH_2Cl \rightarrow CH_2\!\!-\!\!CH_2 + Cl^- \quad\quad (37)$$
$$\underset{O\_}{|} \quad\quad\quad \underset{O}{\diagdown\diagup}$$

of the activation energy, which in turn shows up in a change in ΔH^0 from a large positive quantity to a small negative one.

The labilizing influence of oxonium complex formation is usually utilized in opening oxide rings unless very basic reagents are involved. Under acid conditions there is some evidence that the displacement mechanism may not be working. For unsymmetrical oxides the direction of ring opening is frequently different in acid and in base.[41] The changes observed can be rationalized in terms of a carbonium ion mechanism in

acid, the direction of ring opening being such that the most stable carbonium ion is formed, as shown in (38). For example, styrene oxide is

$$\underset{\Large\diagdown}{\overset{\Large H}{\underset{}{\overset{O^+}{|}}}}\text{C}\text{---}\text{C} \longrightarrow \overset{H}{\underset{}{\overset{O}{|}}}\text{C-C}^+ \xrightarrow{\text{BH}} \overset{H}{\underset{\underset{B}{|}}{\overset{O}{|}}}\text{C-C} + \text{H}^+ \qquad (38)$$

opened at the β position under basic or neutral conditions, and at the α
position in acid, as shown in (39) and (40). The most stable carbonium ion

$$\bigcirc\!\!\!-\overset{H}{\underset{\diagdown\;\;\diagup}{\underset{O}{C}}}\!\!\!-\text{CH}_2 + \text{BH} \xrightarrow{\text{basic}} \bigcirc\!\!\!-\overset{H}{\underset{\underset{H}{\overset{O}{|}}}{\overset{|}{C}}}\!\!\!-\text{CH}_2\text{B} \qquad (39)$$

$$\bigcirc\!\!\!-\overset{H}{\underset{\diagdown\;\;\diagup}{\underset{O}{C}}}\!\!\!-\text{CH}_2 + \text{BH} \xrightarrow{\text{acid}} \bigcirc\!\!\!-\overset{H}{\underset{\underset{B}{|}}{\overset{|}{C}}}\!\!\!-\text{CH}_2\text{OH} \qquad (40)$$

would be the one in which the positive charge is on the α carbon where it
can resonate with the ring.

Actually the lack of rearrangement and racemization observed under
acid conditions for the opening of those oxides where such changes could
occur is evidence against a carbonium ion mechanism (Winstein and
Henderson[1]). The effect of oxonium complex formation in acid seems
rather to be a greater weakening of one bond than of the other, so that
reaction occurs by breaking the most-weakened carbon-oxygen bond.

Kinetic studies have been made[42] for the acid hydrolysis of ten simple
epoxides in aqueous solutions of perchloric acid at $0°C$. Table 2 shows the
results for the pseudo-first-order constant at a pH of 0.12. In seven of the
cases the studies were extended to concentrations of acid above one molar.

There are two points of interest from these studies. There is obviously
an increase in rate when ethylene oxide is substituted by electron-donating
groups such as methyl, and a decrease in rate for electron-attracting groups
such as chloromethyl. This result is perhaps expected on the basis of the
mechanism which requires an addition of a proton prior to reaction. The
rates are well predicted by the use of the Taft σ^* values (Chapter 9, p. 228),
which are simple inductive constants, without any corrections for steric
factors. This lack of a steric factor is surprising if the reaction involves
nucleophilic attack at the most substituted carbon atom. It is reasonable
if the carbonium ion mechanism is operating, however, in which case the

effect of the alkyl groups is to stabilize the carbonium ion and increase the rate as observed.[42]

Also the studies in concentrated acid reveal a dependence of the rate on a function of the acidity more complex than the concentration of the hydrogen ion. This H_0 function will be described in detail in the section on hydrolysis of lactones, but for the moment it can be said that the observed dependence is also strong evidence for the carbonium ion mechanism of reaction 38. The stereochemical results mentioned above

Table 2

HYDROLYSIS RATES OF EPOXIDES IN AQUEOUS SOLUTIONS OF
PERCHLORIC ACID AT 0° C

(Long and Pritchard)

Epoxide	Concentration Range of $HClO_4$, moles/liter	$10^5 k_1$, sec^{-1} at pH = 0.12
Epibromohydrin	0.2–3.5	3.8
Epichlorohydrin	1.0–3.3	4.8
β-Methylepichlorohydrin	0.4–2.6	11.5
Glycidol	0.6–3.0	26.6
β-Methylglycidol	0.1–1.6	75
Ethylene oxide	0.5–2.2	63
Propylene oxide	0.8–2.5	350
trans-2,3-Epoxybutane	0.01–0.5	770^a
cis-2,3-Epoxybutane	3×10^{-3}–0.1	1,480^a
Isobutylene oxide	5×10^{-5}–2×10^{-3}	33,000^a

^{a} Extrapolated value.

include not only no racemization in the case of optically active oxides, but also complete inversion of configuration. This is explained by Pritchard and Long[42] as a consequence of "shielding" of one side of the carbonium ion by the liberated hydroxyl group.

REFERENCES

1. R. C. Elderfield, *Heterocyclic Compounds*, Vol. I, John Wiley & Sons, New York, 1950, Chapter on Ethylene and Trimethylene Oxides by S. Winstein and R. B. Henderson; R. E. Parker and N. S. Isaacs, *Chem. Revs.*, *59*, 737 (1959).
2. D. Porret, *Helv. Chim. Acta*, *24*, 80E (1941); *27*, 1321 (1944).
3. J. N. Brönsted, M. Kilpatrick, and M. Kilpatrick, *J. Am. Chem. Soc.*, *51*, 428 (1929).

4. W. P. Evans, *Z. physik. Chem.*, 7, 337 (1891); L. Smith, *ibid.*, *81*, 339 (1912); L. O. Winstrom and J. C. Warner, *J. Am. Chem. Soc.*, *61*, 1205 (1941); J. E. Stevens, C. L. McCabe, and J. C. Warner, *ibid.*, *70*, 2449 (1947).

5. T. Bergkvist, *Svensk Kem. Tid.*, *59*, 194, 244 (1947).

6. C. L. McCabe and J. C. Warner, *J. Am. Chem. Soc.*, *70*, 4031 (1948).

7. T. Bergkvist, *Svensk Kem. Tid.*, *59*, 215 (1947).

8. P. D. Bartlett and R. H. Rosenwald, *J. Am. Chem. Soc.*, *56*, 1990 (1934); C. M. Suter and G. A. Lutz, *ibid.*, *60*, 1360 (1938).

9. S. Winstein and H. J. Lucas, *J. Am. Chem. Soc.*, *61*, 1576 (1939).

10. (*a*) J. Hine and M. Hine, *J. Am. Chem. Soc.*, *74*, 5267 (1952); (*b*) A. Unmack, *Z. physik. Chem.*, *133*, 45 (1927); (*c*) P. Ballinger and F. A. Long, *J. Am. Chem. Soc.*, *82*, 795 (1960).

11. W. E. Grigsby, J. H. Hind, J. Chanley, and F. H. Westheimer, *J. Am. Chem. Soc.*, *64*, 2606 (1942).

12. G. H. Twigg and H. J. Lichtenstein, *Trans. Faraday Soc.*, *44*, 905 (1948).

13. W. C. J. Ross, *J. Chem. Soc.*, 2257 (1950).

14. J. L. Gleave, E. D. Hughes, and C. K. Ingold, *J. Chem. Soc.*, 236 (1935).

15. F. A. Long and J. G. Pritchard, *J. Am. Chem. Soc.*, *78*, 2663 (1958).

16. (*a*) P. Ballinger and F. A. Long, *J. Am. Chem. Soc.*, *81*, 2347 (1959); (*b*) C. G. Swain, A. D. Ketley, and R. W. F. Bader, *ibid.*, 2353.

17. (*a*) For a review of deuterium isotope effects see K. Wiberg, *Chem. Revs.*, *55*, 713 (1955); (*b*) R. P. Bell, *Acid-Base Catalysis*, Oxford University Press, London, 1941, p. 145; (*c*) E. L. Purlee, *J. Am. Chem. Soc.*, *81*, 263 (1959).

18. (*a*) C. K. Rule and V. K. LaMer, *J. Am. Chem. Soc.*, *60*, 1981 (1938); (*b*) E. Högfeldt and J. Bigeleisen, *ibid.*, *82*, 15 (1960).

19. V. K. LaMer and R. W. Kingerley, *J. Am. Chem. Soc.*, *63*, 3256 (1941).

20. R. G. Pearson, N. C. Stellwagen, and F. Basolo, *82*, 1077 (1960).

21. V. K. LaMer and S. H. Maron, *J. Am. Chem. Soc.*, *60*, 2588 (1938).

22. G. H. Grant and C. N. Hinshelwood, *J. Chem. Soc.*, 258 (1933).

23. H. Freundlich and G. Salomon, *Z. physik. Chem.*, *166*, 161 (1933).

24. G. Salomon, *Helv. Chim. Acta*, *16*, 1361 (1933).

25. H. S. Taylor and H. W. Close, *J. Phys. Chem.*, *29*, 1085 (1925); A. Kailan, *Z. physik. Chem.*, *101*, 63 (1922).

26. C. N. Hinshelwood and A. R. Legard, *J. Chem. Soc.*, 587 (1935).

27. S. Winstein and E. Grunwald, *J. Am. Chem. Soc.*, *70*, 828 (1948).

28. S. Winstein, C. R. Lindegren, H. Marshall, and L. L. Ingraham, *J. Am. Chem. Soc.*, *75*, 147 (1953).

29. C. K. Ingold, *Structure and Mechanism in Organic Chemistry*, Cornell University Press, Ithaca, 1953, p. 511.

30. For a review see M. L. Bender, *Chem. Revs.*, *60*, 53 (1960), and reference 32.

31. M. L. Bender, Y. L. Chow, and F. Chloupek, *J. Am. Chem. Soc.*, *80*, 5380 (1958).

32. M. L. Bender and M. C. Neveu, *J. Am. Chem. Soc.*, *80*, 5388 (1958); T. C. Bruice, *ibid.*, *81*, 5444 (1959); E. Gaetjens and H. Morawetz, *ibid.*, *82*, 5328 (1960).

33. G. Salomon, *Trans. Faraday Soc.*, *32*, 153 (1936).

34. J. F. Lane and H. W. Heine, *J. Am. Chem. Soc.*, *73*, 1348 (1951).

35. H. D. Cowan, C. L. McCabe, and J. C. Warner, *J. Am. Chem. Soc.*, *72*, 1194 (1950).

36. R. Kuhn and F. Ebel, *Ber.*, *58*, 919 (1925).

37. S. Winstein and R. E. Buckles, *J. Am. Chem. Soc.*, *64*, 2780 (1942); S. Winstein and E. Grunwald, *ibid.*, *70*, 828 (1948).

38. H. Nilsson and L. Smith, *Z. physik. Chemie*, *166A*, 136 (1933).

39. References and a discussion are given by (a) F. G. Bordwell, C. E. Osborne, and R. D. Chapman, *J. Am. Chem. Soc.*, *81*, 2698 (1959); (b) G. S. Hammond, *Steric Effects in Organic Chemistry*, M. S. Newman, editor, John Wiley and Sons, New York, 1956, pp. 460–470.
40. Reference 39b and N. L. Allinger and V. Zalkow, *J. Org. Chem.*, *25*, 701 (1960).
41. R. G. Kadesch, *J. Am. Chem. Soc.*, *68*, 41 (1946).
42. J. G. Pritchard and F. A. Long, *J. Am. Chem. Soc.*, *78*, 2667 (1956).

B. The Ammonium Cyanate-Urea Conversion

The conversion of ammonium cyanate into urea in aqueous or alcoholic solutions has been extensively studied by kinetic methods:

$$NH_4^+ + CNO^- = (NH_2)_2CO \tag{1}$$

The reaction is reversible but the equilibrium constant is about 10^4 so that the reverse reaction may usually be neglected. Most of the more recent work on this reaction was done on the assumption that the reactants are ammonium ion and cyanate ion, and the object was to test various theories as to the effect of changing ionic strength and dielectric constant. Since it will be shown that the mechanism probably involves reaction between neutral ammonia and cyanic acid, this might seem to be misspent labor. Actually it turns out that the data are still useful in checking the theories but not quite in the sense that was originally intended.

The first kinetic study of the reaction was made by Walker and Hambly[1] who found the reaction to be second-order and the rate expression to be given by:

$$\text{rate} = k[NH_4^+][CNO^-] \tag{2}$$

The rate was increased if potassium cyanate or ammonium sulfate was added to an ammonium cyanate solution, but the addition of ammonia had very little effect on the rate. From this it was concluded that the reaction was between the ions and not between cyanic acid and ammonia. F. D. Chattaway,[2] among others, pointed out that because of the mobile equilibrium between the possible pairs of reactants

$$NH_4^+ + CNO^- \rightleftharpoons NH_3 + HCNO \tag{3}$$

the two mechanisms were kinetically indistinguishable, the experimental rate equation being also interpretable as

$$\text{rate} = k'[NH_3][HCNO] = (k'K_w/K_aK_b)[NH_4^+][CNO^-] \tag{4}$$

The ratio of ionization constants, K_w/K_aK_b, is the equilibrium constant for (3). Since K_a is 2×10^{-4}, K_w is 10^{-14}, and K_b is 2×10^{-5}, this constant is

2.5×10^{-6} at 25° so that the concentrations of ammonia and cyanic acid are very small unless one or the other is added to a reacting solution. Addition of ammonia, however, will not increase the rate since the concentration of cyanic acid will be reduced a proportionate amount.

Walker and Appleyard studied the analogous reaction for several alkyl ammonium cyanates,[3] as shown in (5). The relative rates at 60° in

$$R_2NH_2{}^+ + CNO^- = R_2N - \overset{\overset{\displaystyle O}{\|}}{C} - NH_2 \tag{5}$$

water for ammonium, methyl ammonium, dimethyl ammonium, ethyl ammonium, and diethyl ammonium ion were found to be 1, 1, 1.8, 0.6, and 0.6 so that all the rate constants are the same within a factor of 3. The ammonium ion, methyl ammonium ion, and ethyl ammonium ion have also been compared in 98 per cent ethanol,[4] the relative rates being 1, 12, and 7 at 10°, so that the ammonia derivative is relatively more slow in alcohol by a factor of 10. All these reactions were much faster in alcohol than in water, and the ammonium ion reaction is also faster in a number of mixed solvents than in water alone.

It had been observed by early workers that the second-order rate constant for the ammonium cyanate reaction tended to decrease with increasing concentration. Warner and Stitt[5] correctly attributed this to a salt effect and made a detailed study of the effect of ionic strength on the rate constant. A substantial negative salt effect was found, and in particular if $\log k$ were plotted against the square root of the ionic strength a straight line was approached at high dilution. The slope of this line agreed with the theoretical value -2α predicted by the Brönsted-Bjerrum-Christiansen equation for a reaction between singly charged ions of opposite sign (Chapter 7, equation 52). Warner and Stitt concluded that this was proof for a mechanism involving the two ions as reactants, since the alternative mechanism involving the neutral molecules would be expected to show a negligible salt effect.

On this basis several other studies were made on the rate of the ammonium cyanate reaction in a number of mixed solvents including water-dioxane and water with various alcohols.[6] The data were used to check the effect of ionic strength against the theoretical predictions of the Brönsted-Bjerrum-Christiansen equation. Also the effect of the dielectric constant on the rate constant as given in the equation (Chapter 7, equation 42)

$$\log k_0 = \log k_0' - \frac{N Z_A Z_B e^2}{2 \cdot 3 D R T r_{\pm}} \tag{6}$$

was determined by plotting $\log k$ versus $1/D$ for a given set of mixtures. The results were always good enough to generally confirm the theories. Reasonable values of $r_{\pm}$ were found.

Another point that was studied had to do with the difference in activation energy in a solvent of fixed composition and one of fixed dielectric constant. If it is postulated that the rate constant is a function of only the temperature, dielectric constant, and ionic strength, it is possible to write:

$$\frac{d \log k_0}{dT} = \left(\frac{\partial \log k_0}{\partial T}\right)_D + \left(\frac{\partial \log k_0}{\partial D}\right)_T \frac{dD}{dT} \qquad (7)$$

where the k_0 refers to the rate constant at zero ionic strength (Svirbely and Warner[6]). From equation 6

$$\left(\frac{\partial \log k_0}{\partial D}\right)_T = \frac{NZ_A Z_B e^2}{2 \cdot 3 RTD^2 r_{\pm}} \qquad (8)$$

so that (7) becomes

$$E_{\text{fc}}{}^0 = E_D{}^0 + \frac{NZ_A Z_B e^2 T}{D^2 r_{\pm}} \left(\frac{dD}{dT}\right)_{\text{fc}} \qquad (9)$$

where $E_{\text{fc}}{}^0$ is the experimental activation energy in a solvent of fixed composition and $E_D{}^0$ is the experimental activation energy in a series of solvents selected to give the same fixed value of the dielectric constant at all the temperatures used. In general, good agreement between theory and experiment was obtained, using values of $r_{\pm}$ of about 2×10^{-8} cm which changed somewhat with the type of solvent mixture. Amis and his co-workers[7] have also used the data of Warner, Svirbely, et al., to check successfully other points of electrostatic theory applied to reaction rates.

All of this seems overwhelming proof that the reaction must go by way of some collision process involving the ammonium ion and the cyanate ion. Actually, because of the equilibrium indicated in (3), it is no more proof for a mechanism involving the two ions than for one involving ammonia and cyanic acid! Weil and Morris[8] have recently emphasized this kinetic ambiguity.

If activity coefficients are included, the rate equations 2 and 4 and the equilibrium expression for (3) will become

$$\text{rate} = k[\text{NH}_4{}^+][\text{CNO}^-]\frac{f_+ f_-}{f^{\ddagger}} \quad \text{ionic reactants} \qquad (10)$$

$$[\text{NH}_3][\text{HCNO}] = \frac{K_w}{K_a K_b} [\text{NH}_4{}^+][\text{CNO}^-] \frac{f_+ f_-}{f_{\text{NH}_3} f_{\text{HCNO}}} \qquad (11)$$

$$\text{rate} = k'[\text{NH}_3][\text{HCNO}]\frac{f_{\text{NH}_3} f_{\text{HCNO}}}{f^{\ddagger}} = \frac{k' K_w}{K_a K_b} [\text{NH}_4{}^+][\text{CNO}^-]\frac{f_+ f_-}{f^{\ddagger}}$$

$$\text{neutral reactants} \quad (12)$$

Clearly (10), the ionic mechanism, and (12), the molecular mechanism, are indistinguishable as far as the effect of ionic strength, etc., on activity coefficients is concerned. Also, the rate constant k includes an equilibrium constant for the formation of a neutral-activated complex from two oppositely charged ions, and the apparent rate constant $k'K_w/K_aK_b$ includes an equilibrium constant for the formation of two neutral molecules from two oppositely charged ions. Therefore, the theoretical effect of changing dielectric constant and ionic strength on the rate constant, the experimental energy of activation, the frequency factor, etc., will be the same for both mechanisms. The only difference is that a critical distance $r_{\ddagger}$ in the activated complex is replaced by a distance of closest approach r_0 for the two ions.

Consequently we have two alternative mechanisms, each of which is equally supported by the available kinetic evidence, and which seem to be indistinguishable by any kinetic method. An instructive procedure at this point is to consider the detailed, stereochemical mechanism of the reaction. As organic chemists have several times pointed out,[9] there is an easy reaction path for the ammonia-cyanic acid mechanism. This involves addition of the nitrogen atom's unshared pair of electrons to the positively polarized carbon atom of the acid (written as the tautomeric and indistinguishable isocyanic acid), as shown in reaction 13. A simple proton shift

$$H—N{=}C{=}O \rightarrow H—N{=}C—O^- \rightarrow HN—C{=}O \qquad (13)$$

$$\begin{matrix} H\ddot{N}H & & HNH^+ & & NH \\ H & & H & & H \end{matrix}$$

can now complete the transformation to urea. This reaction sequence is exactly analogous to what must happen when an alkyl isocyanate reacts with ammonia to give a substituted urea, as shown in (14). It is similar to

$$R—N{=}C{=}O \rightarrow R—N{=}C—O^- \rightarrow RN—C{=}O \qquad (14)$$

$$\begin{matrix} H\ddot{N}H & & HNH^+ & & NH \\ H & & H & & H \end{matrix}$$

a host of reactions involving addition of basic reagents to carbonyl groups.

On the other hand, the reaction of the ammonium ion with cyanate ion to give urea in any simple way is difficult to imagine. The carbon-nitrogen bond which must be established is blocked out by the four hydrogens around the nitrogen. The only thing the ions seem capable of doing is to form a hydrogen-bonded complex

$$\begin{matrix} H \\ HNH—N{=}C{=}O \\ H \end{matrix}$$

which then rearranges into the same addition product shown in (13). The rearrangement seems much less likely than a dissociation into ammonia and isocyanic acid or ammonium and cyanate ions.

A consideration of the true rate constants, activation energies, and frequency factors on the basis of each of the two mechanisms does not lead to any clear decision. If the ionic mechanism is operating, each of the above quantities is equal to that experimentally observed. If the molecular mechanism is operating, the true quantities are obtained from the observed by the following operations

$$k' = k_{obs}K_aK_b/K_w \qquad E_{true} = E_{obs} + \Delta H^0 \qquad (15)$$

$$\log A_{true} = \log A_{obs} + \Delta S^0/2.3R$$

where ΔH^0 and ΔS^0 are the standard changes in heat content and entropy for the reaction

$$HCNO + NH_3 \rightleftharpoons NH_4^+ + CNO^- \qquad (16)$$

Cyanic acid has K_a equal to 3.4×10^{-4} at $25°$ C. The value of the heat of ionization is 2.5 kcal and -7.4 E.U. is the entropy of ionization.[10] For ammonia the corresponding figures are 1.3 kcal and -18.7 E.U., and for the ion product of water the heat is 13.8 kcal and the entropy -19.2 E.U. This makes $\Delta H^0 = -10$ kcal and $\Delta S^0 = -7$ E.U.

Experimentally it is found that the rate constant for the ammonium cyanate reaction in water is given by $4 \times 10^{12}e^{-23,200/RT}$ at an ionic strength of 0.0376 (Svirbely and Warner;[6] Wyatt and Kornberg[11] report this value to be about 40 per cent too large at $70°$ because of the concurrent hydrolysis of cyanate ion to bicarbonate ion and ammonia), the units being liters/mole-second. This represents a reasonable activation energy and frequency factor for a reaction between two oppositely charged ions of unit charge. The probability factor greater than unity, assuming $Z = 10^{11}$, in particular is just what would be expected (see Chapter 7). For the molecular mechanism we have $k' \simeq 10^{11}e^{-13,200/RT}$ which represents not improbable values of A and E_a for a reaction of the kind shown in (13). It would be very desirable to have available the corresponding quantities for the reaction of simple alkyl iocyanates and amines measured under similar conditions for comparison. However, such data would be difficult to obtain since in water or alcohol the reactions would be very fast and the competition of solvent would be serious.

A comparison of the rate constants for different alkyl ammonium cyanates according to each mechanism is of some value. Experimentally the rate constants are the same within a factor of 2 for ammonium ion and ethyl ammonium ion in spite of a thirtyfold difference in base strength

$(K_b$ for ethylamine $= 5 \times 10^{-4})$. On the molecular basis the ratio of k' for ammonia and ethylamine becomes 1 to 18. The other amines fall into a rough order with base strength also except, as will be shown below, base strength is not the dominant factor. In a study involving a larger number of amines a correlation is found with nucleophilic reactivity of the amines towards alkyl halides.[12] A maximum in rate is found for dimethylamine and a minimum for t-butylamine.

The best correlation, and powerful evidence for the assumption of molecular rather than ionic reactants, comes from a study of the reactions of amines with carbon dioxide to form carbamates.[13]

$$RNH_2 + CO_2 \rightarrow RNHCO_2H \qquad (17)$$

It is found that there is a very simple linear free-energy relationship between the rate constants for reaction 17 with a number of amines and the rate constants for the urea reaction calculated on the molecular basis.

$$RNH_2 + HCNO \rightarrow RNHCONH_2 \qquad (18)$$

The relationship is (at $18°$ C)[13]

$$\log k_{18} = \log k_{17} - 1.08 \qquad (19)$$

Thus structural changes in the amine produce identical changes in rate for both reactions. Furthermore, the equation holds for bases other than amines, including water and hydroxide ion.

$$OH^- + CO_2 \rightarrow HCO_3^- \qquad (20)$$

$$OH^- + HCNO \rightarrow NH_2CO_2^- \qquad (21)$$

The reaction with CO_2 is studied in aqueous solution and the question may arise as to whether H_2CO_3 is the reagent or possibly RNH_3^+ and HCO_3^- are the reactants. Fortunately a definite answer is possible in this case. The hydration of CO_2 to form H_2CO_3 or HCO_3^- is a slow enough process so that it can be eliminated as a prior step in the carbamate reaction. The technique used is to dissolve gaseous carbon dioxide quickly in an aqueous solution of the amine and hydroxide ion.[14] The reaction is rapid, a mixture of carbonate and carbamate being formed. The ratio of products then gives the ratio of rate constants.

$$\frac{\% \text{ carbonate}}{\% \text{ carbamate}} = \frac{k_{20}[OH^-]}{k_{18}[\text{amine}]} \qquad (22)$$

The rate constant k_{20} is known from flow studies[15] and hence k_{18} can be calculated.

The mechanism for the carbon dioxide reaction certainly is

$$O{=}C{\overset{\curvearrowright}{=}}O \rightarrow O{=}C{-}O^- \rightarrow O{=}C{-}OH \qquad (23)$$

$$\underset{\underset{H}{H\ddot{N}H}}{} \qquad \underset{\underset{H}{HNH^+}}{} \qquad \underset{\underset{H}{NH}}{}$$

and thus equation 19 is powerful support for the molecular mechanism for the urea reaction. It is found that the two rate constants for hydroxide ion are less than the rate constants for several of the amines, even though hydroxide ion is the most powerful base by far. Hence base strength is not the dominant factor for either the CO_2 or HCNO reaction.

Shaw and Bordeaux[16] have studied the reverse of reaction 1, the thermal decomposition of urea in water. This can conveniently be done at higher temperatures, 60 to 100° C, but only if acid is added to drive the reaction to completion. The products are then ammonium ion and carbon dioxide.

$$(NH_2)_2CO + H_2O + 2H^+ \rightarrow 2NH_4^+ + CO_2 \qquad (24)$$

In acid solution cyanic acid is rapidly hydrolyzed to ammonium ion and carbon dioxide.[17]

The urea decomposition has sometimes erroneously been considered an acid-catalyzed reaction, but this is not so. The rate is independent of the acidity once sufficient acid is added to destroy the cyanate ion.[16] The overall rate constant is given by $k = 5 \times 10^{14} e^{32,700/RT}$ sec^{-1} and is for a simple first-order process. The rate-determining step is the conversion of urea to ammonia and cyanic acid; by the principle of microscopic reversibility we need only to reverse reaction 13 for the details. From the ratio of rate constants the value of ΔH^0 and ΔS^0 for the equilibrium

$$(NH_2)_2CO \rightleftharpoons NH_4^+ + CNO^- \qquad (25)$$

can be calculated as 9.5 kcal and 10.0 E.U. respectively. These numbers are in poor agreement with the thermodynamic data of 11.1 kcal and 16.5 E.U. (see p. 134), but the differences can be blamed on errors in the activation energies and on neglect of likely non-zero values for ΔC_p^0 and $\Delta C_p^{\ddagger}$. The equilibrium constant at several temperatures is in much better agreement when calculated from both the thermodynamic and kinetic data.

A study has also been made of the related thermal decomposition of thiourea and the methyl thioureas.[18] The mechanism

$$(NH_2)_2CS \rightarrow NH_4^+ + CNS^- \qquad (26)$$

is probably identical with that for urea.

Surprisingly, urea is 240 times as reactive as thiourea towards decomposition. This seems to indicate that for the structures

$$\overset{+}{H-N}=\overset{-}{C}-\overset{-}{O} \qquad\qquad \overset{+}{H-N}=\overset{-}{C}-\overset{-}{S}$$

$$H-N=C=O \qquad\qquad H-N=C=S$$

the one involving sulfur in the polar form is relatively more stable. It might be thought the polar form of oxygen is the more stable because of its greater electronegativity. Urea is a stronger base (towards the proton) than thiourea is.[19] The well-known weakness of double bonds between atoms of other than first row elements may account for the anomaly. There is some evidence[20] that thiourea is normally the thiol form $NH_2CSH{=}NH$. As expected, tetramethylthiourea does not decompose into ions, but only hydrolysis into dimethylamine and carbonoxysulfide occurs.[18] The mechanism for forming a thiocyanate is not open to this compound because there is no labile hydrogen.

In the ammonium cyanate reaction, as in any reaction involving a weak acid and base, there should be a change in the observed rate constant with pH. If k_{obs} is defined in terms of initially added ammonium ion and cyanate ion and if no ammonia or cyanic acid is added, the following equation can be derived:

$$k_{obs} = \frac{kK_aK_b[H^+]}{K_aK_b[H^+] + K_aK_w + K_b[H^+]^2} \qquad (27)$$

where k is the rate constant defined by the equation

$$\text{rate} = k[NH_4^+][CNO^-] \qquad (28)$$

the concentrations of the ions here referring to those actually present in a solution whose pH is fixed by added buffers. Equation 27 gives a maximum in the observed rate constant at a hydrogen ion concentration equal to $(K_aK_w/K_b)^{1/2}$. Since a solution of ammonium cyanate is buffered at a hydrogen ion concentration equal to $(K_aK_w/K_b)^{1/2}$, it is easy to see that the experimental rate constants have been the maximum ones and equal to k. From (27) it would take a pH change of two units in either direction to change the rate constant appreciably. No experiments appear to have been done to check this predicted effect of pH. Unfortunately, the results if positive would still not distinguish between the ionic and nonionic mechanisms since (27) holds true for both of them. A negative result would, of course, cast doubt on both mechanisms since a result of changing pH other than that given by (27) would indicate an unexpected catalysis by acids or bases. However, the numerous experiments done to show that there is a neutral activated complex in direct or indirect equilibrium with two oppositely charged ions render such a result very unlikely.

The problem involved in deciding the mechanism of the urea conversion is a fairly frequent one which can arise whenever there is a mobile equilibrium before reaction (see Chapter 8). A very similar case is found in the reaction studied by Weil and Morris[8] between hypochlorous acid and amines to form N-chloramines:

$$RNH_2 + HOCl \rightarrow RNHCl + H_2O \qquad (29)$$

Here the possible reactants are hypochlorous acid and the amine as shown in (29), or ammonium ion and hypochlorite ion which are in equilibrium with them.

$$RNH_2 + HOCl \rightleftharpoons RNH_3^+ + OCl^- \qquad (30)$$

Weil and Morris concluded that the reaction involving the neutral molecules was much more plausible than the ionic reaction, for the same reasons as those given in the preceding discussion. A reasonable transition state can be pictured:

$$\begin{array}{c} H \\ | \\ R\text{---}N\text{---}Cl\text{---}OH \\ | \\ H \end{array}$$

which, with the help of several water molecules to remove the hydrogen and hydroxyl ions, or by a proton shift, becomes the chloramine.

To summarize the situation with regard to the ammonium cyanate-urea reaction, all the kinetic evidence supports equally a mechanism involving ammonium and cyanate ions or a mechanism involving ammonia and cyanic acid. The latter, however, is far more likely from a closer stereochemical analysis of the way in which the reaction must proceed. If this is the mechanism, the extensive work done to check the application of the Debye-Hückel theory of ionic strength and general electrostatic theory to reaction kinetics is still useful, but what has been checked is a thermodynamic equilibrium and not a kinetic process.

The important lesson to be learned from this particular example is that kinetic methods will not distinguish between a particular group of reactants and some other group in rapid equilibrium with the first. That is, kinetic methods will give the overall composition of the activated complex but will not reveal, at least in our present state of knowledge, the exact arrangement of the components in the activated complex. For such information it is necessary to use other experimental methods and fields of knowledge.

REFERENCES

1. J. Walker and F. J. Hambly, *J. Chem. Soc.*, *67*, 746 (1895).
2. F. D. Chattaway, *J. Chem. Soc.*, *101*, 170 (1912).
3. J. Walker and J. R. Appleyard, *J. Chem. Soc.*, *69*, 193 (1896).
4. C. C. Miller and J. R. Nicholson, *Proc. Roy. Soc.*, *A168*, 206 (1938).
5. J. C. Warner and F. B. Stitt, *J. Am. Chem. Soc.*, *55*, 4807 (1933).
6. J. C. Warner and E. L. Warrick, *J. Am. Chem. Soc.*, *57*, 1491 (1935); W. J. Svirbely and J. C. Warner, *ibid.*, *57*, 1883 (1935); W. J. Svirbely and A. Schramm, *ibid.*, *60*, 330 (1938); J. Lander and W. J. Svirbely, *ibid.*, *60*, 1613 (1938); C. C. Miller, *Proc. Roy. Soc.*, *A145*, 288 (1934), *A151*, 188 (1935).
7. E. S. Amis, *Kinetics of Chemical Change in Solution*, The Macmillan Co., New York, 1949, Chapters IV, V, and VI.
8. I. Weil and J. C. Morris, *J. Am. Chem. Soc.*, *71*, 1664 (1949).
9. T. M. Lowry, *Trans. Faraday Soc.*, *30*, 375 (1934).
10. R. Caramazza, *Gazz. chim. ital.*, *88*, 308 (1958).
11. P. A. H. Wyatt and H. L. Kornberg, *Trans. Faraday Soc.*, *48*, 454 (1952).
12. P. Johncock, G. Kohnstam, and D. Speight, *J. Chem. Soc.*, 2544 (1958).
13. M. B. Jensen, *Acta Chem. Scand.*, *13*, 289 (1959).
14. A. Jensen and C. Faurholt, *Acta Chem. Scand.*, *6*, 385 (1952).
15. R. Brinkman, R. Margaria and F. J. W. Roughton, *Phil. Trans.*, *232*, 65 (1933).
16. W. H. R. Shaw and J. J. Bordeaux, *J. Am. Chem. Soc.*, *77*, 4729 (1955).
17. A. R. Amell, *J. Am. Chem. Soc.*, *78*, 6234 (1956).
18. W. H. R. Shaw and D. G. Walker, *J. Am. Chem. Soc.*, *80*, 5337 (1958).
19. J. Walker, *J. Chem. Soc.*, *67*, 576 (1895).
20. N. V. Sidgwick, *The Organic Chemistry of Nitrogen*, Oxford University Press, London, 1936, p. 291.

C. The Hydrolysis of Lactones. The H_0 Function

Mechanisms for Ester Hydrolysis

In Chapter 1 a variety of evidence was mentioned to indicate the probable mechanism of alkaline hydrolysis of simple esters. An equally impressive amount of evidence has been accumulated on the acid-catalyzed hydrolysis of similar esters.† The indicated mechanism in acid-catalyzed hydrolysis is shown in (1), (2), and (3). Step 2 is written to show the formation of an

$$
\begin{array}{ccc}
\text{O} & & \text{O H} \\
\parallel & & \parallel \ | \\
\text{R—C—OR}' + \text{H}^+ & \rightleftharpoons & \text{R—C—OR}' \\
& & +
\end{array}
\qquad \text{fast} \qquad (1)
$$

† For a review of work on ester hydrolysis see references 1 and 29.

intermediate which can decompose to give either the products or the reactants. The existence of such an intermediate in the case of both acid and basic hydrolysis is strongly indicated by the work of Bender,[2] who

$$H_2O + R-\overset{\overset{O}{\|}}{\underset{+}{C}}-\overset{\overset{H}{|}}{O}R' \rightleftharpoons R-\overset{\overset{\bar{O}}{|}}{\underset{\underset{+\dot{O}H_2}{|}}{C}}-\overset{\overset{H}{|}}{\underset{+}{O}}R' \rightarrow R-\overset{\overset{O}{\|}}{C}-\underset{+}{O}H_2 + R'OH \quad \text{slow} \quad (2)$$

$$R-\overset{\overset{O}{\|}}{C}-\underset{+}{O}H_2 \rightleftharpoons R-\overset{\overset{O}{\|}}{C}-OH + H^+ \quad \text{fast} \quad (3)$$

measured the rates of hydrolysis and the rates of oxygen exchange for several esters labeled in the carbonyl group with O^{18}. Since oxygen exchange occurred before hydrolysis, some symmetrical intermediate such as the ester hydrate, $R-C(OH)_2OR'$, which can be derived from the dipolar form shown in (2), is needed.

It might be expected that the subject of ester hydrolysis is well understood and that the mechanism of hydrolysis of any ester under any conditions could be deduced by analogy from the examples which have been mentioned. Actually this is far from the truth. The mechanisms mentioned in Chapter 1 and in the previous paragraph can be relied on only for esters closely related to the model compounds which have been studied and for a limited set of experimental conditions. For other esters and other conditions a variety of behaviors becomes possible. Day and Ingold[1] list no less than six mechanisms of ester hydrolysis for which some experimental evidence can be cited. To this list two others, not as yet observed, can be added, making eight in all, four for acid hydrolysis and four for basic hydrolysis. Furthermore, this listing for acid conditions considers only hydrogen ion catalysis and makes no mention of possible variants involving general acid catalysis. Such general acid catalysis undoubtedly occurs in some nonaqueous solvents, though it has not been convincingly demonstrated in aqueous solution. General base catalysis for ester hydrolysis definitely exists in aqueous solution.

Following Day and Ingold the eight mechanisms can be summarized as follows: let A stand for acid hydrolysis and B for basic, let 1 stand for an unimolecular process and 2 for a bimolecular process, and let the symbol ' refer to acyl oxygen scission and " to alkyl oxygen scission. The rate-determining steps for each mechanism are shown in (4)–(11).

$$R-\overset{\overset{O}{\|}}{C}-\overset{\overset{H}{|}}{\underset{+}{O}}R' \rightleftharpoons R-\overset{\overset{O}{\|}}{C}^+ + R'OH \quad A'1 \quad (4)$$

Of these $A'2$ and $B'2$ will be recognized as the mechanisms found for simple esters under acid and basic conditions. Mechanisms $B'1$ and $A''2$ have not as yet been observed. In mechanism $B''2$ a hydroxide ion could

$$H_2O + R\overset{O}{\underset{+}{\overset{\|}{C}}}\overset{H}{\overset{|}{-}}OR' \rightleftharpoons R\overset{O}{\underset{+}{\overset{\|}{C}}}\overset{H}{\overset{|}{-}}OH + R'OH \quad A'2 \tag{5}$$

$$R\overset{O}{\underset{+}{\overset{\|}{C}}}\overset{H}{\overset{|}{-}}OR' \rightleftharpoons R\overset{O}{\overset{\|}{C}}-OH + R'^+ \quad A''1 \tag{6}$$

$$H_2O + R'\overset{H}{\overset{|}{O}}\underset{+}{-}\overset{O}{\overset{\|}{C}}-R \rightleftharpoons R'OH_2^+ + R\overset{O}{\overset{\|}{C}}-OH \quad A''2 \tag{7}$$

$$R\overset{O}{\overset{\|}{C}}-OR' \rightleftharpoons R\overset{O}{\overset{\|}{C}}{}^+ + R'O^- \quad B'1 \tag{8}$$

$$OH^- + R\overset{O}{\overset{\|}{C}}-OR' \rightleftharpoons R\overset{O}{\overset{\|}{C}}-OH + OR'^- \quad B'2 \tag{9}$$

$$R\overset{O}{\overset{\|}{C}}-OR' \rightleftharpoons R\overset{O}{\overset{\|}{C}}-O^- + R'^+ \quad B''1 \tag{10}$$

$$H_2O + R'O\overset{O}{\overset{\|}{C}}-R \rightleftharpoons R'OH_2^+ + R\overset{O}{\overset{\|}{C}}-O^- \quad B''2 \tag{11}$$

conceivably take the place of the water molecule. Table 1 summarizes the properties of all eight mechanisms with respect to kinetics, expected configuration of an asymmetric alkyl group R' after reaction, the effect of increasing the electronegativity of R and R' on the rate, and the effect or lack of effect of steric hindrance in R and R' on the rate.

The evidence for mechanism $A'1$ comes chiefly from experiments in concentrated sulfuric acid where the work of Treffers and Hammett,[3] for example, indicates that the methyl esters of substituted benzoic acids undergo the ionization of equation 4 rapidly and quantitatively. In agreement with this, the rate of hydrolysis of methyl benzoate in concentrated sulfuric acid is independent of the water concentration in regions of acidity where the conversion of methyl benzoate to its conjugate acid,

$C_6H_5CO\overset{+}{\underset{H}{O}}CH_3$, is quantitative.[4]

Mechanism $A''1$ is indicated in the behavior of esters of allylic alcohols under acid conditions. It is found that for esters of unsymmetrical allyl alcohols rearrangements occur readily in the presence of acids:

$$R\text{—}CH\text{=}CH\text{—}CHR'\text{—}O\overset{\displaystyle O}{\overset{\|}{\text{—}C}}\text{—}CH_3 + H_2O \xrightarrow{\text{H}^+}$$

$$R\text{—}CHOH\text{—}CH\text{=}CHR' + CH_3COOH \quad (12)$$

The rearrangement certainly indicates alkyl-oxygen bond breaking and is consistent with what would be expected of an allylic carbonium ion.

Table I

MECHANISMS FOR ESTER HYDROLYSIS

(Day and Ingold)

Mech-anism	Position of Cleavage	Alkyl Configuration	Kinetics	Electron Requirements[a] R	R'	Steric Hin-drance
$A'1$	Acyl	Retention	[H⁺] [E]	~ 0	$+$	No
$A'2$	Acyl	Retention	[H⁺] [E][H₂O]	~ 0	~ 0	Yes
$A''1$	Alkyl	Racemization	[H⁺] [E]	$+$	~ 0	No
$A''2$	Alkyl	Inversion	[H⁺] [E][H₂O]	$+$	~ 0	Yes
$B'1$	Acyl	Retention	[E]	~ 0		No
$B'2$	Acyl	Retention	[OH⁻][E]	$-$	$-$	Yes
$B''1$	Alkyl	Racemization	[E]	$+$	$-$	No
$B''2$	Alkyl	Inversion	[H₂O] [E]	~ 0	$-$	Yes

[a] A $+$ sign means that an electron-donating substituent will favor hydrolysis. A $-$ sign means that an electron-attracting group will be favorable for hydrolysis.

Also in the esterification of optically active octyl alcohol with acetic acid catalyzed by sulfuric acid there is some racemization.[5] If this is not accounted for by the racemization of the octyl alcohol in an independent reaction, it occurs in the esterification step and must also occur in the reverse reaction, which is the hydrolysis. Such racemization is consistent with the carbonium ion postulated in mechanism $A''1$. Not all of the reaction can go by this mechanism, however, since the amount of racemization is not large.

Salomaa[6] has shown that alkoxymethyl esters of acetic and formic acid hydrolyze in acid by an alkyl oxygen cleavage, at least in part. The products in water are acid, alcohol, and formaldehyde, as in equation 13.

$$R\overset{\displaystyle O}{\overset{\|}{\text{—}C}}\text{—}OCH_2OR' + H_2O \rightarrow R\overset{\displaystyle O}{\overset{\|}{\text{—}C}}\text{—}OH + CH_2O + R'OH \quad (13)$$

However, in mixtures of alcohol and water as solvents, a part of the formaldehyde appears as its acetal. This can only be explained by the $A''1$ mechanism, or possibly the $A''2$. The former is preferable since the carbonium ion $R'OCH_2{}^+$ would be stabilized by the additional structure $R'O^+{=}CH_2$.

Mechanism $B''1$ is strongly indicated by the work of Kenyon, Balfe, et al., on the hydrolysis in neutral or slightly alkaline solution of esters in which the radical R is substituted with aryl groups which would stabilize it as a carbonium ion. Such esters when optically active are found to be extensively racemized in dilute alkali in agreement with the predictions for the $B''1$ mechanism.[7] That alkyl oxygen cleavage is involved is shown in (13) by the behavior of these esters in methanol and ethanol, where racemic ethers are formed instead of the expected alcohols. The asymmetric carbon is, of course, the one attached to the oxygen in the ester.

$$R-\overset{O}{\overset{\|}{C}}-O\overset{*}{R'} + CH_3OH \rightarrow R-\overset{O}{\overset{\|}{C}}-OH + \underset{\text{inactive}}{R'-O-CH_3} \qquad (14)$$

Along the same lines it has been shown that, if R' is a tertiary alkyl radical, alkyl oxygen scission is favored. Thus tertiary butyl benzoate refluxed with methanol gives tertiary butyl methyl ether rather than the products of ester interchange.[8] Kinetic evidence for the $B''1$ mechanism has also been obtained.[7b]

Mechanism $B''2$ is perhaps the most unusual of those which have been observed. There is good evidence for it in the work of Olson and Miller[9] on the hydrolysis of optically active β-butyrolactone under neutral or slightly acid conditions. Lactones are, of course, inner esters, and β-lactones are a particularly reactive species which helps to explain the existence of the $B''2$ mechanism in this case. The results obtained can be explained by assuming a reaction of the lactone with a water molecule involving alkyl oxygen scission and inversion of configuration at the asymmetric carbon, as shown in reaction 15. Because of this interesting

$$CH_3-\overset{\overset{\displaystyle H}{|}}{\underset{\overset{|}{O-C{=}O}}{C^*}}-CH_2 + H_2O \rightarrow CH_3-\overset{\overset{\displaystyle H}{\overset{\displaystyle O}{|}}}{\underset{\overset{|}{H}}{C^*}}-CH_2-COOH \qquad (15)$$

reaction and because considerable work has been done on the mechanisms of hydrolysis of lactones, this subject will be discussed in detail. The experimental results in very strong acid solutions will serve as a good means of introducing the useful H_0, or acidity, function.

The Hydrolysis of β-Lactones

In addition to the usual acid- and base-catalyzed reactions typical of ordinary esters, β-lactones give an added reaction with water.[10] This reaction is the only observable one in the pH range from about 1 to 7. Strong acid is needed to bring in the acid-catalyzed reaction.[11]

It is possible to prepare an optically active β-lactone in which the lactone carbon is asymmetric. Starting with (+) β-bromobutyric acid, (+) butyrolactone can be prepared,[12] as shown in reaction 16. This reaction

$$(+) \ CH_3\overset{Br}{\underset{H}{-C^*-}}CH_2-COO^- \rightarrow (+) \ CH_3\overset{H}{\underset{O-C=O}{-C^*-}}CH_2 + Br^- \quad (16)$$

proceeds with an inversion of configuration at the asymmetric carbon, the negative carboxylate group displacing the bromide ion from the back side.[13] In alkali or in strong acid the product of hydrolysis of this active lactone is the (−) hydroxybutyric acid, but in neutral or slightly acid solution the (+) hydroxybutyric acid is formed (Olson and Miller[9]), as shown in reaction 17. Experiments using water labeled with O^{18}, in which the oxygen of the

$$(+) \ CH_3\overset{H}{\underset{O-C=O}{-C^*-}}CH_2 \xrightarrow{H_2O} CH_3\overset{H}{\underset{\underset{H}{O}}{-C-}}CH_2-COOH \quad (17)$$

$$(-) \ pH \ 3 \ to \ 8$$
$$(+) \ pH \ -2 \ to \ 0 \ or \ 10 \ to \ 12$$

hydroxyl group of the acid after hydrolysis was removed as water by heating, were carried out to determine the point of cleavage under various conditions.[14] In neutral or slightly acid solution the hydroxyl group of the acid contained the equilibrium concentration of O^{18}. In very strong acid and in base solutions the hydroxyl group contained only minor amounts of O^{18}. This indicates alkyl oxygen scission in neutral solution, and the more usual acyl oxygen scission in base or strong acid, as shown in reactions 18 and 19. Thus it appears that the reaction leading to formation of (+) hydroxybutyric acid from the (+) lactone is an inversion reaction in which

$$CH_3\overset{}{\underset{O-C=O}{-CH-}}CH_2 + H_2O^{18} \rightarrow CH_3\overset{}{\underset{\underset{H}{O^{18}}}{-CH-}}CH_2-COOH \quad (18)$$

$$(neutral \ or \ dilute \ acid)$$

a water molecule, uncatalyzed by acid or base, displaces the lactone oxygen from the asymmetric carbon (Olson and Miller[9]). The base and strong acid reactions involve acyl oxygen cleavage which would leave the configuration of the asymmetric carbon unchanged, as shown in reaction 21. Consequently the (+) bromobutyric acid and the (+) hydroxybutyric

$$CH_3—CH—CH_2 \ + \ H_2O^{18} \rightarrow CH_3—CH—CH_2CO\overset{18}{O}H \qquad (19)$$
$$\underset{O——C=O}{|\qquad\qquad|} \qquad\qquad \underset{\underset{H}{O}}{|}$$

(base or strong acid)

$$H_2O + CH_3—CH—CH_2 \ \rightarrow CH_3—\overset{\overset{+OH_2}{|}}{C}H—CH_2—COO^- \qquad (20)$$
$$\underset{O——C=O}{|\qquad\qquad|} \qquad\qquad \underset{\underset{O}{H}}{\downarrow}$$

$$CH_3—CH—CH_2—COOH$$

$$CH_3—CH—CH_2 \ \rightarrow CH_3—CH—CH_2—COOH \qquad (21)$$
$$\underset{O——C=O}{|\qquad\qquad|} \qquad\qquad \underset{\underset{H}{O}}{|}$$

$$(H^+ + H_2O \ or \ OH^-)$$

acid have the same relative configuration, which is opposite to that of the (+) lactone. Similar results have been obtained with (+) malolactonic acid, which in dilute acid hydrolyzes to give (−) malic acid and in strong acid or base solution gives (+) malic acid.[15]

Additional information on the several reactions comes from a study of the products formed from the related β-propiolactone under various conditions. Gresham, Jansen, Shaver, and their co-workers have carried out a study of the behavior of this reactive compound with a number of reagents.[16]

Base-catalyzed alcoholysis gives chiefly esters of hydracrylic acid which is the expected product of acyl oxygen cleavage. For neutral or slightly acid

$$CH_2—CH_2 \ + \ ROH \xrightarrow{\text{base}} HOCH_2—CH_2—COOR \qquad (22)$$
$$\underset{O——C=O}{|\qquad\quad|}$$

conditions the alcoholysis products are chiefly alkoxy derivatives, indicating alkyl oxygen cleavage, as shown in reaction 23. In more concentrated acid

$$CH_2—CH_2 \ + \ ROH \xrightarrow{\text{neutral}} ROCH_2—CH_2—COOH \qquad (23)$$
$$\underset{O——C=O}{|\qquad\quad|}$$

the proportion of alkoxy derivatives decreases and the proportion of hydracrylic esters increases. This indicates an incursion of an acid-catalyzed reaction involving acyl oxygen cleavage again. Considerable amounts of polymeric substances are also formed.

The work of Gresham, Jansen, Shaver, et al., shows that a variety of nucleophilic reagents can take the place of water in the neutral reaction. Some quantitative data on relative reactivity is furnished by the kinetic studies of Bartlett and Small,[17] who measured the second-order rate

Table 2

RATE CONSTANTS AT 25° FOR ATTACK ON β-PROPIOLACTONE AND OTHER SUBSTANCES BY NUCLEOPHILIC REAGENTS IN AQUEOUS SOLUTION

(Bartlett and Small)

Reagent	β-Propio-lactone, k_2, liters/mole-sec	Relative Rate toward		
		Propiolactone	Epichloro-hydrin[b]	β-Chloroethyl-ethylenesulfonium Ion[c]
H_2O	5.6×10^{-5a}	0.31	0.04	0.05
CH_3COO^-	3.0×10^{-4}	1.7	0.54	0.48
Cl^-	1.8×10^{-4}	(1.0)	(1.0)	(1.0)
Br^-	6.0×10^{-4}	3.3	5.2	....
I^-	3.0×10^{-3}	17.0	85.0	31.0
SCN^-	3.8×10^{-3}	21.0	57.0	32.0
$S_2O_3^{-2}$	1.9×10^{-1}	1,050	....	1,286
OH^-	1.2	6,700	100^d	381

[a] First-order, sec^{-1}.
[b] J. N. Brönsted, M. Kilpatrick and M. Kilpatrick, *J. Am. Chem. Soc.*, *51*, 428 (1929).
[c] A. G. Ogston and E. R. Holliday, *Trans. Faraday Soc.*, *44*, 49 (1948).
[d] H. J. Lichtenstein and G. H. Twigg, *ibid*, *44*, 905 (1948).

constants for the reaction of β-propiolactone with a number of nucleophilic reagents. Table 2 shows their data together with the relative rate constants for some of the same reagents with epichlorohydrin ClCH$_2$—CH—CH$_3$ and the β-chloroethylethylenesulfonium ion
            \ /
             O
        CH$_2$—CH$_2$. These last two are included for comparison since
        \ /
ClCH$_2$CH$_2$—S$^+$
their reactions also involve the opening of strained rings by nucleophilic reagents. It can be seen that the order of reactivity is about the same for

all reagents in the three series except for the hydroxide ion reaction of β-propiolactone. This is much faster than would be expected and is evidence that a different mechanism is operating, presumably the normal $B'2$ ester hydrolysis at the carbonyl group.

There appears to be still another way in which bases can react with β-lactones. Thus Olson and Miller[9] and Olson and Youle[18] have shown that there is a general base-catalyzed hydrolysis of β-butyrolactone. Optically active lactone changes its rotation in carbonate, borate, or phosphate buffers much more rapidly than the rates of the combined water and hydroxide ion reactions. There seems to be one reaction which produces inversion of configuration and one which gives retention. It has been suggested (Olson and Miller[9]) that a water molecule coordinated with a carbonate ion or phosphate ion is the reactive species. Attack can occur either at the carbonyl carbon or the asymmetric carbon as in equations 20 and 21. Certainly a water molecule associated with a negative ion would be expected to have enhanced nucleophilic character as well as basicity. However, such an effect becomes difficult to separate from salt effects in general since it appears only in rather concentrated buffers. The rate constants for several basic ions seem to follow the Brönsted relationship, which may rule out an ionic strength explanation.

Long and Purchase[11] have carefully studied the rate of hydrolysis of β-propiolactone under various conditions. For basic conditions the reaction is second-order, dependent on the concentration of hydroxide ion. The rate constant is given by $k_b = 1.43 \times 10^{10} e^{-13,400/RT}$ liters/mole-sec. In the pH range from 1 to 7 there is a first-order reaction independent of pH. This is the water reaction with a rate constant $k_w = 1.1 \times 10^{10} e^{-19,500/RT}$ sec^{-1}. In strong acid another reaction sets in which is second-order in the sense that it depends in some way on the concentration of hydrogen ion in solution. The data can be represented by

$$k_{obs} = k_w a_{H_2O} + k_a F(H_3O^+) \tag{24}$$

where k_{obs} is the observed first-order rate constant, a_{H_2O} is the activity of water in these concentrated solutions, k_a is the acid-catalyzed rate constant, and $F(H_3O^+)$ is a function of the hydrogen ion concentration, whose nature will be discussed shortly. Table 3 shows the experimental results obtained in concentrated acid solutions for β-propiolactone. It is evident at once that there is no simple relationship between the observed rate constant and the hydrogen ion concentration. A change in perchloric acid concentration from 2.75 molar to 5.36 molar causes the observed rate constant to change by a factor of 13 and the part of the rate constant due to acid to change by a factor of 30.

Table 3

HYDROLYSIS OF β-PROPIOLACTONE AT 25° IN STRONG ACID SOLUTIONS

(Long and Purchase)

Acid	Conc., M	a_{H_2O}	k_{obs} $\times 10^3$	$k_a F(H_3O^+)$ $\times 10^3$	$-H_0$
$HClO_4$	1.83	0.916	3.65	0.61	0.58
$HClO_4$	2.60	0.862	5.38	2.52	0.94
$HClO_4$	2.75	0.850	5.21	2.39	1.01
$HClO_4$	3.26	0.804	8.89	6.22	1.24
$HClO_4$	3.93	0.734	14.41	11.97	1.55
$HClO_4$	4.53	0.665	32.2	30.1	1.84
$HClO_4$	4.58	0.659	32.6	30.4	1.85
$HClO_4$	5.36	0.567	69.9	68.0	2.28
$NaClO_4$	4.0 ⎫	0.5	17.6	16.4	1.72
$HClO_4$	2.0 ⎭				
H_2SO_4	2.01	0.903	5.39	2.39	0.73
H_2SO_4	3.14	0.810	9.52	6.83	1.28
H_2SO_4	3.77	0.744	13.1	10.6	1.57
H_2SO_4	3.92	0.728	18.9	16.4	1.64
H_2SO_4	5.07	0.595	57.3	55.3	? ??

$$k_w = 3.32 \times 10^{-3}\,\text{min}^{-1}$$

The H_0 Function†

These changes, however, become intelligible if the acidity of the solutions is expressed in another way. Hammett and his co-workers[19] have shown how it is possible to define an acidity function which accurately expresses the tendency of a given solution to transfer a proton to a neutral base under conditions where pH or hydrogen ion concentration lose their significance. This function, H_0, is defined by the equation

$$H_0 = -\log \left(\frac{a_{H^+} f_B}{f_{BH^+}} \right) \tag{25}$$

where a_{H^+} is the total activity of the hydrogen ion in the solution, and (f_B/f_{BH^+}) is the ratio of the activity coefficients of a neutral base B and its conjugate acid BH^+. There is much evidence to indicate that this ratio has the same value for all bases in a given solvent.[20] Hence the acidity function

† For a review see reference 21.

H_0 is a property of the solvent and is independent of the base used. If we take the equation for the ionization constant of an acid BH^+

$$pK_a = -\log \frac{a_{H^+} a_B}{a_{BH^+}} \qquad (26)$$

and combine it with (24) we obtain the useful relation

$$H_0 = pK_a + \log \frac{C_B}{C_{BH^+}} \qquad (27)$$

since the activities are related to the concentrations and activity co-efficients by $a_B = C_B f_B$, etc. Equation 27 shows that the fraction of a neutral base B which is converted to its conjugate acid in any solution depends on two factors only, the base strength of B as determined by the thermodynamic quantity pK_a, independent of solvent, and the H_0 value for the solvent. Equation 27 also gives a method for measuring H_0 since an indicator whose pK_a value is known from data in dilute aqueous solution (which is the standard state) may be used in another solution and the ratio C_B/C_{BH^+} determined spectrometrically or colorimetrically. Values of H_0 for mixtures of various acids in water and other solvents are given by Paul and Long.[21]

The acidity function H_0 is not the same as the pH, but it does approach it in dilute aqueous solution where the activity coefficient ratio approaches unity. If negatively charged bases such as bisulfate ion or the picrate ion are involved, then the H_0 function loses its significance and must be replaced by a function H_- defined by

$$H_- = -\log \left(\frac{a_{H^+} f_{B^-}}{f_{BH}} \right) \qquad (28)$$

which gives the tendency of a solution to transfer a proton to a singly negative charged base. H_- is a property of the solvent also but is not equal to H_0.

Since H_0 is a logarithmic function it is convenient to define its anti-logarithm by the equation

$$H_0 = -\log h_0 \qquad (29)$$

This new function, h_0, is equal to the activity of the hydrogen ion in dilute aqueous solution. Also in these solutions it becomes equal to the activity or concentration of the hydronium ion, H_3O^+. For other solutions the terms a_{H^+} and $a_{H_3O^+}$ are not synonomous since the proton can be bound to to a variety of donor molecules.

Figure 1 shows the result of plotting $\log (k_{obs} - k_w a_{H_2O})$ against the H_0 function in the acid-catalyzed hydrolysis of β-propiolactone. The solid line is drawn with unit slope, which is the correct value if the acid reaction has a first-order dependence on the acidity as determined by h_0. The agreement with the best straight line that could be drawn through the

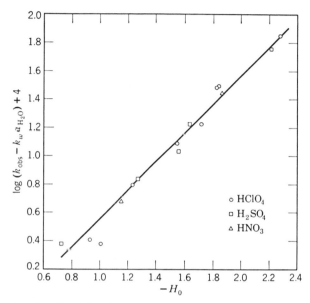

Fig. 1. Rate of acid-catalyzed hydrolysis of β-propiolactone versus H_0 (Long and Purchase).

experimental points is good. It can be concluded then that the rate of the acid-catalyzed reaction is given by

$$\text{rate} = k_a[\text{lactone}]h_0 = k_a[\text{lactone}]\frac{a_{H^+}f_B}{f_{BH^+}} \tag{30}$$

where B is some neutral base. The interpretation of this in terms of mechanism will be postponed until the experimental facts on the hydrolysis of γ-butyrolactone have been presented.

The Hydrolysis of γ-Butyrolactone

The hydrolysis of the γ-lactone, which is an unstrained five-membered ring, differs in several important respects from that of the β-lactone.

$$\text{CH}_2\text{—CH}_2\text{—CH}_2\text{—C}{=}\text{O} + \text{H}_2\text{O} \rightarrow \text{HO—CH}_2\text{—CH}_2\text{—CH}_2\text{—COOH}$$
$$\underline{\qquad\quad\text{O}\qquad\quad}$$

$$\tag{31}$$

There is a base-catalyzed reaction and an acid-catalyzed reaction but no detectable water reaction. Also the acid-catalyzed reaction shows up at low values of the hydrogen ion concentration and is pseudo-first-order with a rate constant directly proportional to the hydrogen ion concentration for moderate concentrations.[22] The base-catalyzed reaction is second-order, and the rate constant can be expressed by the equation $k = 2.2 \times 10^8 e^{-11,300/RT}$ liters/mole-sec.[23] A comparison of this figure with that for the β-lactone basic hydrolysis, $k = 1.43 \times 10^{10} e^{-13,400/RT}$ and that of ethyl acetate, $k = 1.7 \times 10^7 e^{-11,300/RT}$, shows that the γ-lactone more closely resembles the simple ester. It is perhaps surprising that the activation energy for breaking the strained ring is higher than for the unstrained compounds. Evidently the activation energy reflects the repulsive energy of adding an hydroxide ion even more than the energy required to rupture the ring.

For the γ-lactone Long and Friedman[24] have shown by the use of O^{18} labeled water that acyl oxygen scission occurs in both the acid- and base-catalyzed reactions. Here also the reactions resemble those of simple aliphatic esters. The γ-lactone is like simple esters also in not displaying the unusual reactivity towards nucleophilic reagents shown by the β-lactone.

Long, McDevit, and Dunkle[25] have studied carefully the kinetics of the acid-catalyzed hydrolysis of γ-butyrolactone, paying particular attention to salt effects and to the reaction in concentrated acid solutions. The activity coefficients of the lactone in various solutions were found from distribution experiments (see Chapter 7). Also the variation in the concentration equilibrium constant between the lactone and its hydrolysis product, γ-hydroxybutyric acid, was studied in strong acid and salt solutions.

$$CH_2—CH_2—CH_2—C{=}O + H_2O \rightleftharpoons HOCH_2—CH_2—CH_2—COOH$$
$$\underset{\hspace{2em}O\hspace{2em}}{\rule{4em}{0.4pt}} \tag{32}$$

$$K_c = \frac{C_L}{C_A} = K_{eq} \frac{f_A}{f_L a_{H_2O}} \tag{33}$$

where the subscripts L and A stand for lactone and acid, respectively, and K_{eq} is the thermodynamic equilibrium constant. The value of K_{eq} at $25°$ is 2.60 (determined in dilute aqueous solution under slightly acid conditions).

The rate equation for the acid-catalyzed hydrolysis is

$$\text{rate} = k_h' C_L - k_l' C_A \tag{34}$$

Since the reaction does not go to completion but comes to equilibrium, the reverse reaction must be considered. The pseudo-first-order constants for hydrolysis and lactonization, k_h' and k_l', are functions of the hydrogen ion concentration and can be converted to second-order constants k_h and k_l by dividing by the hydrogen ion concentration. Since $k_l' = K_c k_h'$, where K_c is the concentration equilibrium constant, the above equation can be integrated to give

$$k_h'(1 + K_c) = \frac{1}{t} \ln \frac{x_e}{x - x_e} \tag{35}$$

where x and x_e are the hydroxyacid concentrations at $t = t$, and $t = \infty$, respectively. (See Chapter 8 for reversible first-order reactions.) For any solution used the value of K_c must be independently determined.

Table 4

HYDROLYSIS IN HYDROCHLORIC ACID SOLUTIONS, $0°$ C, LACTONE CONCENTRATION APPROXIMATELY $0.7M$
$k \times 10^2$, time in minutes

(Long, McDevit, and Dunkle)

C_{acid}	K_c	k_h'	k_h	k_l' (calc.)	k_l (calc.)
0.485	2.63	0.0507	0.105	0.133	0.274
0.996	2.97	0.106	0.106	0.314	0.315
1.46	3.32	0.153	0.105	0.508	0.348
1.96	3.77	0.214	0.109	0.807	0.412
2.46	4.27	0.283	0.115	1.208	0.491
2.815	4.68	0.328	0.116	1.54	0.544
3.45	5.51	0.410	0.119	2.26	0.652
3.94	6.27	0.494	0.125	3.10	0.786

To determine the effect of concentrated acids a number of experiments were run in perchloric acid and hydrochloric acid solutions. Table 4 shows some of the data collected in hydrochloric acid at $0°$. The values of k_h' were calculated from equation 35 and the values of k_l' from the concentration equilibrium constant and corresponding k_h'. It is seen that k_h, the second-order constant for hydrolysis obtained by dividing by the acid concentration, is fairly constant, despite large changes in the ionic strength. The calculated values of k_l, however, show a much wider variation. Figures 2 and 3 show log k_h' plotted against log $C_{\mathrm{H_3O^+}}$, and log k_l' against the H_0 function. A straight line of unit slope can be drawn through the points in each case, showing that the rate equation 33 is closely given by

$$\text{rate} = k_h C_{\mathrm{L}} C_{\mathrm{H_3O^+}} - k_l C_{\mathrm{A}} h_0 \tag{36}$$

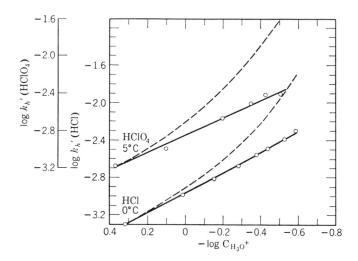

Fig. 2. Rate of hydrolysis of γ-butyrolactone in acid solutions. Dotted lines are predicted curves for dependence of rate on H_0 (Long, McDevit, and Dunkle).

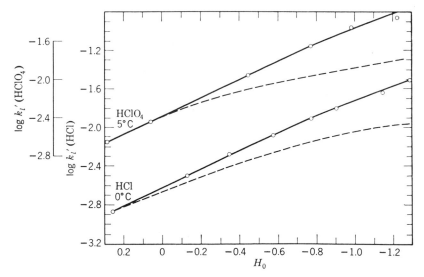

Fig. 3. Rate of lactonization of γ-hydroxybutyric acid in acid solutions. Dotted lines are predicted curves for dependence of rate on $\log C_{H_3O^+}$ (Long, McDevit, and Dunkle).

That is, the rate of the hydrolysis reaction is proportional to the concentration of hydrogen ion, but the rate of the lactonization reaction is proportional to $a_{H^+} f_B / f_{BH^+}$, where B is a neutral molecule which can function as a base. For comparison Figs. 2 and 3 also show the predicted values of log k_h' and log k_l' for dependence on H_0 and log $C_{H_3O^+}$. These lines are quite incompatible with the data.

The difference in behavior of the forward and reverse reaction is explained by Long et al. on the basis of a suggestion made originally by Zucker and Hammett.[26] This was that an acid-catalyzed reaction's dependence on the H_0 function would show whether a water molecule was present in the activated complex. Of the four mechanisms for acid hydrolysis of an ester listed by Day and Ingold, two involve acyl oxygen cleavage. These are the $A'1$ and $A'2$ mechanisms of equations 4 and 5.

For the examples under discussion the $A'1$ mechanism can be written as

$$L + H^+ \rightleftharpoons LH^+ \qquad \text{equilibrium} \tag{37}$$

$$LH^+ \rightarrow M^+ \rightarrow \text{product} \qquad \text{slow} \tag{38}$$

where M^+ is the activated complex. The rate equation would be

$$\text{rate} = k' C_{LH^+} f_{LH^+} / f_{M^+} = k C_L a_{H^+} f_L / f_{M^+} \tag{39}$$

where the constant k includes the rate constant k' for (38) and the equilibrium constant for (37). Since the lactone and the activated complex differ in composition from each other only by a proton, they have the same relationship as B and BH$^+$. Hence $a_{H^+} f_L / f_{M^+}$ may be expected to be equal to h_0. The log of the rate constant will be a linear function of H_0 with a slope of unity.

For the $A'2$ mechanism the first two steps are

$$L + H^+ \rightleftharpoons LH^+ \qquad \text{equilibrium} \tag{40}$$

$$LH^+ + H_2O \rightarrow M^+ \rightarrow \text{product} \qquad \text{slow} \tag{41}$$

$$\text{rate} = k C_L a_{H^+} a_{H_2O} f_L / f_{M^+} \tag{42}$$

Here M^+ differs from L by containing a water molecule as well as a proton. Hence f_L / f_{M^+} is not equal to f_B / f_{BH^+}, and a proportionality between the observed rate constant and h_0 is not expected. Also the activity of water appears in the numerator of the rate equation. Equation 41 may be rewritten as shown in (43). This is done by making use

$$\text{rate} = k C_L C_{H_3O^+} f_L f_{H_3O^+} / f_{M^+} \tag{43}$$

of the equilibrium

$$H_3O^+ \rightleftharpoons H^+ + H_2O \tag{44}$$

and by defining the unknown equilibrium constant for this reaction as unity.

In order for the $A'2$ mechanism to have a rate which is a linear function of the hydrogen ion concentration, it is necessary that the ratio $f_L f_{H_3O^+}/f_{M^+}$ be constant in different solutions. This may be nearly true, since the total charge is the same and the sum of the components is the same in numerator and denominator. We may conclude from this analysis that the hydrolysis of γ-butyrolactone under acid conditions follows the usual $A'2$ mechanism postulated for simple esters. The β-lactone in strong acid, with a rate proportional to h_0, appears to follow an $A'1$ mechanism in which the rate-determining step is a cleavage to an acylium ion, as shown in (45).

$$\begin{array}{c} \text{CH}_2\text{—CH}_2 \\ |\qquad\quad | \\ {}^+\text{O}\text{——}\text{C}\text{=}\text{O} \\ | \\ \text{H} \end{array} \rightarrow \text{HO—CH}_2\text{—CH}_2\text{—C}\underset{+}{=}\text{O} \qquad (45)$$

Undoubtedly the strained four-membered ring makes such a rupture easier than for unstrained esters or lactones.

The above analysis is greatly strengthened by the observations[27] that in concentrated acid solutions simple ester hydrolysis is more nearly proportional to the concentration of hydrogen ion than to h_0. Exact proportionality between rate and concentration is not found, nor is it really expected since the ratio of activity coefficients contained in (43) will not be exactly unity always.

A consideration of the $A'2$ mechanism for the reverse reaction of lactone formation is instructive. The steps would be, from the principle of microscopic reversibility,

$$A + H^+ \rightleftharpoons AH^+ \qquad \text{equilibrium} \qquad (46)$$

$$AH^+ \rightarrow M^+ \rightarrow \text{product} \qquad \text{slow} \qquad (47)$$

The rate equation is

$$\text{rate} = kC_A a_{H^+} f_A/f_{M^+} = kC_A h_0 \qquad (48)$$

Again A and M$^+$ differ only by a proton, so that the rate is a function of h_0 and not of hydrogen ion concentration. This prediction is in agreement with the experimental results for the lactonization reaction of γ-butyrolactone in strong acid (compare Fig. 3). The agreement lends further support to the $A'2$ mechanism for hydrolysis, since the forward and reverse reactions must have the same mechanism.

Liang and Bartlett[28] have studied the hydrolysis of the tertiary lactone, β-isovalerolactone under acid, neutral, and basic conditions. The rate

$$\begin{array}{c} \text{CH}_3 \\ | \\ \text{CH}_3\text{—C—CH}_2 \\ |\qquad\quad | \\ \text{O—C=O} \end{array} + \text{H}_2\text{O} \rightarrow \begin{array}{c} \text{CH}_3 \\ | \\ \text{CH}_3\text{—C—CH}_2\text{—COOH} \\ | \\ \text{O} \\ | \\ \text{H} \end{array} \qquad (49)$$

constant in strong acid follows the H_0 function as expected. Table 5 shows the rate constants for the primary, secondary, and tertiary lactones under the three conditions. The rates for basic conditions seem normal for attack of the hydroxide ion at the carbonyl group. However, the drop in rate in going from primary to secondary, and the sharp increase in rate going from secondary to tertiary, suggest a change in mechanism in neutral or acid solution.

It is reasonable that in the isovalerolactone we are dealing with a case of alkyl oxygen fission because of the stabilizing influence of the alkyl groups

Table 5

RATE CONSTANTS IN WATER AT 25° FOR β-LACTONES

(Bartlett and Liang)

Lactone	k_{H_2O}, sec.$^{-1}$	k_{OH^-}, l/mole sec.	k_a/h_0 l/mole sec.
β-Propio-	5.6×10^{-5}	1.2	5.8×10^{-6}
β-Butyro-	1.4×10^{-5}	0.82	$\sim 2.3 \times 10^{-6}$
β-Isovalero-	1.35×10^{-3}	0.22	2.0×10^{-3}

on the carbonium ion formed.[28] This would be the case under neutral or acid conditions but not alkaline. In acid the mechanism is thus $A''1$ since the H_0 function is followed. The corresponding reaction under neutral

$$
\begin{array}{cc}
(CH_3)_2C\!-\!CH_2 & (CH_3)_2C\!-\!CH_2 \\
| \quad | & + \quad | \\
O\!-\!C\!\!=\!\!\overset{+}{O}H & O\!\!=\!\!C\!-\!OH
\end{array} \rightarrow \tag{50}
$$

conditions would be as in (51). The intermediate in this case could either

$$
\begin{array}{cc}
(CH_3)_2C\!-\!CH_2 & (CH_3)_2C\!-\!CH_2 \\
| \quad | & + \quad | \\
O\!-\!C\!\!=\!\!O & O\!\!=\!\!C\!-\!O^-
\end{array} \rightarrow \tag{51}
$$

react with water to give the hydroxy acid or, very easily, form carbon dioxide and isobutylene. In fact the lactone gives 63 per cent decarboxylation and 37 per cent hydrolysis in neutral solution.[28] Under strongly acid

$$
\begin{array}{c}
(CH_3)_2C\!-\!CH_2 \\
+ \quad | \\
O\!\!=\!\!C\!-\!O^-
\end{array} \rightarrow (CH_3)_2C\!\!=\!\!CH_2 + CO_2 \tag{52}
$$

conditions the amount of decarboxylation is small and hydroxy acid is chiefly formed.

The H_0 function thus appears as a useful tool not only for describing the properties of strong acid solutions but also for investigating reaction mechanisms. The definition of the H_0 function predicts a linear relationship between h_0 and the rate constant for those acid-catalyzed reactions in which the activated complex differs from the reactant only by the addition of a proton. Other examples that may be cited are the inversion of sucrose and the decomposition of trioxane. Long and Paul discuss a number of other cases.[29]

Another acid-catalyzed reaction which, like the hydrolysis of γ-butyro-lactone, does not depend on the H_0 function but on the concentration of hydrogen ion, is the enolization of acetophenone (Zucker and Hammett[26]). Here the rate-determining step is the removal of a proton by a water molecule from an oxonium complex of the ketone, as shown in (53). The

$$C_6H_5-\underset{+}{\overset{\overset{\displaystyle OH}{|}}{C}}-CH_3 + H_2O \rightarrow C_6H_5-\overset{\overset{\displaystyle OH}{|}}{C}=CH_2 + H_3O^+ \tag{53}$$

activated complex differs from acetophenone by not only a proton but also a water molecule. Hence the dependence on acidity is similar to that for the hydrolysis of the γ-lactone.

In the similar case of enolization of acetone under acid conditions, different results are found.[30] After correcting for the amount of acetone bound up as oxonium complex, it is found that the rate of enolization depends on the H_0 function rather than the concentration of acid. Since almost certainly the mechanism is the same as for acetophenone, and a molecule of water is in the transition state, this illustrates the uncertainty of the method in trying to prove the existence of the water molecule. For the case where water is absent in the activated complex, adherence to h_0 seems to be always found. For the case where water is present, dependence on h_0, $C_{H_3O^+}$, or on neither may be found.

REFERENCES

1. S. C. Datta, J. N. E. Day, and C. K. Ingold, *J. Chem. Soc.*, 838 (1939); J. N. E. Day and C. K. Ingold, *Trans. Faraday Soc.*, 37, 686 (1941).
2. M. L. Bender, *J. Am. Chem. Soc.*, 73, 1626 (1951).
3. H. P. Treffers and L. P. Hammett, *J. Am. Chem. Soc.*, 59, 1708 (1937).
4. J. A. Leisten, *J. Chem. Soc.*, 1572 (1956).
5. E. D. Hughes, C. K. Ingold, and S. Masterman, *J. Chem. Soc.*, 840 (1939).
6. P. Salomaa, *Acta Chem. Scand.*, 11, 125, 132, 141, 235, 239 (1957).
7. (a) M. Balfe et al., *J. Chem. Soc.*, 556, 605 (1942), 797, 803, 807 (1946); (b) G. S. Hammond and J. T. Rudesill, *J. Am. Chem. Soc.*, 72, 2769 (1950).
8. S. G. Cohen and A. Schneider, *J. Am. Chem. Soc.*, 63, 3382 (1941).
9. A. R. Olson and R. J. Miller, *J. Am. Chem. Soc.*, 60, 2687 (1938).

10. H. Johansson, *Chem. Zentr.*, *87*, II, 557 (1916).
11. F. A. Long and M. Purchase, *J. Am. Chem. Soc.*, *72*, 3267 (1950).
12. H. Johansson, *Ber.*, *48*, 1256 (1915).
13. E. Grunwald and S. Winstein, *J. Am. Chem. Soc.*, *70*, 841 (1948).
14. A. R. Olson and J. L. Hyde, *J. Am. Chem. Soc.*, *63*, 2459 (1941).
15. B. Holmberg, *J. prakt. Chem.*, *88*, 553 (1913); H. N. K. Rørdam, *J. Chem. Soc.*, 2931 (*1932*).
16. T. L. Gresham, J. E. Jansen, F. W. Shaver, et al., *J. Am. Chem. Soc.*, *70*, 998, 999, 1001, 1004, 4277 (1948); *71*, 661, 2807 (1949); *72*, 72 (1950).
17. P. D. Bartlett and G. Small, *J. Am. Chem. Soc.*, *72*, 4867 (1950).
18. A. R. Olson and P. V. Youle, *J. Am. Chem. Soc.*, *73*, 2468 (1951).
19. L. P. Hammett, *Physical Organic Chemistry*, McGraw-Hill Book Co., New York, 1940, p. 267 ff.
20. L. P. Hammett and M. A. Paul, *J. Am. Chem. Soc.*, *56*, 827 (1934).
21. M. A. Paul and F. A. Long, *Chem. Revs.*, *57*, 1 (1957).
22. P. Henry, *Z. physik. Chem.*, *10*, 96 (1892); A. Kailan, *ibid.*, *94*, 111 (1920); *101*, 63 (1922).
23. D. S. Hegen and J. H. Wolfenden, *J. Chem. Soc.*, 508 (1939).
24. F. A. Long and L. Friedman, *J. Am. Chem. Soc.*, *72*, 3692 (1950).
25. F. A. Long, W. F. McDevit, and F. B. Dunkle, *J. Phys. & Colloid Chem.*, *55*, 829 (1951).
26. L. Zucker and L. P. Hammett, *J. Am. Chem. Soc.*, *61*, 2791 (1939).
27. (a) R. P. Bell, A. L. Dowding and J. A. Noble, *J. Chem. Soc.*, 3106 (1955); (b) C. T. Chmiel and F. A. Long, *J. Am. Chem. Soc.*, *78*, 3326 (1956).
28. H. T. Liang and P. D. Bartlett, *J. Am. Chem. Soc.*, *80*, 3585 (1958).
29. F. A. Long and M. Paul, *Chem. Revs.*, *57*, 935 (1957); for another viewpoint, see E. Whalley, *Trans. Faraday Soc.*, *55*, 798 (1959).
30. G. Archer and R. P. Bell, *J. Chem. Soc.*, 3228 (1959).

D. The Aldol Condensation and the Cleavage of Diacetone Alcohol

The condensation of acetaldehyde to aldol and of acetone to diacetone alcohol are two reactions which may be discussed together since they have almost identical mechanisms as far as the steps involved are concerned. Small differences, however, are sufficient to make the observed kinetics of the two reactions quite dissimilar. The reaction is shown in (1).

$$
\underset{\substack{\parallel \\ O}}{R-C-CH_3} + \underset{\substack{\parallel \\ O}}{R-C-CH_3} \rightleftharpoons R-\underset{\substack{| \\ O \\ | \\ H}}{\overset{\substack{CH_3 \\ |}}{C}}-CH_2-\underset{\substack{\parallel \\ O}}{C}-R \tag{1}
$$

where R is H or CH_3, basic catalysts being used in each case. The reaction is reversible though subsequent reactions in the aldol condensation (dehydration and polycondensation) tend to make it irreversible. The point of equilibrium is well to the right for acetaldehyde, even in dilute aqueous solution, but it is well to the left for acetone even in concentrated acetone solution. For this reason the aldol condensation is conveniently studied only from the acetaldehyde side and the acetone reaction only from the diacetone alcohol side. The reverse process can in each case be reconstructed from the mechanism of the forward process so that a complete comparison of (1) can be made for $R = CH_3$ and $R = H$.

The assumed mechanism is one of great importance in organic chemistry, since it is operative for most condensation reactions. The general sequence can be given as follows (see Chapter 9):

$$HA + B \underset{k_2}{\overset{k_1}{\rightleftharpoons}} BH^+ + A^- \tag{2}$$

$$A^- + C \xrightarrow{k_3} \text{product} \tag{3}$$

where HA is an organic compound with an acidic hydrogen which can be ionized by a base B, forming a carbanion A^-. The carbanion can either regenerate the reactants by pulling a proton from BH^+, or react with a substrate C to form the desired product of the reaction. In the present example, HA and C are the same substance, both being either acetaldehyde or acetone. The kinetics of (2) and (3) are complicated unless, as is usual, the carbanion is reactive enough to come to a kinetic equilibrium so that the steady-state method may be used for its concentration. This method gives

$$[A^-] = \frac{k_1[HA][B]}{k_3[C] + k_2[BH^+]} \tag{4}$$

$$\frac{d[\text{product}]}{dt} = \frac{k_1 k_3[HA][B][C]}{k_3[C] + k_2[BH^+]} \tag{5}$$

If more than one base is present and effective, then (5) can be written as

$$\frac{d[\text{product}]}{dt} = \frac{k_3[HA][C] \sum_j k_{1j}[B_j]}{k_3[C] + \sum_j k_{2j}[BH_j^+]} \tag{6}$$

Two limiting cases are of interest in which $k_3[C]$ is either much greater or much smaller than $k_2[BH^+]$, assuming only one base present. The first case leads to

$$\text{rate} = k_1[HA][B] \tag{7}$$

and the second to

$$\text{rate} = \frac{k_1 k_3[HA][C][B]}{k_2[BH^+]} = \frac{k_1 k_3[HA][C][S^-]}{k_2 K_B} \tag{8}$$

where K_B is the ionization constant of the base and S^- is the anion of the solvent SH. Equation 7 corresponds to general base catalysis, and equation 8 to a specific catalysis by S^-, the usual case being by hydroxide ion in aqueous solution.

For acetone and acetaldehyde equations 2 and 3 are shown in (9)–(11),

$$R-\overset{\overset{\text{O}}{\|}}{C}-CH_3 + B \underset{k_2}{\overset{k_1}{\rightleftharpoons}} R-\overset{\overset{\text{O}}{\|}}{C}-CH_2^- + BH^+ \tag{9}$$

$$R-\overset{\overset{\text{O}}{\|}}{C}-CH_2^- + R-\overset{\overset{\text{O}}{\|}}{C}-CH_3 \underset{k_4}{\overset{k_3}{\rightleftharpoons}} R-\overset{\overset{\text{O}}{\|}}{C}-CH_2-\overset{\overset{\text{R}}{|}}{\underset{\underset{\text{CH}_3}{|}}{C}}-O^- \tag{10}$$

$$R-\overset{\overset{\text{O}}{\|}}{C}-CH_2-\overset{\overset{\text{R}}{|}}{\underset{\underset{\text{CH}_3}{|}}{C}}-O^- + BH^+ \underset{k_6}{\overset{k_5}{\rightleftharpoons}} R-\overset{\overset{\text{O}}{\|}}{C}-CH_2-\overset{\overset{\text{R}}{|}}{\underset{\underset{\text{CH}_3}{|}}{C}}-OH + B \tag{11}$$

the last step being very rapid. The rate equations 7 and 8 become, in water,

$$\text{rate} = k_1 \left[R-\overset{\overset{\text{O}}{\|}}{C}-CH_3 \right] [B] \tag{12}$$

$$\text{rate} = (k_1 k_3 / k_2 K_B) \left[R-\overset{\overset{\text{O}}{\|}}{C}-CH_3 \right]^2 [OH^-] \tag{13}$$

The factor which determines whether (12) or (13) is observed is the relative velocity of the carbanion towards condensing at the carbonyl compared to reaction with BH^+. A high rate of condensation leads to (12), and a high rate of neutralization or a low rate of condensation to (13). The base B is not used up during a reaction so that B, BH^+, and hydroxide ion remain constant during a run. The reaction is thus either first- or second-order in the carbonyl compound, and the rate varies from run to run with either the concentration of the base B or the hydroxide ion. Since in water solution more than one base is usually present, (12) is more correctly written as

$$\text{rate} = \left[R-\overset{\overset{\text{O}}{\|}}{C}-CH_3 \right] \sum_j k_{1j}[B_j] \tag{14}$$

Figure 1 represents schematically how the free energy varies with extent of reaction for the situations given by (12) and (13). The heights of the free-energy barriers which the carbanion A^- has to overcome in going backwards or forwards determine which kinetics will be found, the steps and the intermediates being identical. Figure 1 is useful for considering the mechanism of the reverse reaction, the cleavage of the condensed

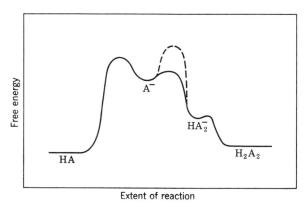

Extent of reaction

Fig. 1. Free-energy barriers in condensation reactions of carbanions. Full line is for easy condensation step. Dotted line is for difficult condensation step.

product. In each case there will be an equilibrium between the dimer and its ion established first, as shown in (15). The ion will then be cleaved to

$$R-\overset{\overset{O}{\|}}{C}-CH_2-\overset{\overset{R}{|}}{\underset{\underset{H}{|}}{\underset{O}{|}}}{C}-CH_3 + B \underset{k_5}{\overset{k_6}{\rightleftharpoons}} R-\overset{\overset{O}{\|}}{C}-CH_2-\overset{\overset{R}{|}}{\underset{\underset{O_-}{|}}{C}}-CH_3 + BH^+ \quad (15)$$

carbanion and carbonyl compound in a slow step, as shown in (16). This

$$R-\overset{\overset{O}{\|}}{C}-CH_2-\overset{\overset{R}{|}}{\underset{\underset{O_-}{|}}{C}}-CH_3 \underset{k_3}{\overset{k_4}{\rightleftharpoons}} R-\overset{\overset{O}{\|}}{C}-CH_2^- + R-\overset{\overset{O}{\|}}{C}-CH_3 \quad (16)$$

reaction is reversible, but another reaction competes with the reverse process, as shown in (17). The reaction with BH^+ is less rapid than the

$$R-\overset{\overset{O}{\|}}{C}-CH_2^- + BH^+ \underset{k_1}{\overset{k_2}{\rightleftharpoons}} R-\overset{\overset{O}{\|}}{C}-CH_3 + B \quad (17)$$

reversal in one case, so that (16) becomes an equilibrium also and the picking up of a proton becomes the rate-determining step. The rate equation is then

$$\text{rate} = k_2[\text{BH}^+][\text{A}^-] = \frac{k_2 k_4 k_6 [\text{H}_2\text{A}_2][\text{B}]}{k_3 k_5 [\text{HA}]} \tag{18}$$

where H_2A_2 is the condensed product, that is, aldol or diacetone alcohol. The ratios k_4/k_3 and k_6/k_5 are the equilibrium constants for (16) and (15). The cleavage reaction is thus general base-catalyzed and inhibited by the presence of the monomer. Equating the rates of the forward reaction given by (12) and the reverse reaction gives the equilibrium condition:

$$[\text{H}_2\text{A}_2]/[\text{HA}]^2 = k_1 k_3 k_5 / k_2 k_4 k_6 = K_{\text{eq}} \tag{19}$$

For the second case the reaction with BH^+ is fast compared to the reversal of (16). Hence the cleavage to carbanion and carbonyl is the rate-determining step. The rate is given by

$$\text{rate} = k_4[\text{HA}_2^-] = \frac{k_4 k_6 [\text{H}_2\text{A}_2][\text{B}]}{k_5[\text{BH}^+]} = \frac{k_4 k_6 [\text{H}_2\text{A}_2][\text{OH}^-]}{k_5 K_{\text{B}}} \tag{20}$$

Setting the rate of the forward reaction, equation 13, equal to the rate to the reverse reaction gives the equilibrium constant

$$[\text{H}_2\text{A}_2]/[\text{HA}]^2 = k_1 k_3 k_5 / k_2 k_4 k_6 = K_{\text{eq}} \tag{21}$$

Turning now to a consideration of the experimental facts for the acetone-diacetone alcohol reaction, Koelichen[1] has measured the equilibrium constant. The equilibrium was established in various mixtures of acetone and water, using potassium hydroxide as a catalyst. The concentration of the catalyst had no effect on the point of equilibrium, but the amount of water had some influence. Table 1 shows the equilibrium constant in liters per mole at 25° for various compositions. A safe extrapolation can be made to find the equilibrium constant in dilute aqueous solution. Koelichen also determined the equilibrium constant at 0° so that the heat and entropy changes can be found. For a solution containing 60 per cent water, these are $\Delta H^0 = -7.8$ kcal and $\Delta S^0 = -32$ E.U. The decrease in entropy, chiefly due to the loss of translational entropy of one mole of acetone, offsets the somewhat greater bond energies of the dimer so that only a few per cent of dimer exists at equilibrium.

Koelichen also measured the rate of cleavage of diacetone alcohol. Somewhat more accurate data have been obtained by French;[2] by Murphy;[3] and by La Mer and Miller.[4] Table 2 shows the results of French on the effect of catalyst concentration on the rate. The reaction is

Table I

EQUILIBRIUM CONSTANT AT 25° FOR
DIACETONE ALCOHOL
(Koelichen)

$$2CH_3-\overset{\overset{\displaystyle O}{\|}}{C}-CH_3 \rightleftharpoons CH_3-\overset{\overset{\displaystyle CH_3}{|}}{\underset{\underset{\displaystyle H}{|}}{\underset{\displaystyle O}{|}}}{C}-CH_2-\overset{\overset{\displaystyle O}{\|}}{C}-CH_3$$

% Water by Weight	K_{eq}, liters/mole
4	0.024
20	0.029
39.5	0.032
60	0.035
80	0.037
100	(0.039)

pseudo-first-order, being dependent on the first power of the diacetone alcohol concentration and on the first power of the hydroxide ion concentration. The latter, however, remains constant during a run. The second-order

Table 2

(French)

Kinetics of $CH_3-\overset{\overset{\displaystyle CH_3}{|}}{\underset{\underset{\displaystyle OH}{|}}{}}{C}-CH_2-\overset{\overset{\displaystyle O}{\|}}{C}-CH_3 \rightarrow 2CH_3-\overset{\overset{\displaystyle O}{\|}}{C}-CH_3$ at 25°

NaOH Concentration, moles/liter	k_1, min^{-1}	k_1/[NaOH]
5×10^{-3}	2.32×10^{-3}	0.465
10×10^{-3}	4.67×10^{-3}	0.467
20×10^{-3}	9.40×10^{-3}	0.470
40×10^{-3}	19.2×10^{-3}	0.479
100×10^{-3}	47.9×10^{-3}	0.479

rate constant can be expressed as a function of temperature by the relation $k = 1.31 \times 10^{11}e^{-18,020/RT}$ liters/mole-sec. The activation energy is not constant but varies somewhat with temperature (La Mer and Miller[4]).

The data of Table 2 can be represented by an equation

$$\text{rate} = k[H_2A_2][OH^-] \tag{22}$$

This does not exclude general base catalyses, since water which is the only other base present in appreciable amounts may have too small an effect to be noticeable. However, by using buffered solutions of phenol and phenolate ion (French[2]), and of tertiary amines with their 'onium salts,[5] it was shown that other bases are not effective as such and that equation 22 correctly represents the rate. Some confusion existed earlier because of a specific effect of ammonia, primary and secondary amines, which accelerate the cleavage.[6] The acceleration was shown to be due to the formation of ketimines which were readily cleaved, presumably because of the stability of the dipolar-ion form analogous to the anion of equation 16 (Westheimer).

$$
\begin{array}{c}
\text{CH}_3 \\
|
\end{array}
\qquad
\begin{array}{c}
\text{CH}_3 \\
|
\end{array}
$$

$$
\text{B} + \text{CH}_3\!-\!\overset{\overset{\displaystyle CH_3}{|}}{\underset{\underset{\displaystyle OH}{|}}{C}}\!-\!\text{CH}_2\!-\!\overset{\overset{}{}}{\underset{\underset{\displaystyle \overset{N^+}{\underset{R_2}{}}}{||}}{C}}\!-\!\text{CH}_3 \rightarrow \text{CH}_3\!-\!\overset{\overset{\displaystyle CH_3}{|}}{\underset{\underset{\displaystyle O_-}{|}}{C}}\!-\!\text{CH}_2\!-\!\overset{}{\underset{\underset{\displaystyle \overset{N^+}{\underset{R_2}{}}}{||}}{C}}\!-\!\text{CH}_3 + \text{BH}^+ \tag{23}
$$

La Mer and Miller[4] have also shown that the rate of cleavage of diacetone alcohol is not affected by the addition of acetone. Hence all the facts agree with equation 20 and not with equation 18. The rate-controlling step is the cleavage of the anion into acetone and carbanion as given in (16). The velocity of condensation, then, should be given by (13) rather than (12). This can be checked indirectly by calculating the rate of condensation from the equilibrium constant and the rate of cleavage. The third-order rate constant for condensation is equal to $0.039 \times 0.465 = 1.8 \times 10^{-2}$ liters^2/mole^2-min. Now k_1 in equation 9 when B is hydroxide ion can be separately measured, since it is the rate of ionization of acetone which is equal to the rate of halogenation of acetone and (within a factor of 40 per cent) to the rate of deuterium uptake of acetone in heavy water. Its value is about 15 liters/mole-min at $25°$.[7] Hence for a one-molar acetone solution the rate of ionization is about 1000 times as great as the rate of condensation. This is in agreement with the ionization being an equilibrium and the rate-determining step being the condensation of the carbanion with a molecule of acetone as in (10).

Although there is no appreciable salt effect in dilute solution for the cleavage of diacetone alcohol (viz., Table 2), in more concentrated salt solutions a decrease in rate is noticed.[8] Some salts with divalent anions, however, increase the rate, as does potassium fluoride. The salt effect for a reaction between an ion and a neutral molecule should be small at low ionic strengths (Chapter 7). At higher salt concentrations the specific nature of the salt is important and no simple theory can be used.

The reaction has been studied in mixed solvents including water, methanol, ethanol, *n*-propanol, *i*-propanol, ethylene glycol, and glycerine.[9]

The rate increases with addition of *i*-propanol, as would be expected for a reaction in which the transition state is an ion of the same charge as the reactant ion but of greater size (Chapter 7). For the other alcohols the rate decreases on adding the alcohol to water. This must be attributed to the reaction

$$OH^- + ROH \rightarrow H_2O + OR^- \tag{24}$$

which reduces the hydroxide ion concentration. (See the similar effect in ethylene-chlorohydrin hydrolysis.) The results are compatible with methanol, glycerine and ethylene glycol being stronger acids than water, and ethanol, and *n*-propanol being acids of about the same strength as water.[7] In the last two cases the rate eventually increases again with increasing alcohol content. The secondary propyl alcohol is evidently a weaker acid than water. An attempt has been made[10] to treat the effect of alcohols as one of changing dielectric constant. Unless (24) is taken into account, such a treatment cannot be valid except, possibly, for *i*-propanol.

A different mechanism for the cleavage of diacetone alcohol has been proposed by Nelson and Butler.[11] Their reaction scheme is shown in (25), (26), and (27). The rate-determining step is the removal of the proton

$$CH_3-\overset{O}{\underset{\|}{C}}-CH_2-\overset{CH_3}{\underset{\underset{CH_3}{|}}{C}}-OH + B \xrightarrow{slow} CH_3-\overset{O}{\underset{\|}{C}}-CH_2-\overset{CH_3}{\underset{\underset{CH_2^-}{|}}{C}}-OH + BH^+ \tag{25}$$

$$CH_3-\overset{O}{\underset{\|}{C}}-CH_2-\overset{CH_3}{\underset{\underset{CH_2^-}{|}}{C}}-OH \xrightarrow{fast} CH_3-\overset{O}{\underset{\|}{C}}-CH_2^- + CH_3-\overset{\overset{H}{O}}{\underset{\|}{C}}=CH_2 \tag{26}$$

$$CH_3-\overset{O}{\underset{\|}{C}}-CH_2^- + BH^+ \xrightarrow{fast} CH_3-\overset{O}{\underset{\|}{C}}-CH_3 + B \tag{27}$$

from a carbon atom. A rapid cleavage to the enol and anion of acetone then occurs. Both the anion and the enol are then converted to the keto form in rapid steps. The chief evidence for this mechanism is that a deuterated diacetone alcohol sample (60 per cent of the hydrogen atoms replaced by deuterium) cleaved 25 per cent more slowly than ordinary diacetone alcohol.

It is known that the removal of the deuteron will be slower than the removal of a proton in a step such as (25). This depends upon the difference in zero point energies in the bonds of hydrogen and deuterium to other atoms.[12] The zero point energy is the half-quantum of vibrational energy that a molecular vibration has even at the absolute zero of temperature. Its value is $\frac{1}{2}h\nu$ where ν is the vibrational frequency. Since the frequency depends inversely on the square root of a reduced mass for the vibration, it will be different for different isotopes even though the force constants are very closely the same.

The situation is illustrated in Fig. 2 for a diatomic hydride, X—H and X—D. Since the frequencies are different by nearly the square root of two

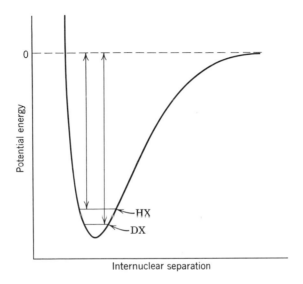

Fig. 2. Potential energy diagram for diatomic molecules HX and DX. The lengths of the arrows represent the dissociation energies of the two isotopic molecules.

in this case, when X is heavy, it can be seen that the dissociation energy into X and D atoms requires more energy than into X and H atoms. Since the frequencies also enter into the vibrational partition function (p. 86), there will also be some entropy effects in the equilibrium constants between the molecules and the free atoms. These are of lesser importance as a rule except at high temperatures.

In a chemical rate, it is not dissociation into a free atom which is to be considered, since the transition state will not correspond to complete dissociation. Further, in an acid-base reaction the proton is transferred from one base to another and the transition state involves at least three atoms. It is the vibrational properties of the activated complex which now must be considered and compared to the vibrational properties of the reactants. Using transition-state theory, it is possible to calculate the isotope effect on the rate if assumptions are made about the several vibrations involving the isotopic atom in the transition state.[13] A safe rule is that if the bonding is weaker in the transition state (usually it will be), then the reaction rate of hydrogen will be greater than that of deuterium.

A loosening of the bonds to hydrogen reduces their force constants and, hence, the vibrational frequencies. For smaller frequencies the differences in zero point energy for hydrogen and deuterium become smaller. The situation becomes similar to Fig. 2 since in the completely dissociated state there is no vibration to consider and no difference between the isotopes. A simplified theory[13b,c] valid at temperatures below 100° C gives the ratio of rate constants as shown in (28)

$$k_H/k_D = \prod_i e^{-h(\nu_{Hi}{}^{\ddagger}-\nu_{Di}{}^{\ddagger}-\nu_{Hi}+\nu_{Di})/2kT} \tag{28}$$

where $2\pi\nu_{Hi} = \sqrt{f_i/m_H}$, f_i being the force constant for the ith vibration. The isotope effect for tritium can be calculated if that for deuterium is known.[13b]

For acid-base reactions of hydrogen bound to carbon, the ratio k_H/k_D has always been found[14] to be as large as 2.7 and sometimes as large as 10. Such large effects can be used as evidence that a hydrogen atom transfer is rate-determining. For bonds to oxygen and nitrogen, the isotope effect is not well established as yet but seems to be of the same order of magnitude as for bonds to carbon. Because of the polarity of bonds to oxygen there can be quite large isotope effects even in reactions in which the proton or hydrogen atom is not directly transferred.[15] Accordingly, kinetic isotope effects are not as valuable as diagnoses of mechanism in these cases.

In any event the observed ratio of 1.25 in the case of diacetone alcohol seems too small to be consistent with a proton transfer as rate-determining. It is a perfectly reasonable result for a secondary isotope effect such as the breaking of bonds $CH_2-C(CH_3)_2$ and $CD_2-C(CD_3)_2$. Isotope factors of this order of magnitude have been observed for such less direct effects.[16] The cleavage of diacetone alcohol also shows another secondary isotope effect in that it goes 45 per cent more rapidly in D_2O as a solvent than in H_2O.[17] However, this is just what is predicted on the basis of a mechanism with a preliminary acid-base equilibrium as in (15).

There are other objections to the mechanism of Nelson and Butler. If (25) is the rate-determining step, then the reaction should be general base-catalyzed, whereas, as we have seen, it appears to be catalyzed specifically by hydroxide ion. However, this criterion is not rigorous since, as mentioned in Chapter 9, a value of β close to unity in the Brönsted equation would make catalysis by other bases than hydroxyl ion difficult to detect. In fact, it is fairly certain that β for a proton as difficult to remove as the methyl hydrogen of equation 25 would be close to unity.

A more serious objection comes from a consideration of the mechanism of the reverse reaction, formation of diacetone alcohol from acetone based on a reversal of reactions 25 to 27. Equation 27 would be a mobile equilibrium, as would the formation of the enol form (29). The rate-

$$CH_3-\overset{\overset{\textstyle O}{\|}}{C}-CH_2^- + BH^+ \rightleftharpoons CH_3-\overset{\overset{\textstyle H}{|}{\underset{}{O}}}{C}=CH_2 + B \qquad (29)$$

determining step could be either the condensation of the enol with the carbanion (reverse of equation 26) or the picking up of the proton (reverse of equation 25), depending on the shape of the free-energy diagram. In either case the reverse of (26) involving addition of a negative ion to a carbon-carbon double bond is certainly less likely than the competing reaction involving addition of the same negative ion to the highly polar carbon-oxygen double bond, as in (30). However, this more likely process

$$CH_3-\overset{\overset{\textstyle O}{\|}}{C}-CH_2^- + CH_3-\overset{\overset{\textstyle O}{\|}}{C}-CH_3 \rightarrow CH_3-\overset{\overset{\textstyle O}{\|}}{C}-CH_2-\overset{\overset{\textstyle CH_3}{|}}{\underset{\underset{\textstyle CH_3}{|}}{C}}-O^- \qquad (30)$$

is just the step proposed in our original mechanism (see equation 10). Hence from the principle of microscopic reversibility the mechanism embodied in equations 25 to 27 is ruled out, at least in the sense that it contributes less to the overall reaction than the mechanism of equations 9 to 11.

To turn now to a consideration of the aldol condensation of acetaldehyde, the equilibrium here is too far to the right to be determined.[18] The kinetics of (31) have been studied by Bell, using sodium hydroxide and other basic

$$2CH_3-\overset{\overset{\textstyle H}{|}}{C}=O \rightleftharpoons CH_3-\overset{\overset{\textstyle H}{|}}{\underset{\underset{\textstyle H}{\overset{|}{O}}}{C}}-CH_2-\overset{\overset{\textstyle H}{|}}{C}=O \qquad (31)$$

catalysts.[19] The results are not entirely clear since polycondensation also occurs. There is also the complication that acetaldehyde exists, at least in part, as a hydrate in water solution. The formation of free aldehyde from the hydrate is not instantaneous, and probably the free aldehyde only is involved in the condensation.[20] The amount of hydrate at equilibrium

$$CH_3-\overset{\overset{\textstyle H}{|}}{C}(-OH)_2 \rightleftharpoons CH_3-\overset{\overset{\textstyle H}{|}}{C}=O + H_2O \qquad (32)$$

appears to be about 50 per cent, and the dehydration is catalyzed by acids and bases (Bell and Higginson).

The aldolization was studied by Bell, using a dilatometric method. Empirically the volume changes could be represented by the equation

$$V_0 - V = A(1 - e^{-k_1 t} + Bt) \qquad (33)$$

where V_0 is the volume at $t = 0$ and A, k_1, and B are constants. This is what would be expected if two consecutive first-order reactions existed, the second being much slower than the first:

$$M_1 \xrightarrow{k_1} M_2 \xrightarrow{k_2} M_3 \qquad (34)$$

The solution to this system for $k_1 \gg k_2$ is (see Chapter 8)

$$M_2 + 2M_3 = M_0(1 - e^{-k_1 t} + k_2 t) \qquad (35)$$

Bell identified k_1 with the rate constant for aldolization and k_2 with a second reaction, probably polycondensation, which is relatively unimportant.

The rate constant k_1 depended on the concentration of sodium hydroxide, the rate equation being given in terms of seconds and moles per liter at 25° by

$$\text{rate} = (2.6 + 10^{-4} + 0.111[OH^-])[CH_3CHO] \qquad (36)$$

except for solutions containing less than $0.003M$ base where the first term disappeared. Accordingly, the rate went to zero at zero concentration of sodium hydroxide. Sodium carbonate was a catalyst but only to the extent given by the hydroxide ion formed by its hydrolysis. A 0.5 molar sodium acetate solution buffered to a pH of 5.5 by acetic acid had no catalytic effect. Also slightly acid solutions of acetaldehyde in water were stable for months with no sign of aldol formation.

One other piece of evidence is available in that Bonhoeffer and Walters[7] formed aldol from acetaldehyde in the presence of heavy water and found that no uptake of deuterium occurred except in the hydroxyl group of the aldol. Their procedure involved treating 10 grams of acetaldehyde with 10 grams of deuterium oxide at 0° with potassium carbonate. After 6 hours the aldol was formed in good yield free of any deuterium attached to carbon.

Thus at high concentrations of acetaldehyde the aldolization is a first-order reaction in acetaldehyde catalyzed by hydroxide ions and possibly by water. It might be considered that the dehydration of the aldehyde hydrate is the rate-determining step since this might obey the experimental kinetics. However, the data on the rate of this reaction show it is much too fast to be a factor.[21]

As Bonhoeffer and Walters pointed out, their data and those of Bell are best explained by assuming that the slow step in forming aldol is the ionization of acetaldehyde to give a carbanion which then rapidly condenses with another acetaldehyde molecule. This would explain the failure of deuterium uptake to occur, since the reaction

$$\overset{-}{C}H_2\overset{\overset{\displaystyle H}{|}}{C}=O + D_2O \rightarrow CH_2D\overset{\overset{\displaystyle H}{|}}{C}=O + OH^- \qquad (37)$$

is slower than the condensation. The rate of aldolization is thus given by (12) or (14), the latter being preferred since general base catalysis should occur.

It is tempting to try to attribute the constant term in equation 35 to catalysis by water molecules. This would help satisfy the above requirement. However, it can then be calculated that the value of β in the Brönsted equation is 0.16 (by setting $2.6 \times 10^{-4}/1.11 \times 10^{-1} = (10^{-14}/55)^\beta$). For organic compounds of the same acidity as acetaldehyde (acetone, acetonylacetone, etc.) β is nearly 0.9.[22] A value of β equal to 0.16 would be expected only for a rather strong acid whose rate of ionization would show little dependence on the strength of the base. Also if β were 0.16, catalysis by carbonate ion would be easily observed in a sodium carbonate solution. A value of 0.9 for β, on the other hand, predicts that catalysis by other bases than hydroxide ion would not be observed until the hydroxide-ion concentration had fallen to a very low value. In such a solution, however, say in acetic acid-acetate buffer, the rate at which the carbanion picks up a proton could readily be so large that the rate would be imperceptible. That is, any carbanion formed in an acidic solution would be converted back to aldehyde again before being able to condense.

The rate constant for hydroxide ion acting as a catalyst, 0.111 M^{-1} sec^{-1} or 7 M^{-1} min^{-1}, should be the rate constant for ionization of acetaldehyde by hydroxide ion. This value is very close to that observed for similar compounds, acetone being 15 M^{-1} min^{-1}. The constant term is inexplicable unless it is due to side reactions or experimental complications.

Thus it seems that two very similar reactions show different kinetics because of the relative rates of the steps involved, the steps themselves being the same. For acetaldehyde in which the carbonyl group is very reactive, the condensation step is very rapid. For acetone with one more methyl group, the rate of condensation is reduced to the point where the carbanion is reverted to acetone a thousand times before condensation finally occurs. The alkyl group probably exerts both a steric effect and an inductive effect in slowing down reaction at the carbonyl carbon. For aldehydes of greater branching than acetaldehyde something intermediate in kinetics might be observed.

It is not necessary to choose other compounds to find such intermediate behavior. It is a necessary consequence of the mechanism postulated for the aldol condensation that changes in the kinetic order will occur at low aldehyde concentration. An examination of equation 5 shows that only when $k_3[C]$ is greater than $k_2[BH^+]$ will first-order kinetics be observed. Here C is acetaldehyde itself and BH^+ is water. As C goes to zero, the rate must become second-order in acetaldehyde. This change in kinetics has been observed by Broche and Gibert.[23] Between 0.1 and $0.5M$ aldehyde, the order of the reaction is between one and two.

Similarly it is predicted that acetaldehyde treated with base in heavy water will pick up deuterium before condensing if the concentration of aldehyde is low. This has also been observed if the aldehyde concentration goes below $1.4M$.[24] From the experiments on the change in the kinetic order with concentration and from the amount of deuterium taken up in D_2O, it is possible to calculate the ratio of k_2/k_3 in equation 5.[24] This ratio is equal to 0.10 in light water and 0.023 in heavy water. If k_3 is nearly the same in both solvents, then k_2, a proton transfer from water to a carbanion, shows an isotope effect of 4.3.

Results similar to the acetaldehyde case have been obtained for the condensation of glyceraldehyde to hexoses catalyzed by base as shown in (38). d-Glyceraldehyde gives a mixture of approximately equal amounts of d-fructose and d-sorbose, the total yield being about 90 per cent.[25] If d,l-glyceraldehyde is used, then d,l-fructose and d,l-sorbose are obtained.

(38)

d-glyceraldehyde   d-fructose   d-sorbose

Dihydroxyacetone can be used also as a starting material, but the yields of straight-chain hexoses are decreased and branched-chain carbohydrates are obtained.[26] It was observed by Bonhoeffer and Walters[27] that the condensation of either gylceraldehyde or a mixture of glyceraldehyde and dihydroxyacetone in heavy water led to the formation of hexoses without deuterium exchange of any of the hydrogens bound to carbon. The kinetics of condensation have been studied and found to be first-order in glyceraldehyde and first-order in hydroxide ion.[28] Addition of equal amounts of dihydroxyacetone increases the rate of formation of the straight-chain hexoses by a factor of 2 or 3, but the kinetics remained first-order with respect to the total triose concentration. Dihydroxyacetone by itself condenses at a considerably slower rate.

All of this is consistent with a mechanism analogous to that proposed for acetaldehyde. Ionization of glyceraldehyde (or dihydroxyacetone) is the rate-determining step. The carbanions of dihydroxyacetone and

$$
\begin{array}{c}
\text{CHO} \\
| \\
\text{HCOH} \\
| \\
\text{CH}_2\text{OH}
\end{array}
+ \text{OH}^- \rightarrow
\begin{array}{c}
\text{CHO} \\
| \\
{}^-\text{COH} \\
| \\
\text{CH}_2\text{OH}
\end{array}
+ \text{H}_2\text{O}
\qquad (39)
$$

glyceraldehyde can be assumed to be in mobile equilibrium with each other, as shown in (46). The proton shift from oxygen to oxygen is very rapid.

$$
\begin{array}{c}
\text{CH}_2\text{—}\overset{-}{\text{C}}\text{—C—H} \\
| \quad\;\; | \;\;\; \| \\
\text{O} \quad \text{O} \;\; \text{O} \\
| \quad\;\; | \\
\text{H} \quad \text{H}
\end{array}
\rightleftharpoons
\begin{array}{c}
\text{CH}_2\text{—C—}\overset{-}{\text{C}}\text{H} \\
| \quad\;\; \| \;\;\; | \\
\text{O} \quad \text{O} \;\; \text{O} \\
| \quad\;\;\;\;\;\; | \\
\text{H} \quad\;\;\;\;\;\; \text{H}
\end{array}
\qquad (40)
$$

Next a rapid condensation of the carbanion of dihydroxyacetone with a glyceraldehyde molecule leads to the oxy-anion of the ketohexose, as shown in (41). If the starting materials were inactive, then there are eight

$$
\begin{array}{c}
\text{CH}_2\text{—C—}\overset{-}{\text{C}}\text{H} \\
| \quad\;\; \| \;\;\; | \\
\text{O} \quad \text{O} \;\; \text{O} \\
| \quad\;\;\;\;\;\; | \\
\text{H} \quad\;\;\;\;\;\; \text{H}
\end{array}
+
\begin{array}{c}
\text{H—C—CH—CH}_2 \\
\;\; \| \;\;\; | \quad\;\; | \\
\;\; \text{O} \;\; \text{O} \quad \text{O} \\
\;\;\;\;\;\;\;\; | \quad\;\; | \\
\;\;\;\;\;\;\;\; \text{H} \quad\;\; \text{H}
\end{array}
\rightarrow
\begin{array}{c}
\text{CH}_2\text{—C—CH—CH—CH—CH}_2 \\
| \quad\;\; \| \;\;\; | \quad\;\; | \quad\;\; | \quad\;\; | \\
\text{O} \quad \text{O} \;\; \text{O} \;\; \text{O}_- \;\; \text{O} \;\; \text{O} \\
| \quad\;\;\;\;\;\; | \quad\;\;\;\;\;\;\;\;\;\;\; | \quad\;\; | \\
\text{H} \quad\;\;\;\;\;\; \text{H} \quad\;\;\;\;\;\;\;\;\;\;\; \text{H} \quad\;\; \text{H}
\end{array}
$$
$$(41)$$

possible stereoisomeric products, or four d,l pairs. Actually only two pairs are isolated, d,l-fructose and -sorbose. The other possible isomers need not be formed since, being diastereomers, different activation energies may be necessary for their formation. Any isomer formed will, of course, always have equal amounts of the d and l forms so that the product is inactive.

The rate constant for the condensation of glyceraldehyde catalyzed by sodium hydroxide is about 9 liters/mole-min at 20° after correcting for the lowering of the hydroxide-ion concentration by the acidic hydroxyl groups of glyceraldehyde. This is the expected magnitude for a carbon-hydrogen ionization of an aldehyde when hydroxide ion is the base. The increase in rate when dihydroxyacetone is added must be attributed to a greater rate of ionization of the ketone than the aldehyde since the change from the aldehyde carbanion to the ketone carbanion is assumed to be very fast. If all the glyceraldehyde were replaced, however, it would be expected that the condensation step would be slowed down since the carbanion would then have to add to a ketonic carbonyl. If the condensation step became a factor in the rate, a switch to second-order dependence on the carbonyl compound would be predicted.

REFERENCES

1. K. Koelichen, *Z. physik. Chem.*, *33*, 129 (1900).
2. C. C. French, *J. Am. Chem. Soc.*, *51*, 3215 (1929).
3. G. M. Murphy, *J. Am. Chem. Soc.*, *53*, 977 (1931).
4. V. K. La Mer and M. L. Miller, *J. Am. Chem. Soc.*, *57*, 2674 (1935).
5. F. H. Westheimer and H. Cohen, *J. Am. Chem. Soc.*, *60*, 90 (1938).
6. J. G. Miller and M. Kilpatrick, *J. Am. Chem. Soc.*, *53*, 3217 (1931).
7. K. F. Bonhoeffer and W. D. Walters, *Z. physik. Chem.*, *A181*, 441 (1938); R. P. Bell and O. M. Lidwell, *Proc. Roy. Soc.*, *A176*, 88 (1940).
8. G. Åkerlof, *J. Am. Chem. Soc.*, *48*, 3046 (1926); *49*, 2960 (1927).
9. G. Åkerlof, *J. Am. Chem. Soc.*, *50*, 1272 (1928).
10. E. S. Amis, G. Jaffé, and R. T. Overman, *J. Am. Chem. Soc.*, *66*, 1823 (1944).
11. W. E. Nelson and J. A. V. Butler, *J. Chem. Soc.*, 957 (1938).
12. H. C. Urey and G. K. Teal, *Rev. Mod. Phys.*, *7*, 34 (1935).
13. (*a*) J. Bigeleisen, *J. Chem. Phys.*, *17*, 675 (1949); (*b*) C. G. Swain, E. C. Stivers, J. F. Reuwer, and L. J. Schaad, *J. Am. Chem. Soc.*, *80*, 5885 (1958); (*c*) A. Streitwieser, Jr., R. H. Jagow, R. C. Fahey, and S. Suzuki, *ibid.*, 2328 (1958).
14. R. P. Bell, *The Proton in Chemistry*, Cornell University Press, Ithaca, 1959, Chapter 11; see especially Y. Pocker, *Chem. and Ind.*, 89, 599, 1383 (1959).
15. J. Hudis and R. W. Dodson, *J. Am. Chem. Soc.*, *78*, 911 (1956); A. Zwickel and H. Taube, *ibid.*, *81*, 1288 (1959); Y. Pocker, *Proc. Chem. Soc.*, Jan., (1960), p. 17.
16. K. Wiberg, *Chem. Revs.*, *55*, 713 (1955); P. M. Laughton and R. E. Robertson, *Can. J. Chem.*, *34*, 1714 (1956).
17. W. E. Nelson and J. A. V. Butler, *J. Chem. Soc.*, 2019 (1938).
18. E. H. Usherwood, *J. Chem. Soc.*, *123*, 1717 (1923).
19. R. P. Bell, *J. Chem. Soc.*, 1637 (1937).
20. J. B. M. Herbert and I. Lauder, *Trans. Faraday Soc.*, *34*, 432 (1938); I. Lauder, *Trans. Faraday Soc.*, *44*, 729 (1948); R. P. Bell and W. C. E. Higginson, *Proc. Roy. Soc.*, *A197*, 141 (1947); R. P. Bell and J. C. Clunie, *Trans. Faraday Soc.*, *48*, 439 (1952).
21. R. P. Bell and B. deB. Darwent, *Trans. Faraday Soc.*, *46*, 34 (1950).
22. R. P. Bell, *Trans. Faraday Soc.*, *39*, 253 (1943).
23. A. Broche and R. Gilbert, *Bull. Soc. chim. France*, 131 (1955).
24. R. P. Bell and W. E. Smith, *J. Chem. Soc.*, 1691 (1958).
25. H. O. L. Fischer and E. Baer, *Helv. Chim. Acta*, *19*, 519 (1936).
26. L. M. Utkin, *Doklady Akad. Nauk U.S.S.R.*, *67*, 301 (1950).
27. K. F. Bonhoeffer and W. D. Walters, *Z. physik. Chem.*, *A181*, 447 (1938).
28. W. G. Berl and C. E. Feazel, *J. Am. Chem. Soc.*, *73*, 2054 (1951).

E. The Nitration of Aromatic Compounds

The nitration reaction has been the most widely studied of

$$ArH + HNO_3 = ArNO_2 + H_2O \qquad (1)$$

all reactions of aromatic compounds, and in the last few years a rather complete picture of the mechanism has been worked out. This is due largely to the work of Ingold, Hughes, and their collaborators. A number of earlier papers were followed by a veritable avalanche of articles appearing together in the *Journal of the Chemical Society* (pp. 2400–2684, 1950). Two review articles summarize most of the important information.[1]

The kinetics of nitration was studied quite early by other workers, but little of value was obtained since the importance of maintaining a nearly constant medium was not realized. However, Martinsen[2] was able to show that simple kinetics were obtained when concentrated sulfuric acid was used as a solvent. The nitration of nitrobenzene, for example, was second-order, first-order in nitric acid, and first-order in nitrobenzene. Similar results were found for other of the less reactive aromatic compounds. Benzene itself would be nitrated too rapidly in this solvent for kinetic measurements.

An important step was made when Benford and Ingold[3] showed that in organic solvents, such as glacial acetic acid, dioxane, and nitromethane, nitration was zero-order for the more reactive compounds, benzene, toluene, and ethylbenzene, if an excess of nitric acid was used. For a given solvent composition the same rate was obtained for all three aromatics, independent of their concentration. For aromatic compounds of somewhat lower reactivity such as chlorobenzene there was a dependence on concentration, the order being between zero and one. Hughes, Ingold, and Reed[1] showed that for trichlorobenzene and ethylbenzoate the order did become one. The same result was found for nitrobenzene in concentrated nitric acid solvent, the rate equation being

$$\text{rate} = k[\text{ArH}] \qquad (2)$$

The relative reactivities of the aromatic compounds were based not only on qualitative observations, such as would distinguish benzene from nitrobenzene, but also on an elegant competitive method developed by Ingold and Shaw.[5] This consisted of allowing a mixture of two aromatic compounds to react with a limited quantity of nitrating reagent and then analyzing the products to determine the amounts of each compound undergoing nitration. In this way the following relative reactivities were found: toluene 24, benzene 1, fluoro- and iodobenzene $\frac{1}{6}$, chloro- and bromobenzene $\frac{1}{30}$, and ethylbenzoate $\frac{1}{300}$.

The solvent had a great effect on the reaction rate, as well as on the reaction order. Some common solvents can be arranged qualitatively in the following order of decreasing effectiveness as a medium: $H_2SO_4 >$ $HNO_3 \gg CH_3NO_2 > CH_3COOH > H_2O$. In a mixture, say of H_2SO_4 and H_2O, the relative amounts can have a large influence on the reaction rate, as will be brought out in detail later. In general it can be said that the more acidic the medium, or the less basic, the better the medium is for nitration.

Additional data to be considered are the earlier work on the ratios of *meta* nitration to *ortho-para* nitration for various substituted benzenes. As is well known, substituents which activate the benzene nucleus cause reactions to occur chiefly *ortho* and *para*, and substituents which strongly deactivate the benzene ring cause reaction at the *meta* positions. Modern electronic theory proposes that activation occurs because of a donation of electrons to the benzene ring by the substituent, and deactivation occurs because of withdrawal of electrons from the ring.†

All this information leads to a tentative mechanism for nitration, at least if essentially the same thing is going on in the different solvents for each of the aromatic compounds. First, there must be at least two steps in the reaction, the first of which involves only nitric acid or the solvent and nitric acid, and the second of which involves the aromatic compound and some intermediate produced in the first step. Either of these may be the slower or rate-determining step. For the more reactive compounds such as benzene, the first step is the slower, at least in organic solvents, and the second step is fast. This leads to a zero-order dependence of the rate on the concentration of the aromatic compound. For the less reactive compounds the second step is the slow one. From the kinetics in concentrated sulfuric acid this slow step involves one molecule of the aromatic compound and a reactive intermediate derived from one molecule of nitric acid. The overall scheme is

$$HNO_3 + \text{solvent} \rightleftharpoons \text{intermediate} \qquad (3)$$

$$\text{intermediate} + ArH \rightarrow ArNO_2 \qquad (4)$$

with the certainty that (3) is reversible and the possibility that (4) may consist of several steps and may involve the solvent.

Furthermore, the active intermediate which reacts with the benzene ring must be a positively charged particle (or a very strong dipole). This is indicated by the relative reactivities since electron withdrawal from the

† For a discussion of inductive and resonance effects see reference 6.

ring, as in ethyl benzoate or nitrobenzene, reduces the reactivity. Also the facts on orientation by substituents are consistent with a positive-ion attack since, for example, Ri and Eyring[7] were able to calculate with reasonable accuracy the *ortho-para* to *meta* ratio for the nitration of several benzene derivatives. They assumed a positive-ion reagent and used dipole moment and bond distance data only and made calculations based on electrostatic interactions.

To get more information on this positive ion it is necessary to consider some physical properties of nitric acid under nitrating conditions. Hantzsch[8] made a number of studies of nitric acid dissolved in concentrated sulfuric acid. He concluded, from freezing-point depressions, conductivity data, and ultraviolet absorption data, that nitric acid is extensively ionized in sulfuric acid to give positive ions containing nitrogen. He incorrectly concluded that the ions present were $H_2NO_3^+$ and $H_3NO_3^{+2}$ arising from the reactions

$$HNO_3 + H_2SO_4 \rightleftharpoons H_2NO_3^+ + HSO_4^- \qquad (5)$$

$$H_2NO_3^+ + H_2SO_4 \rightleftharpoons H_3NO_3^{+2} + HSO_4^- \qquad (6)$$

This conclusion was based chiefly on an inaccurate value of the van't Hoff *i* factor which he believed to be about 3. The above reactions give 2 and 3 particles per molecule of nitric acid, respectively. Bennett, Brand, and Williams[9] pointed out that a value of 4 was more probable, and this was confirmed by Gillespie, Graham, Hughes, Ingold, and Peeling,[10] who in a very accurate study showed that *i* was 3.77. This agrees with the reaction

$$HNO_3 + 2H_2SO_4 \rightleftharpoons NO_2^+ + H_3O^+ + 2HSO_4^- \qquad (7)$$

almost exactly, if allowance is made for the side reaction

$$H_3O^+ + HSO_4^- \rightleftharpoons H_2O + H_2SO_4 \qquad (8)$$

which does not go completely to the left even in concentrated sulfuric acid. That the NO_2^+ or nitronium ion is the reactive species in nitration had been proposed as early as 1903 by Euler.[11]

Additional evidence for the nitronium ion comes from the Raman spectra of nitric acid dissolved in sulfuric and other strong acids. Chedin[12] found that when nitric acid was dissolved in sulfuric acid, two strong nonmixture lines were produced in the Raman spectrum at 1050 cm^{-1} and 1400 cm^{-1}. Of these two only the 1400 cm^{-1} line was characteristic of the positive ion produced since it was the only one that persisted when nitric acid was dissolved in perchloric acid and selenic acid.[13] The line at 1050 cm^{-1} was attributed to the bisulfate ion.

The only line in the Raman spectrum of the positive ion formed from nitric acid by strong acids is the 1400 cm^{-1} one. Now to have such a simple spectrum, the ion must be very simple, being either a diatomic ion or a linear, symmetric, triatomic ion (Ingold, Millen, and Poole[13]). Since the spectra of all possible diatomic species containing the elements of nitric acid are known, and since the azide ion is eliminated by its spectrum and chemical behavior, the only possibility left is the centrosymmetric NO_2^+ ion. Like the carbon dioxide molecule, this ion would have only one line in the Raman spectrum.

Final proof for the nitronium ion lies in the preparation of a number of its stable salts,[14] such as NO_2ClO_4, NO_2SO_3F, and $(NO_2)_2S_2O_7$. Only the salts of strong acids appear to be stable. Surprisingly enough, there is a well-known nitronium salt. The familiar nitrogen pentoxide, N_2O_5, which is usually thought of as a gas, is a solid of high vapor pressure. Millen[15] has identified the solid by Raman spectra as ionic nitronium nitrate, NO_2NO_3. The stable salt NO_2BF_4 is a good nitrating agent.[14b]

It is reasonable now to assume that the nitronium ion is the active nitrating agent in those solutions where it is known to form. These would usually be solutions of nitric acid in strong acids. Nitric acid itself is a weaker acid than sulfuric and perchloric, but it shows a weak line at 1400 cm^{-1} so that some nitronium ion is present from the reaction

$$2HNO_3 \rightleftharpoons NO_2^+ + NO_3^- + H_2O \qquad (9)$$

In general the ease of formation of NO_2^+ will parallel the acidity of the solvent, since the nitronium ion is a very powerful generalized acid and can be produced only from another strong acid.

From what has been said it is evident that the kinetics of nitration is complex and that two extreme cases can be (a) that in which rate of formation of NO_2^+ is rate-determining, and (b) that in which rate of reaction of NO_2^+ with the aromatic compound is rate-determining. It is now necessary to account for each of these and for intermediate cases. Since equations 7 and 9 represent steps which may be rate-determining in the nitration of the reactive benzene derivative, it is necessary to consider in more detail the process whereby NO_2^+ is formed.

Equations 7 and 9 represent stoichiometry and not mechanism. This is shown by the observations[16] that, in the zero- and first-order nitrations run in organic solvents, added nitrate ion represses the rate, added sulfuric acid increases the rate, and, for zero-order reactions, small amounts of added water have no effect. The effect of strong acid is to produce a linear increase in the observed rate constant for small additions. The observed rate constant in the case of added ionized nitrates follows a hyperbolic law as follows:

$$k_{obs} = \frac{k_0}{1 + k'[NO_3^-]} \qquad (10)$$

so that the reciprocal of the observed rate constant is a linear function of the added nitrate ion. In every case the kinetic order is not affected, zero-order reactions remaining zero-order and first-order reactions remaining first-order. Nitric acid is, of course, always in excess in these experiments. Water has little influence on the rate or on the zero-order when added in concentrations at which nitrate ion has a four- or five-fold influence on the rate (ca., 0.05 molar). Larger amounts of water definitely begin to depress the rate.

To explain these effects consider equation 9. If this were rate-determining and irreversible, then added nitrate ion should have no effect. On the other hand, if it were reversible, then nitrate ion could have a depressing effect but the kinetic order would change from zero to one for the reactive aromatics. This would be on the basis of a competition between nitrate ion and the aromatic compound to capture the nitronium ion, and the observed rate would become first-order in the aromatic compound. Also there would be a decelerating effect of water comparable to that for nitrate ion. The same arguments would still be true even if (9) were followed by the reaction

$$H_2O + HNO_3 \rightleftharpoons H_3O^+ + NO_3^- \tag{11}$$

A process which does satisfy the experimental facts, however, is the two-stage one:

$$2HNO_3 \rightleftharpoons H_2NO_3^+ + NO_3^- \quad \text{fast} \tag{12}$$

$$H_2NO_3^+ \rightleftharpoons H_2O + NO_2^+ \quad \text{slow} \tag{13}$$

where (12) is a rapid equilibrium and (13) is rate-determining. The effect of added nitrate ion is to repress the equilibrium in (12). As long as the reaction of the nitronium ion with the aromatic compound is fast compared to the reversal of (13), added water will have no effect. Added strong acid increases the concentration of $H_2NO_3^+$ by the process

$$H_2SO_4 + HNO_3 \rightleftharpoons H_2NO_3^+ + HSO_4^- \tag{14}$$

Higher concentrations of water would be expected to reduce the rate by a competition between water and aromatic compound for the nitronium ion. The order would also be expected to change slowly over from zero to one. The latter effect has been observed, adding one molar water to nitromethane solvent will cause the expected shift to first-order in the case of benzene.

However, the reactivity of the aromatic compound determines the effect of water since there is a competition between the two for NO_2^+. The anion 2-mesitylethenesulfonate is almost zero-order even in 60 mole per cent water and 40 mole per cent nitric acid.[1b] The similarly very active aromatic compound mesitylene-α-sulfonate is 1500 times as reactive as water in trapping nitronium ion.[1b]

To investigate the details of the way in which NO_2^+ attacks the benzene ring and replaces a proton

$$NO_2^+ + ArH \rightarrow ArNO_2 + H^+ \tag{15}$$

it is necessary to consider the second-order nitrations of the less reactive compounds in concentrated sulfuric acid. Evidently in these cases the

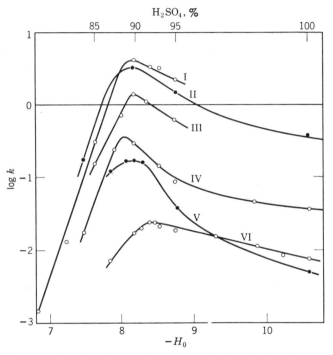

Fig. I. Rates of nitration of aromatic compounds in sulfuric acid-water mixtures. [Gillespie and Millen, *Quart. Rev. Chem. Soc.*, *2*, 227 (1948).]

formation of NO_2^+ is not rate-determining but the attack on the aromatic compound is. Presumably an equilibrium concentration of nitronium ion is formed in the solution, the magnitude of which depends on the concentration of nitric acid and the composition of the solvent. The observed rate constant will depend upon this equilibrium concentration, and hence upon the composition, and upon the reactivity of the aromatic compound. Figure 1 and Table 1 show data collected by several workers which illustrate these points. The maximum rate obtained at about 90 per cent sulfuric acid is quite characteristic and has been found for all compounds whose rates have been studied.

The increase in rate up to the maximum seems to be readily explained in terms of an increasing NO_2^+ concentration with decreasing water content.

Table I

SECOND-ORDER RATE CONSTANTS FOR THE NITRATION OF SOME
AROMATIC COMPOUNDS IN SULFURIC ACID MEDIA

[Gillespie and Millen, *Quart. Rev. Chem. Soc.*, 2, 277 (1948)]

Compound	Temperature, °C	k, liters mol^{-1} min^{-1} 90% H_2SO_4	95% H_2SO_4	100% H_2SO_4	Authors[a]
Nitrobenzene	25	3.22	1.50	0.37	M
Nitrobenzene	25	4.1	2.0	. . . . .	W and K
4:6-Dinitro-*m*-xylene	25	. . . . .	0.0040	0.0014	M
4:6-Dinitro-*m*-xylene	50	0.085	0.046	0.035	W and K
Dinitromesitylene	25	1.4	0.62	. . . . .	W and K
2:4-Dinitrotoluene	90	0.017	0.019	0.0074	B
2:4-Dinitrotoluene	100	0.037	0.043	. . . . .	B
2:4-Dinitrotoluene	100	0.021	0.023	. . . . .	W and K
p-Chloronitrobenzene	25	. . . . .	0.18	0.05	M
m-Chloronitrobenzene	25	. . . . .	0.39	0.14	M
o-Chloronitrobenzene	25	. . . . .	7.15	2.18	M
2:4-Dinitroanisole	25	. . . . .	0.17	0.053	M
2:4-Dinitrophenol	25	. . . . .	0.85	0.39	M
Anthraquinone	25	0.248	0.037	0.0053	L and O
Benzenesulfonic acid[b]	25	. . . . .	~26	2.3	M
Benzoic acid[b]	25	. . . . .	>100	~5.4	M

[a] Authors: M, Martinsen; W and K, Westheimer and Kharasch; B, Bennett et al.; L and O, Lauer and Oda.

[b] The values for these compounds are only approximate.

(Since these constants vary with the inital concentration of the reactant, those for the smallest reactant concentrations investigated are given in this table.)

In fact there is a fair quantitative agreement between the rate of nitration and the NO_2^+ concentration as determined spectroscopically[17] up to the maximum. The limit of detection of NO_2^+ spectroscopically is at 85 per cent sulfuric acid by weight, which is the composition corresponding to $H_2SO_4 \cdot H_2O$. However, the nitronium ion apparently is still present in 75 per cent sulfuric acid in amounts large enough for nitration.[18]

Westheimer and Kharasch[19] also demonstrated the linear relationship between the nitronium-ion concentration and the rate of nitration in an ingenious manner. They showed that the rate of nitration of nitro-benzene paralleled the degree of ionization of the dye trinitrotriphenyl-carbinol. This dye ionizes in sulfuric acid in the same way that nitric acid does:

$$R_3COH + 2H_2SO_4 \rightleftharpoons H_3O^+ + R_3C^+ + 2HSO_4^- \qquad (16)$$

(compare equation 7). On the other hand, there was no parallelism between rate of nitration and the ionization of anthraquinone which ionizes in a different fashion:

$$R_2CO + H_2SO_4 \rightleftharpoons R_2COH^+ + HSO_4^- \qquad (17)$$

The ionization of the dyes was determined spectrophotometrically.

The maximum in the rate occurs at a concentration of sulfuric acid where ionization to NO_2^+ is virtually complete. We can understand then why an increase in sulfuric acid causes no further increase in the rate. The observed decrease (Fig. 1), however, requires an explanation and has a bearing on the mechanism. The question arises whether the removal of the proton from the aromatic ring precedes, coincides with, or follows the addition of NO_2^+. That it precedes cannot be allowed, since then the rate would depend upon the acid-strengthening properties of substituents in a way which is opposite to that observed. Furthermore, it would then be expected that isotopic exchange could occur before nitration in a medium containing deuterium or tritium. This does not occur.[20] The two possibilities left are:

$$HSO_4^- + ArH + NO_2^+ \rightarrow ArNO_2 + H_2SO_4 \quad \text{slow} \qquad (18)$$

and $\qquad\qquad ArH + NO_2^+ \rightarrow ArNO_2H^+ \quad \text{slow} \qquad (19)$

$$ArNO_2H^+ + HSO_4^- \rightarrow ArNO_2 + H_2SO_4 \quad \text{fast} \qquad (20)$$

where bisulfate ion is taken as the most likely base present in the system though any other base would serve.

Bennett et al.[21] assumed that the removal of the proton was part of the rate step as in (18). Hence the rate should depend upon the concentration of bisulfate ion as well as the nitronium ion. As the composition changes from 90 per cent to 100 per cent sulfuric acid the concentration of HSO_4^- drops sharply because it is produced chiefly by the reaction

$$H_2O + H_2SO_4 \rightleftharpoons H_3O^+ + HSO_4^- \qquad (21)$$

Thus the decrease in rate is explained. Adding bisulfate ion as sodium bisulfate to a mixture on the left-hand side of the maximum in Fig. 1 should then increase the rate. Bisulfate ion added to mixtures on the right-hand side of the maximum, however, will decrease the rate of nitration by reducing the NO_2^+ concentration. Both of these effects of added bisulfate ion are actually observed (Westheimer and Kharasch[19]). Another way of stating the observed effect of added bisulfate ion is to say that the rate depends upon a definite value of the H_0 function rather than upon a definite concentration of sulfuric acid.

There are several objections to the concept that the removal of the proton is part of the rate-determining step. It would predict, for one thing, that the maximum in the rate would be a function only of the solvent and independent of the aromatic compound used. This is nearly, but not exactly, true, the differences being greater than the experimental errors. Furthermore, the ratio of rate constants for two compounds should be constant and independent of the solvent composition. This is not true, the ratio of rates for nitrobenzene and anthraquinone being 16.5 in 90 per cent acid and 54 in 95 per cent acid.

The correct explanation for the decrease in rate in stronger sulfuric acid solutions appears to be related in part to a conversion of the aromatic compound to a positively charged, and hence unreactive, species.[1,22] These may be positive ions such as $C_6H_5NO_2H^+$ or simply hydrogen-bonded complexes $C_6H_5NO_2 \cdot H_2SO_4$ with a strong positive pole on the aromatic compound. In Fig. 1, for example, there appears to be a correlation between the steepness of the drop in rate after the maximum and the tendency of the organic molecule to add protons, that is, to its basicity. Nitrobenzene has a steeper maximum than dinitrobenzene because the latter compound is a poorer base. Additional support for this hypothesis comes from the work of Baker and Hey,[23] who showed that the percentage of *meta* nitration for benzaldehyde, acetophenone, and ethyl benzoate increases with the concentration of sulfuric acid in the nitrating mixture. This increase was explained as due to the conversion of these substances into positively charged ions which would, of course, increase the tendency for *meta* orientation.

Another factor in the decrease in rate may be the change in polarity of the solvent. Sulfuric acid is more polar than water; hence in going from 90 per cent acid to 100 per cent acid the polarity is actually increasing. Now, if the rate-determining step is between a neutral molecule and a positive ion as indicated in equation 19, there should be a small decrease in rate on going to a more polar solvent. This is because of the spreading out of charge in the activated complex (see Chapter 7). Reaction 18 would show a large decrease in rate in going to a more polar solvent because it involves the destruction of charges.

More direct evidence that the removal of the proton is not part of the rate-determining step comes from the work of Melander,[20] who used aromatic compounds labeled in the ring with tritium, the radioactive hydrogen isotope of mass three. He showed that in nitration the tritium was removed from the ring as readily as hydrogen. Now, if removal of the proton or tritium ion were part of the slow step, a difference in rates would be expected, tritium being removed less readily. In a similar comparison of hydrogen and deuterium, differences of two- to ten-fold are found, deuterium being removed more slowly. Also if there were a prior equilibrium involving removal of a proton or tritium ion, a difference in the equilibrium constant would be expected. The conclusion then is that in nitration the removal of the proton is a rapid reaction following the slow step.

To summarize what has been said, the following mechanism seems consistent with the facts:[24]

$$HNO_3 + HA \underset{k_2}{\overset{k_1}{\rightleftharpoons}} H_2NO_3^+ + A^- \quad fast \tag{22}$$

$$H_2NO_3^+ \underset{k_4}{\overset{k_3}{\rightleftharpoons}} H_2O + NO_2^+ \quad slow \tag{23}$$

$$NO_2^+ + ArH \overset{k_5}{\longrightarrow} ArNO_2H^+ \quad slow \tag{24}$$

$$ArNO_2H^+ + A^- \overset{k_6}{\longrightarrow} ArNO_2 + HA \quad fast \tag{25}$$

where HA is any strong acid and A^- is any base. The exact solution of the rate expressions corresponding to equations 22 to 25 is not possible. However, the kinetics for two limiting cases of interest can be worked out. These correspond to one of the steps in (23) or (24) being much slower than the other.

If k_5, for example, is much less than k_3 and k_4, we shall have the case of an equilibrium concentration of nitronium ion being established. This will be given by

$$[NO_2^+] = \frac{k_1 k_3 [HNO_3][HA] f_{HNO_3} f_{HA}}{k_2 k_4 [H_2O][A^-] f_{H_2O} f_{A^-} f_{NO_2^+}} \tag{26}$$

If the solvent is of fixed composition, that is, dilute in the reactants, then the concentration of NO_2^+ is directly proportional to the concentration of molecular nitric acid at any time. In kinetic terms, if $(a - x)$ is the total concentration of nitric acid left unreacted, regardless of its state in the solution,

$$[NO_2^+] = K(a - x) \tag{27}$$

where K is a constant dependent on the solvent. (In concentrated sulfuric acid K would be essentially unity.) Now since (24) is the slow step

$$\frac{dx}{dt} = k_5 [NO_2^+][ArH] \frac{f_{NO_2^+} f_{ArH}}{f^{\ddagger}} = k_5 K'(a - x)(b - x) \tag{28}$$

where b is the initial concentration of aromatic compound and K' absorbs the activity coefficients. The reaction is thus second-order in agreement with what is observed for the less active compounds in sulfuric acid.

If k_5, on the other hand, is much greater than k_3 or k_4, then the formation of the nitronium ion will be rate-determining. The equation will be

$$\frac{dx}{dt} = k_3[H_2NO_3^+]\frac{f_{H_2NO_3^+}}{f^{\ddagger}} = \frac{k_1k_3[HNO_3][HA]f_{HNO^3}f_{HA}}{k_2[A^-]f_{A^-}f^{\ddagger}} \qquad (29)$$

For a solvent of constant composition, including nitric acid in excess, the reaction will be of zero-order. It the concentration of HA or of HNO_3 is increased, there will be an increase in the zero-order constant. If HA is sulfuric acid there will be a linear increase in the rate constant, providing that the self-ionization of the nitric acid is small. If HA is nitric acid and A^- is added in the form of ionized nitrates, there will be a decrease in the zero-order rate constant. If the concentration of nitrate ion due to self-ionization, $[NO_3^-]_0$, is small, it is possible to write

$$[NO_3^-]_{total} = [NO_3^-]_0[1 + k'[NO_3^-]_{added}] \qquad (30)$$

Now if the zero-order rate constant in the absence of added nitrate ion is

$$k_0 = \frac{k_1k_3[HNO_3]^2f^2{}_{HNO_3}}{k_2[NO_3^-]_0 f_{NO_3^-}f^{\ddagger}} \qquad (31)$$

then it will become for added nitrate ion

$$k_0' = \frac{k_0}{1 + k'[NO_3^-]_{added}} \qquad (32)$$

so that the reciprocal of the zero-order constant will be a linear function of the added nitrate ion as is observed. This treatment is only approximate and will break down particularly when the added nitrate ion begins to appreciably affect the ratio of activity coefficients occurring in (31) (Hughes, Ingold, and Reed[16]).

It should be mentioned also that, if the concentration of nitric acid is continuously increased, the rate will not increase in proportion. In fact, the zero-order kinetics will be replaced by first-order kinetics eventually. This is because the complete kinetics depends on whether or not the nitronium ion is produced more rapidly than the aromatic compound can remove it. The formation of nitronium ion as shown in the overall equation

$$2HNO_3 \rightleftharpoons H_2O + NO_2^+ + NO_3^- \qquad (33)$$

depends strongly on the concentration of nitric acid and also on the polarity of the medium. Since nitric acid is more polar than the organic solvents used, eventually nitronium ion will be formed more rapidly than it can be removed and the order will change over from zero to first as the nitric acid concentration is increased.

This analysis is not intended to give a complete picture of aromatic nitration under all conditions. It seems to fit the facts for the systems which have been investigated and for compounds which undergo nitration without side reactions. For many details of anomalous nitration and nitration by agents other than nitric acid the papers of Ingold, Hughes, and their collaborators should be consulted. There are some specific effects of nitrous acid which must be taken into account in any system studied.

On the other hand, the nitronium-ion mechanism seems to be a general one for other nitrations, including those of amines and alcohols.[25]

$$ROH + NO_2^+ \rightarrow RONO_2 + H^+ \tag{34}$$

$$RNH_2 + NO_2^+ \rightarrow RNHNO_2 + H^+ \tag{35}$$

For example, in nitromethane solvent containing nitric acid the compounds methylpicramide, methyl alcohol, glycerol (primary hydroxyls only), benzene, and toluene all undergo nitration at the same rate and with no dependence on the concentration of the substrate being nitrated. In a very convincing set of experiments using water labeled with oxygen-18, it has been shown that isotope exchange between water and nitric acid occurs at the same rate as the zero-order rate of nitration of aromatic compounds in the same medium.[26] In this case NO_2^+ is formed

$$NO_2\text{---}OH + H_2O^{18} \rightarrow NO_2\text{---}O^{18}H + H_2O \tag{36}$$

and can only react with water. In the process the isotope label is exchanged.

There is no stereochemical evidence available on the details of the reaction between NO_2^+ and ArH. It is not known, for example, whether the nitronium ion adds to a single carbon atom of the benzene ring, to a double bond, or to the entire ring. The facts on orientation and the effects of substituents on rates seem to point to the first of these. Coupled with valence-bond theory, a plausible transition-state configuration can be inferred in which the nitronium ion is above the plane of the ring and the

proton is below. There are then several resonance structures contributing to the electronic distribution.

Although the symmetric structure above may represent the transition state, it is still possible that the initial attack of the electrophilic reagent is at a double bond to form a π-complex[27] or charge-transfer complex.[28]

This addition may be fast and the rate step the rearrangement to a new π-complex containing the proton to be eliminated. †

$$Ar-H \rightleftharpoons Ar \begin{matrix} H \\ / \\ \\ \backslash \\ E \end{matrix} + \overset{H^+}{\underset{\vdots}{Ar-E}} \rightleftharpoons Ar-E \qquad (37)$$

The similar equilibrium formation of a π-complex and the slow rearrangement to a carbonium ion has been proposed as the mechanism of hydration of olefins by strong acid.[29]

$$\underset{\diagup}{\overset{\diagdown}{C}} = C \overset{\diagup}{\underset{\diagdown}{}} + H^+ \rightleftharpoons \overset{H^+}{\underset{\diagup}{\overset{\diagdown}{C}} = C \overset{\diagup}{\underset{\diagdown}{}}} \qquad \text{fast} \qquad (38)$$

$$\overset{H^+}{\underset{\diagup}{\overset{\diagdown}{C}} = C \overset{\diagup}{\underset{\diagdown}{}}} \rightarrow \overset{H}{\underset{\diagup}{\overset{\diagdown}{C}} - \overset{+}{C} \overset{\diagup}{\underset{\diagdown}{}}} \qquad \text{slow} \qquad (39)$$

$$\overset{H}{\underset{\diagup}{\overset{\diagdown}{C}} - \overset{+}{C} \overset{\diagup}{\underset{\diagdown}{}}} + H_2O \rightarrow \overset{H \ O}{\underset{\diagup}{\overset{\diagdown}{C}} - C \overset{\diagup}{\underset{\diagdown}{}}} + H^+ \qquad \text{fast} \qquad (40)$$

† The symmetric species is a stable intermediate in some cases, as shown by G. A. Olah and S. J. Kuhn, *J. Am. Chem. Soc.*, *80*, 6535 (1958).

REFERENCES

1. (a) R. J. Gillespie and D. J. Millen, *Quart. Rev. Chem. Soc.*, II, 277 (1948); (b) E. D. Hughes, in *Theoretical Organic Chemistry*, the Kekulé Symposium, Butterworths Scientific Publications, London, 1959, p. 209.
2. H. Martinsen, *Z. physik. Chem.*, *50*, 385 (1905); *59*, 605 (1907).
3. G. A. Benford and C. K. Ingold, *J. Chem. Soc.*, 929 (1938).
4. E. D. Hughes, C. K. Ingold, and R. I. Reed, *Nature*, *158*, 448 (1946).
5. C. K. Ingold and F. R. Shaw, *J. Chem. Soc.*, 2918 (1927).
6. A. E. Remick, *Electronic Interpretations of Organic Chemistry*, 2nd edition, John Wiley & Sons, New York, 1949, Chapter V. [Out of print.]
7. T. Ri and H. Eyring, *J. Chem. Phys.*, *8*, 433 (1940).
8. A. Hantzsch, *Z. physik. Chem.*, *61*, 257 (1907); *65*, 41 (1908); *Ber.*, *58*, 941 (1925).
9. G. M. Bennett, J. C. Brand, and G. Williams, *J. Chem. Soc.*, 869 (1946).
10. R. J. Gillespie, J. Graham, E. D. Hughes, C. K. Ingold, and E. R. A. Peeling, *J. Chem. Soc.*, 2504 (1950).
11. H. Euler, *Ann.*, *330*, 280 (1903).
12. J. Chedin, *Compt. rend.*, *200*, 1397 (1935).

13. C. K. Ingold, D. J. Millen, and H. G. Poole, *J. Chem. Soc.*, 2576 (1950).
14. (*a*) D. R. Goddard, E. D. Hughes, C. K. Ingold, *J. Chem. Soc.*, 2559 (1950); (*b*) G. Olah, S. Kuhn, and A. Mlinko, *J. Chem. Soc.*, 4257 (1956).
15. D. J. Millen, *J. Chem. Soc.*, 2606 (1950).
16. E. D. Hughes, C. K. Ingold, and R. I. Reed, *J. Chem. Soc.*, 2400 (1950).
17. J. Chedin, *Ann. chim.*, *8*, 295 (1937).
18. G. Williams and A. M. Lowen, *J. Chem. Soc.*, 3312 (1950); A. M. Lowen, M. A. Murray, and G. Williams, *ibid.*, 3318 (1950).
19. F. Westheimer and M. S. Kharasch, *J. Am. Chem. Soc.*, *68*, 1871 (1946).
20. L. Melander, *Nature*, *163*, 599 (1949); *Arkiv Kemi*, *2*, 211 (1950).
21. G. M. Bennett, J. C. D. Brand, D. M. James, T. G. Saunders, and G. Williams, *J. Chem. Soc.*, 474 (1947).
22. R. J. Gillespie, *J. Chem. Soc.*, 2542 (1950).
23. J. W. Baker and L. Hey, *J. Chem. Soc.*, 1226, 2917 (1932).
24. R. J. Gillespie, E. D. Hughes, C. K. Ingold, and R. I. Reed, *Nature*, *163*, 599 (1949).
25. E. L. Blackall, E. D. Hughes, C. K. Ingold, and R. B. Pearson, *J. Chem. Soc.*, 4357, 4366 (*1958*).
26. C. A. Bunton, E. A. Halevi, and D. R. Llewellyn, *J. Chem. Soc.*, 4913, 4917 (1952); C. A. Bunton and G. Stedman, *ibid.*, 2420 (1958).
27. M. J. S. Dewar, *The Electronic Theory of Organic Chemistry*, Oxford University Press, London, 1949, p. 168.
28. V. Gold and D. P. N. Satchell, *J. Chem. Soc.*, 3609, 3619 (1955); R. D. Brown, *ibid.*, 2224, 2232 (1958).
29. R. W. Taft, Jr., *J. Am. Chem. Soc.*, *74*, 5372 (1952); R. W. Taft, Jr., E. L. Purlee, P. Riesz, and C. A. De Fazio, *ibid.*, *77*, 1584 (1955); R H. Boyd, R. W. Taft, Jr., A. P. Wolf and D. R. Christman, *ibid.*, *82*, 4729 (1960); for an alternative see P. B. D. De la Mare, E. D. Hughes, C. K. Ingold, and Y. Pocker, *J. Chem. Soc.*, 2930 (1954).

F. The Decomposition of Di-*t*-Butyl Peroxide

It is difficult to find examples of reactions in the gas phase with clean-cut kinetics and where the mechanism is well worked out. A suitable case seems to be the vapor-phase decomposition of di-*t*-butyl peroxide in the range of 110 to 280°. In the presence of a large amount of glass surface the stoichiometry is given quite accurately by

$$(CH_3)_3COOC(CH_3)_3 = 2(CH_3)_2CO + C_2H_6 \tag{1}$$

two moles of acetone and one of ethane being formed from one mole of peroxide.[1] In larger reaction vessels, however, methyl ethyl and higher ketones can be formed as well as methane.

The peroxide is rather stable considering its nature, being resistant to strong base and to concentrated hydrochloric acid and not giving the usual tests for peroxides. It is definitely more stable thermally than the lower homologs, dimethyl, diethyl, and di-*n*-propyl peroxide (Milas and Surgenor[1]). While undergoing thermal decomposition it can function as an efficient polymerization catalyst.

Raley, Rust, and Vaughan have carried out an extensive study of the kinetics of decomposition of di-*t*-butyl peroxide under a variety of conditions.[2] In the gas phase and in several liquid solvents the reaction is cleanly first-order with very nearly the same rates and activation energies, indicated in Table 1.

Table I

(Raley, Rust, and Vaughan)

Solvent	A, sec^{-1}	E_a, kcal	Moles *t*-Butyl Alcohol	Moles Acetone
Vapor	3.2×10^{16}	39.1 ± 0.5	0	2.0
i-Propylbenzene	0.6×10^{16}	37.5	1.61	0.39
t-Butyl benzene	1.1×10^{16}	38	0.75	1.25
Tri-*n*-butyl amine	0.35×10^{16}	37	ca. 1.9	ca. 0.1

The gas-phase reaction is homogeneous as shown by the constancy of the rate constant obtained in packed and unpacked reaction vessels. A change of more than tenfold in the surface/volume ratio leaves the constant almost unchanged. Thus a wall reaction for the rate-determining step is excluded.

The identified products of the gaseous reaction were acetone and methyl ethyl ketone in the ratio of about 13 to 1 and ethane and methane in the ratio of about 10 to 1. The sum of all ketones and of all paraffins added up to very nearly 100 per cent product yield. The kinetics were followed by observing the increase in pressure. According to (1) the final pressure of the products should be three times as great as the initial pressure of peroxide. Actually the final pressure was found to be 2.88 times as great, so that something like 5 per cent of the peroxide must have given products other than those identified, or the peroxide was impure. The rate constants were the same calculated from either the observed or the theoretical value of the final pressure (see Chapter 3 for details).

Several substances recognized as being inhibitors of chain reactions were added to reaction mixtures, and the rates were studied. Thus it was found that the addition of large amounts of nitric oxide and of propylene had no more effect on the rate than the addition of equal amounts of an inert gas such as nitrogen. Oxygen had an accelerating effect on the rate of pressure increase, oxygen being at the same time consumed. However, the rate of formation of ketones as determined by analysis was unchanged. An eightfold increase in the initial pressure of the peroxide caused an increase of only 6 per cent in the rate constant, and the first-order nature of the reaction was unchanged.

When nitric oxide was added, it was converted in substantial yields into formaldoxime. Propylene gave rise, in addition to ketones, methane, and ethane, to appreciable amounts of isopentane, n-butane, 1-butene, 2-butene, isobutane, 3,4-dimethyloctane, 2,4-dimethyloctane, and heptenes. Addition of 2-butene and isobutene gave rise to a mixture of hydro-carbons also.[3]

The pyrolysis of di-t-amyl peroxide at 130–150° was also studied with similar results except that there was some evidence for a higher-order reaction. The products were chiefly acetone and n-butane with smaller amounts of methane, ethane, ethylene, propane, and methyl ethyl and higher ketones. The activation energy was estimated as 37–41 kcal. The rate constants were about the same as for the t-butyl derivative.

The reactions in solution were followed by an infrared analysis of the liquid products. These products were found to be t-butyl alcohol and acetone (with ca. 5 per cent methyl ethyl ketone) in varying amounts, depending on the solvent and temperature. The sum of the moles of t-butyl alcohol and ketone was equal to the moles of peroxide decomposed. Table 1 shows the moles of alcohol and of ketone formed per mole of peroxide for each solvent at 125°.

The decomposition has also been studied in a semi-quantitative way in pure liquid di-t-butyl peroxide.[4] The rate of decomposition is several times greater than in any other solvent. There is a major change in the reaction product in that isobutylene oxide is formed to the extent of 70 moles per 100 moles of peroxide decomposed. Also 50 moles of t-butyl alcohol, 66 moles of acetone, 63 moles of methane, and only 2 moles of ethane are formed per 100 moles of peroxide. Essentially the same products are formed in the photochemical decomposition at room temperature.

In the vapor-phase hydrogen chloride has a large accelerating effect on the decomposition of the peroxide. A major product is isobutylene chlorohydrin which could result from the reaction of hydrogen chloride on isobutylene oxide. Hydrogen bromide is inert and has no accelerating influence.[2]

Possible mechanisms for the first-order decomposition include a direct unimolecular decomposition into the products as in equation 1, or a decomposition into free radicals which then react further. Such free radicals could be formed in several ways, viz.:

$$(CH_3)_3COOC(CH_3)_3 \rightarrow CH_3 \cdot + \cdot (CH_3)_2COOC(CH_3)_3 \qquad (2)$$

$$(CH_3)_3COOC(CH_3)_3 \rightarrow 2(CH_3)_3CO \cdot \qquad (3)$$

$$(CH_3)_3COOC(CH_3)_3 \rightarrow (CH_3)_3C \cdot + \cdot OOC(CH_3)_3 \qquad (4)$$

The free radicals formed could react with each other (or with the solvent) to give the observed products, or the free radicals could attack other molecules of peroxide, giving rise to a chain reaction in which a number of molecules of peroxide react for each molecule which initially dissociates.

The direct intramolecular decomposition can be ruled out as a major reaction path because of the wide variety of products observed under different conditions. If the rate-determining step is the same in the vapor phase as in solvents, as the similar rates and activation energies strongly suggest, then the direct reaction is eliminated except as a very minor side reaction.

The products that are observed can all be explained on the basis of a preliminary dissociation into *t*-butoxyl radicals as indicated in (3).[25,] These can then either decompose to give acetone and a methyl free radical

$$(CH_3)_3CO \cdot \rightarrow (CH_3)_2CO + CH_3 \cdot \tag{5}$$

or the *t*-butoxyl radical can abstract a hydrogen atom from a hydrogen-bearing solvent, forming *t*-butyl alcohol:

$$(CH_3)_3CO \cdot + RH \rightarrow (CH_3)_3COH + R \cdot \tag{6}$$

The fate of the methyl and R· radicals depends on their nature and environment. Methyl free radicals formed in the vapor phase and with large glass surface areas available will combine to give ethane almost exclusively. In a solvent, or with a reduction of surface area, attack by

$$2CH_3 \cdot \rightarrow C_2H_6 \qquad \text{surface} \tag{7}$$

methyl radical on a hydrogen bearer can also occur.

$$CH_3 \cdot + RH \rightarrow CH_4 + R \cdot \tag{8}$$

Reactions 3, 5, and 7 constitute the important processes in the gas phase, accounting for the products almost quantitatively. Methyl ethyl ketone and methane which are formed in small amounts would arise from the sequence

$$CH_3 \cdot + (CH_3)_2CO \rightarrow CH_4 + \cdot CH_2COCH_3 \tag{9}$$

$$CH_3 \cdot + \cdot CH_2COCH_3 \rightarrow CH_3CH_2COCH_3 \tag{10}$$

The existence of free methyl radicals is convincingly demonstrated by the isolation of formaldoxime which can only come from the trapping of methyl radicals by nitric oxide:

$$CH_3 \cdot + NO \cdot \rightarrow CH_3NO \rightarrow CH_2 {=} NOH \tag{11}$$

The reaction products in the presence of propylene (and other unsaturates) can be adequately accounted for by an addition of a methyl radical to the double bond followed by plausible reactions of the butyl radical thus formed:

$$CH_3\cdot + CH_2{=}CHCH_3 \rightarrow CH_3CH_2{-}CHCH_3 \qquad (12)$$

$$\text{or } \cdot CH_2{-}CH(CH_3)_2$$

Both *sec*-butyl (about 90 per cent) and *iso*-butyl radicals (about 10 per cent) must be formed to account for the reaction products observed.

The proof for the *t*-butoxyl radical lies in the formation of *t*-butyl alcohol in solvents, the amount of alcohol formed compared to the amount of acetone being a function of the concentration of hydrogen-bearing molecules and of the ease with which hydrogen can be abstracted from such a molecule. The ratio of alcohol to ketone also depends on temperature in such a way that the decomposition of the *t*-butoxyl radical to acetone as in (5) must have a higher activation energy than the abstraction of hydrogen as in (6) by about 10 kcal. By using a competition method over a range of temperature, it has been estimated that reaction 5 has an activation energy of 11.2 kcal.[6] This estimate is based on the assumption that the alternative addition reaction of the *t*-butoxyl radical to butadiene has an activation energy of 5.4 kcal. The latter figure is the activation energy for the propagation step in the polymerization of butadiene in the gaseous phase and refers, of course, to the addition of a polymeric radical to butadiene.

The products in pure liquid peroxide are of interest in that they show that the peroxide itself at high concentrations can act as a hydrogen donor:

$$(CH_3)_3CO\cdot + (CH_3)_3COOC(CH_3)_3 \rightarrow$$

$$(\text{or } CH_3\cdot)$$

$$(CH_3)_3COH + \cdot CH_2(CH_3)_2COOC(CH_3)_3 \quad (13)$$

$$(CH_4)$$

The intermediate free radical formed is assumed to cleave to a *t*-butoxyl radical and isobutylene oxide:

$$\cdot CH_2(CH_3)_2COOC(CH_3)_3 \rightarrow (CH_3)_2C{-}{-}{-}CH_2 + (CH_3)_3CO\cdot \quad (14)$$
$$\underset{O}{\diagdown \diagup}$$

A chain reaction is thus set up which accounts for the higher rate of decomposition found with pure peroxide than with other liquid solvents.

The vapor-phase reaction involving hydrogen chloride can be explained similarly as involving a chain set off by attack of a chlorine atom on the peroxide

$$Cl\cdot + (CH_3)_3COOC(CH_3)_3 \rightarrow HCl + \cdot CH_2(CH_3)_2COOC(CH_3)_3 \text{ etc.} \quad (15)$$

The isobutylene oxide is not stable in the presence of hydrogen chloride and is converted into a mixture of α and β chlorohydrins as shown in (16).

$$(CH_3)_2C\underset{O}{\overset{}{\diagdown\diagup}}CH_2 + HCl \rightarrow \begin{array}{c} (CH_3)_2COHCH_2Cl \\ \text{and} \\ (CH_3)_2CClCH_2OH \end{array} \qquad (16)$$

Apparently a bromine atom is too inert to initiate a chain by removing hydrogen from the peroxide. Oxygen also is not a chain initiator in the vapor state.

Of all these reactions, the slowest is the initial decomposition into two t-butoxyl radicals as shown in (3). Hence this is rate-determining, and the overall rate, except where chains are formed, is equal to the rate of this reaction. The experimental activation energy is also equal to the activation energy of this process:

$$(CH_3)_3COOC(CH_3)_3 \overset{k_{exp}}{\rightleftharpoons} 2(CH_3)_3CO\cdot \qquad (17)$$

If the assumption can be made that the activation energy for the reverse reaction is zero since it involves a recombination of two free radicals, then the activation energy of the forward reaction should be equal to ΔE^0 for the reaction. In spite of a low activation energy, the reverse reaction is unimportant

It is instructive to compare the experimental value of the activation energy with ΔE^0 as calculated from the thermal data. The activation energy is 39.1 kcal, and ΔE^0 is essentially the energy of dissociation of the oxygen-oxygen bond in the peroxide. The bond energy of a single oxygen-oxygen bond is estimated as about 34 kcal, but the value, as is generally true for bond energies, varies from compound to compound. A better estimate can be made from thermal data given by Raley, Rust, and Vaughan.[2] The heats of combustion of di-t-butyl peroxide and of t-butyl alcohol lead to the equation

$$\underset{1282.6}{C_4H_9O_2C_4H_9\ (g)} + \underset{68.4}{H_2\ (g)} = \underset{2(640.7)}{2C_4H_9OH\ (g)} \qquad (18)$$

$$\Delta H = -69.6 \text{ kcal} \qquad \Delta E = -69.6$$

This can be written in terms of bond energies as

$$D_{O\text{-}O}\ (peroxide) = 2D_{O\text{-}H} - D_{H\text{-}H} - 69.6 \text{ kcal} \qquad (19)$$

The energy of dissociation of the oxygen-hydrogen bond in t-butyl alcohol is computed from the heats of combustion of isobutane and t-butyl alcohol,

$$\underset{686.3}{C_4H_{10}\ (g)} + \underset{}{\tfrac{1}{2}O_2\ (g)} = \underset{640.7}{C_4H_9OH} \qquad (20)$$

$$\Delta H = -45.6 \text{ kcal} \qquad \Delta E = \Delta H + \tfrac{1}{2}RT = -45.3 \text{ kcal}$$

which in terms of bond energies becomes

$$D_{O-H} = D_{C-H} + \tfrac{1}{2}D_{O-O} \text{ (in oxygen) } - D_{C-O} + 45.3 \qquad (21)$$

Taking the values D_{C-H} (in isobutane, tertiary hydrogen) as 86.0 kcal, D_{C-O} (*t*-butyl alcohol) as 85.0 kcal, and D_{O-O} (in oxygen, normal atoms) as 118 kcal, a value of D_{O-H} can be calculated as 105.2 kcal. This with $D_{H-H} = 103$ kcal makes D_{O-O} (peroxide) equal to 38 kcal, which is in excellent agreement with the experimental activation energy.

Raley, Rust, and Vaughan believe that the vapor-phase decomposition of di-*t*-butyl peroxide is a nonchain process, reactions 3, 5, and 7 being the important ones, and attacks on the peroxide such as (13) being very minor. This is borne out by the kinetics in the presence of nitric oxide and propylene. Nitric oxide, being itself an odd electron molecule, is an efficient trap for free radicals, as in equation 11. Hence chain reactions which require an uninterrupted series of free-radical reactions should be strongly inhibited by nitric oxide.[7] Propylene is also efficient in reacting with free radicals, as in equation 12, but also by losing a hydrogen atom to form allyl radical. Since this radical is relatively unreactive, the chain is interrupted. The constancy of rate observed with di-*t*-butyl peroxide in the presence of these inhibitors indicates that a chain mechanism is not involved in the gaseous decomposition. It is true that there is evidence that nitric oxide does not completely suppress chain reactions[8] and it has been pointed out that nitric oxide may start chains as well as stop them.[9] However, it seems unlikely that these effects would just cancel each other.

The work of Raley, Rust, and Vaughan has been extended by Szwarc and his co-workers.[10] Szwarc has developed a useful technique for studying reactions involving free radicals, the "toluene carrier" technique.[11] The method involves mixing the substance whose decomposition is to be studied with an excess of toluene vapor. The bond energy in the substance to be investigated must be less than the carbon-hydrogen bond energy in toluene. If free radicals are formed from the substance, they react more or less readily with toluene, forming a benzyl free radical.

$$R\cdot + C_6H_5CH_3 \rightarrow RH + C_6H_5CH_2\cdot \qquad (22)$$

Benzyl free radicals being stabilized by resonance to a considerable degree will be relatively inert, and will dimerize to form dibenzyl exclusively:

$$2C_6H_5CH_2\cdot \rightarrow C_6H_5CH_2CH_2C_6H_5 \qquad (23)$$

The presence of dibenzyl in the reaction products proves the existence of free radicals. A 1 to 2 relation between the amounts of dibenzyl and RH also proves the identity of the free radical. Furthermore, if a chain reaction exists in the absence of toluene, the addition of toluene will cause a reduction in the rate since the chains will be broken.

The kinetics of decomposition of the peroxide were studied in the vapor state in the presence and absence of toluene. The first-order, homogeneous nature of the reaction was confirmed down to pressures of 0.05 mm of mercury. This does not argue against a unimolecular process since the peroxide is a complex molecule with many degrees of freedom. The rates were followed by measuring the amounts of methane and ethane formed as a function of time, the assumption being made that two methyl radicals were formed from the decomposition of one peroxide molecule. The relative amounts of methane and ethane depended on the ratio of toluene pressure to peroxide pressure which varied from 5 to 1 to 25 to 1.

It was found that toluene had no influence on the rate of decomposition of di-*t*-butyl peroxide.[12] By measurements of the rate over a range from 116 to 350° C, a better value for the rate constant of $7 \times 10^{15} e^{-38,000/RT}$ sec^{-1} in the vapor state has been selected.[13] The frequency factor corresponds to a positive entropy of activation of about twelve units, which means a loosening of structure in the transition state.

The mechanism of this reaction is well enough understood, and the reaction goes at such a convenient rate above 100° C that the peroxide is frequently used as a known source of free radicals. As an example of its utility, it is possible to get relative values for the rate of the reaction[14]

$$(CH_3)_3CO \cdot + RH \rightarrow (CH_3)_3COH + R \cdot \tag{24}$$

for a number of hydrocarbons, RH. It is only necessary to analyze for the amounts of *t*-butyl alcohol and of acetone, the latter arising from the competing reaction 5.

An interesting study of the effect of very high pressures on the rate of the peroxide decomposition has been made by Walling and Metzger.[15] According to transition-state theory,[16] the effect of pressure on a reaction rate is given by

$$\frac{d \ln k}{dP} = - \frac{\Delta V^{\ddagger}}{RT} \tag{25}$$

where $\Delta V^{\ddagger}$ is the difference in molar volume between the transition state and reactants. Pressures of thousands of atmospheres are needed to get substantial changes in rate.

For a reaction in which a single bond is stretching and then breaking it is expected that $\Delta V^{\ddagger}$ would be positive. Experimentally this is so since high pressures cause a decrease in the rate for di-*t*-butyl peroxide in several solvents. The values of $\Delta V^{\ddagger}$ run from 5.4 ml/mole in toluene to 13.3 ml/mole in carbon tetrachloride.

The unexpected variations in $\Delta V^{\ddagger}$ with solvent have been explained in terms of the requirement that the newly formed radicals must escape from the solvent cage in which they find themselves.[16] That is, in solution the unimolecular decomposition must be regarded as a two-step process to be complete.

$$ROOR \rightleftharpoons (2RO\cdot) \tag{26}$$

$$(2RO\cdot) \rightarrow 2RO\cdot \tag{27}$$

The symbol (2RO·) refers to two alkoxy free radicals formed by bond breaking of the peroxide, but still neighboring each other and surrounded by a collection of solvent molecules which holds them together as in a cage.[17]   In this condition they can readily recombine to form the peroxide again.   Only through an escape from the cage by means of diffusion can the radicals get far enough apart so that their chance of recombining with each other is negligible.[18]   Alternatively, a radical can effectively escape from the solvent cage by reacting with a solvent molecule and hence passing the radical characteristic to a new site.[19]

Table 2

QUANTUM YIELDS FOR PHOTODISSOCIATION OF IODINE

(F. W. Lampe and R. M. Noyes)

Solvent	Temperature, °C	ϕ
Hexane	25	0.66
Carbon tetrachloride	25	0.14
Hexachlorobutadiene	25	0.075
Hexachlorobutadiene	35	0.15

Clearly both (26) and either (27) or reaction with the solvent must occur to produce the products of reaction.   Hence the effect of pressure includes not only the effect on (26), which may be considered an equilibrium, but also its effect on (27) or on the solvent reaction.   The diffusion process and the solvent reaction can vary from one solvent to another as is found.

Table 2 shows the nature of the cage effect for the photochemical dissociation of iodine molecules into atoms.[20]

$$I_2 \rightleftharpoons (2I\cdot) \rightarrow 2I\cdot \tag{28}$$

The quantum yield is reduced because of the reversal of the primary dissociation into caged atoms.   The reduction depends on the viscosity of the solvent, which determines the rate of escape from the solvent cage.

REFERENCES

1. N. A. Milas and D. M. Surgenor, *J. Am. Chem. Soc.*, *68*, 205 (1946).
2. J. H. Raley, F. F. Rust, and W. E. Vaughan, *J. Am. Chem. Soc.*, *70*, 88, 1336, 2767 (1948).

3. F. F. Rust, F. H. Seubold, and W. E. Vaughan, *J. Am. Chem. Soc.*, 70, 95 (1948).
4. E. R. Bell, F. F. Rust, and W. E. Vaughan, *J. Am. Chem. Soc.*, 72, 337 (1950).
5. P. George and A. D. Walsh, *Trans. Faraday Soc.*, 42, 94 (1946).
6. D. H. Volman and W. M. Graven, *J. Am. Chem. Soc.*, 75, 3111 (1953).
7. L. A. K. Stavely and C. N. Hinshelwood, *Proc. Roy. Soc.*, A154, 335 (1936).
8. E. W. R. Steacie and H. O. Falkins, *Can. J. Res.*, B17, 105 (1939).
9. L. A. Wall and W. J. Moore, *J. Am. Chem. Soc.*, 73, 2840 (1951).
10. M. Szwarc and J. S. Roberts, *J. Chem. Phys.*, 18, 561 (1950).
11. M. Szwarc, *Chem. Revs.*, 47, 75 (1950).
12. J. Murawski, J. S. Roberts, and M. Szwarc, *J. Chem. Phys.*, 19, 698 (1951).
13. F. Lossing and A. W. Tickner, *J. Chem. Phys.*, 20, 907 (1952).
14. A. L. Waters, E. A. Oberright, and J. W. Brooks, *J. Am. Chem. Soc.*, 78, 1190 (1956).
15. C. Walling and G. Metzger, *J. Am. Chem. Soc.*, 81, 5365 (1959).
16. M. G. Evans and M. Polanyi, *Trans. Faraday Soc.*, 31, 875 (1935).
17. J. Franck and E. Rabinowitch, *Trans. Faraday Soc.*, 30, 120 (1934).
18. R. M. Noyes, *J. Am. Chem. Soc.*, 77, 2042 (1955).
19. N. N. Semenov, *Some Problems of Chemical Kinetics and Reactivity*, Vol. I, Pergamon Press, New York, 1958, p. 160.
20. F. W. Lampe and R. M. Noyes, *J. Am. Chem. Soc.*, 76, 2140 (1954).

G.   The Thermal Isomerization of α- and β-Pinene

When optically active α-pinene I is heated above 200° in the vapor or

I

liquid, it becomes optically inactive. Smith[1] measured the change in rotation with time and found that it obeyed a first-order law:

$$\ln \frac{\alpha_0}{\alpha} = k't \tag{1}$$

Under the impression that the only reaction occurring was the racemization of pinene according to the equation

$$d\text{-pinene} \underset{k}{\overset{k}{\rightleftharpoons}} l\text{-pinene} \tag{2}$$

where the rate constants for the forward and reverse reactions are necessarily equal, Smith set the observed first-order rate constant k' equal to $2k$.

This gave a value of k in the gas phase expressible as $4.6 \times 10^{11}e^{-43,700/RT}$ sec^{-1}. Smith also measured the rate of loss of activity in several nonpolar, high-boiling solvents and found the rates to be about 50 per cent greater in solution and with about the same activation energies. Conant and Carlson[2] pointed out that dipentene II was formed when α-pinene is heated. Since dipentene is inactive (being actually *dl*-limonene) these

II

authors concluded that no racemization took place and that the thermal reaction leading to loss of activity was isomerization to *dl*-limonene.

$$d\text{-pinene} \xrightarrow{k_1} dl\text{-limonene} \tag{3}$$

Hence the observed rate constant k' was equal to k_1 which would then be $9.2 \times 10^{11}e^{-43,700/RT}$ sec^{-1}.

Thurber and Johnson[3] showed that the recovered α-pinene after heating was of reduced optical activity so that some actual racemization had occurred. Also another reaction was present which led to unidentified products. Fuguitt and Hawkins[4] identified these products as allo-ocimene III, a dimer of allo-ocimene, and α- and β-pyronene, the latter three being formed from allo-ocimene. Fuguitt and Hawkins[5] succeeded in measuring

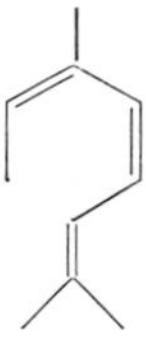

III

the rates of all three primary processes in the pure liquid-phase pyrolysis of α-pinene:

$$d\text{-pinene} \xrightarrow{k_1} dl\text{-limonene} \tag{4}$$

$$d\text{-pinene} \xrightarrow{k_2} \text{allo-ocimene} \tag{5}$$

$$d\text{-pinene} \xrightarrow{k_3} dl\text{-pinene} \tag{6}$$

The procedure used was to fractionate a reaction mixture after a known reaction time and to determine the amounts of each of the products. All three reactions were first-order, the rate constants being given by

$k_1 = 3 \times 10^{11} e^{-37,000/RT}$ sec^{-1}, $k_2 = 6 \times 10^{13} e^{-42,700/RT}$ and $k_3 = 1 \times 10^{14} e^{-44,200/RT}$. The relation between these constants and k' the overall rate constant for loss in optical activity as found by Smith is that $k' = k_1 + k_2 + k_3$. The overall activation energy is related to the activation energies of the several reactions by the equation

$$E' = \frac{k_1 E_1 + k_2 E_2 + k_3 E_3}{k_1 + k_2 + k_3} \tag{7}$$

which can be derived by the application of the Arrhenius equation

$$\frac{d \ln k'}{dT} = \frac{E'}{RT^2} = \frac{d \ln (k_1 + k_2 + k_3)}{dT} = \frac{d(k_1 + k_2 + k_3)}{(k_1 + k_2 + k_3)dT}$$

$$= \frac{E_1 k_1 + E_2 k_2 + E_3 k_3}{(k_1 + k_2 + k_3)RT^2} \tag{8}$$

At 190° the relative values of k_1, k_2, and k_3 are 6, 3, and 1 which leads to a value of 40 kcal for E'. This compares with Smith's value of 43.7 kcal, which is probably more accurate because of the difficulty of analysis of the separate reactions.

Since all three reactions appear to be clearly first-order, racemization cannot involve *dl*-limonene or allo-ocimene as an intermediate.[6] That is, a sequence of the form

$$d\text{-pinene} \rightarrow dl\text{-limonene} \rightarrow dl\text{-pinene} \tag{9}$$

would not show simple first-order kinetics for the formation of *dl*-pinene. Furthermore, *dl*-limonene is formed as such since *l*-limonene resists racemization under the reaction conditions. The similarity of the overall reaction rate in the gas phase and in solution indicates that a chain mechanism cannot play an important part nor can a wall reaction be important. In the liquid phase the rates are the same in the presence of added quinoline, benzoic acid, and hydroquinone.[7] The last substance being an antioxidant would have an effect if peroxides due to air contamination of the olefin played a part.

An attractive mechanism in agreement with all the above facts is one involving a biradical intermediate IV.[8] One electron of the biradical is

IV

part of an allylic resonance system which would help to stabilize it. This biradical is symmetrical and any product formed from it would necessarily be optically inactive. The biradical is formed from α-pinene by breaking the strained four-membered ring, as shown in (10). The biradical could react in several ways. If the four-membered ring were reformed *dl*-pinene

$$\tag{10}$$

would result. If hydrogen transfer from the isopropyl residue to either end of the allylic system occurred *d*- and *l*-limonene would be formed with equal probability. If the six-membered ring were broken, the triolefin (ocimene) V would be formed, as shown in (11). This compound V being unconjugated would be expected to rearrange readily under the experimental conditions to the conjugated allo-ocimene III.[9]

$$\tag{11}$$

IV   V   III

The overall activation energy is assumed to be the energy required to form the biradical. This is the rate-determining step for all three processes. The relative amounts of the primary products, however, depend on the rate constants of the three possible follow-up reactions.

$$\text{IV} \xrightarrow{k_1'} \text{II} \tag{12}$$

$$\text{IV} \xrightarrow{k_2'} \text{V} \rightarrow \text{III} \tag{13}$$

$$\text{IV} \xrightarrow{k_3'} dl \ \ \text{I} \tag{14}$$

The magnitudes of these rate constants cannot be found (except that they are larger than k'), but their ratios can be. Also from the temperature variation of the ratio of products, differences in the activation energies of the reactions 12 to 14 can be calculated. The appropriate equations are

$$k_1 = k' \frac{k_1'}{k_1' + k_2' + k_3'} \qquad (15)$$

$$E_1 = E' + E_1' - \frac{k_1'E_1' + k_2'E_2' + k_3'E_3'}{k_1' + k_2' + k_3'} \quad \text{etc.} \qquad (16)$$

where E_1 is the apparent activation energy for the formation of dl-limonene, E' is the activation energy for the formation of the diradical, and E_1' is the activation energy for the formation of dl-limonene from the diradical. If we put in either 40 or 43.7 kcal for E', the difference in E_1' and E_2', for example, is 5.7 kcal and between E_1' and E_3' is 7.2 kcal from the data of Fuguitt and Hawkins. Thus the formation of dl-limonene requires less energy than either the racemization or the formation of the triolefin V. The overall activation energy is consistent with the energy required to form the biradical from α-pinene. Thus Szwarc and Sheon[10] find an activation energy of 62 kcal for the process

$$CH_2 = CH - CH_2 - CH_3 \rightarrow CH_2 = CH - CH_2\cdot + CH_3\cdot \qquad (17)$$

This value is lower than the normal carbon-carbon bond energy of ca. 80 kcal by about the resonance energy of the allyl radical. If in α-pinene we make additional allowance for the bond strain in the cyclobutane ring (about 6 kcal per methylene group, though the thermal data are not very precise), the activation energy for formation of the biradical could easily be as low as the observed value (40–44 kcal). The differences in the activation energies of the follow-up steps are difficult to explain, but the fact that differences exist at all is consistent with some stability for the biradical intermediate. That is, the biradical is not an activated complex but corresponds to a shallow minimum in the diagram of energy versus extent of reaction shown in Fig. 1. The structural formula for IV does not necessarily give the geometry of the biradical which may have a coiled configuration so that the isopropyl residue is close to the allylic system:

This would agree with the ease of hydrogen transfer to form dl-limonene.

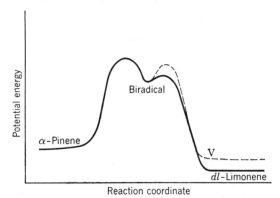

Fig. 1. Thermal isomerization of α-pinene. Full line for formation of *dl*-limonene (forward reaction) or of *dl*-pinene (reverse reaction). Dotted line for formation of compound V, triolefin.

The application of the same theory of a biradical intermediate to the pyrolysis of β-pinene VI, as shown in (18), is interesting (Burwell[8]).

$$ \text{VI} \longrightarrow \quad \longleftrightarrow \quad \tag{18}$$

        VI   VII

Here the biradical VII is also stabilized by an allylic resonance but is not symmetrical. Therefore, if we started with optically active β-pinene, the biradical would also be optically active. If hydrogen transfer occurred between the isopropyl residue and the two ends of the allylic system, then limonene, which is optically active, and the compound VIII would be

active   active   inactive
biradical   limonene   VIII

formed, as shown in (19). Formation of limonene would be favored since the internal double bond is more stable than the *exo*-double bond (β-pinene readily rearranges to α-pinene in the presence of acid, for example). If bond breaking between carbon atoms occurred, the product would be the triolefin myrcene IX which is necessarily inactive. Furthermore, if

(19)

IX

bond closure back to β-pinene occurred, this could happen in only one way so that the same optical isomer as the starting material would be regenerated. Thus no racemization of the β-pinene would result.

The experimental facts bear out these predictions as far as they go. Goldblatt and Palkin[11] report that at 400° the vapor-phase pyrolysis of *l*-β-pinene leads to about 70 per cent of myrcene and to 13 per cent of *l*-limonene of very high rotation (about 10 per cent under the maximum possible). No VIII was reported. There was no mention of any loss of activity of the unreacted β-pinene. It could not, of course, be determined whether the recovered pinene had passed through the biradical intermediate and returned or whether it had simply never reacted.

Hunt and Hawkins[7] measured the rates of formation of *l*-limonene and of myrcene from the pyrolysis of β-pinene in the pure liquid and diluted with limonene. Both reactions are first-order. The rate constants are expressible as $1.6 \times 10^{16} e^{-50,000/RT}$ sec^{-1} for isomerization to limonene and $2.5 \times 10^{15} e^{-47,000/RT}$ for isomerization to myrcene (actually found as its polymer). The relative rates are thus about 1 to 2 at 235°. The overall activation energy is about 48 kcal. This would be approximately the energy of activation for forming the biradical, according to the theory. Why the energy required in this case should be 4–8 kcal higher than for α-pinene is not clear unless the biradical has the coiled structure which might permit additional resonance structures with the internal allylic system of α-pinene. This would lower the energy of formation. The activation energy data (obtained over a 15-degree range) may not be sufficiently accurate to indicate a real difference, however. The biradical theory predicts that *l*-β-pinene and *l*-limonene have the same relative configuration.

REFERENCES

1. D. F. Smith, *J. Am. Chem. Soc.*, *49*, 43 (1927).
2. J. B. Conant and G. H. Carlson, *J. Am. Chem. Soc.*, *51*, 3464 (1929).
3. F. H. Thurber and C. H. Johnson, *J. Am. Chem. Soc.*, *52*, 786 (1930).
4. R. E. Fuguitt and J. E. Hawkins, *J. Am. Chem. Soc.*, *67*, 242 (1945).
5. R. E. Fuguitt and J. E. Hawkins, *J. Am. Chem. Soc.*, *69*, 319 (1947).
6. R. L. Burwell, *J. Am. Chem. Soc.*, *70*, 2865 (1948).
7. H. G. Hunt and J. E. Hawkins, *J. Am. Chem. Soc.*, *72*, 5618 (1950).
8. R. L. Burwell, *J. Am. Chem. Soc.*, *73*, 4461 (1951).
9. J. L. Simonsen, *The Terpenes*, 2nd edition, University Press, Cambridge, 1947, Vol. I, p. 20.
10. M. Szwarc and A. H. Sheon, *J. Chem. Phys.*, *18*, 237 (1950).
11. L. A. Goldblatt and S. Palkin, *J. Am. Chem. Soc.*, *63*, 3517 (1941).

H.   The Decomposition of Nitrogen Pentoxide

The thermal decomposition of nitrogen pentoxide to nitrogen tetroxide (or dioxide) and oxygen

$$2N_2O_5 = 2N_2O_4 + O_2 \tag{1}$$

$$N_2O_4 \rightleftharpoons 2NO_2 \tag{2}$$

is historically important as the first homogeneous first-order reaction in the gas phase to be reported.[1] It was consequently used as a proving ground for various theories of unimolecular reactions. Ogg's work[2] indicates, however, that the reaction, in spite of being first-order, is not unimolecular and has a more complex mechanism.

Experimentally the rate of decomposition is not changed by surface effects, is uncatalyzed, and is independent of the presence of a great many foreign substances. The rate of the reaction is almost the same in the gas phase as in a variety of solvents, and the activation energies are almost the same.[3] The first-order nature of the reaction holds down to very low pressures. At pressures below 0.05 mm the first-order rate constant begins to fall off, and below 0.004 mm the reaction appears to be second-order.[4]

This falling off of the first-order rate constant at low pressures is an expected feature of unimolecular reactions (see Chapter 4). However, its appearance in the case of nitrogen pentoxide does not confirm the unimolecular nature of this reaction since it occurs at far too low a pressure. Various detailed theories of unimolecular activation have been proposed.[5] These permit a calculation of the change in the experimental first-order rate constant with pressure if the diameter of the molecule and the number of fully excited vibrational degrees of freedom are known. No reasonable values can be assigned to these quantities to fit the experimental data on nitrogen pentoxide.

If the reaction is a unimolecular one, the immediate products of the decomposition must react further since the stoichiometry of equation 1 refers to two molecules of the pentoxide. Several possibilities exist for the initial decomposition:

$$N_2O_5 \rightarrow N_2O_3 + O_2 \tag{3}$$

$$N_2O_5 \rightarrow N_2O_4 + O \tag{4}$$

$$N_2O_5 \rightarrow NO + NO_2 + O_2 \tag{5}$$

$$N_2O_5 \rightleftharpoons NO_2 + NO_3 \tag{6}$$

The first of these was proposed by Bodenstein,[6] the nitrogen trioxide rapidly breaking down into nitric oxide and nitrogen dioxide. The nitric oxide would then be oxidized to dioxide to give the final products. This mechanism is ruled out, however, because it involves a change in the electronic multiplicity in the initial step. Nitrogen pentoxide and nitrogen trioxide are diamagnetic and have all their electrons paired off. Oxygen is paramagnetic and has two unpaired electrons. For reaction 3 to occur would necessitate changing the spin of one electron. Such a process is possible but occurs with very low probability †

The next possibility, equation 4, involving an oxygen atom is eliminated because of the high energy involved. From heats of formation it can be calculated that this reaction is endothermic by 61 kcal. The experimental activation energy for the nitrogen pentoxide decomposition is only 24.6 kcal, and the overall energy change in the rate-determining step cannot exceed this value.

The dissociation given in (5) was suggested by Busse and Daniels.[8] From thermochemical data given by Ogg[9] it is possible to calculate the ΔE^0 for this reaction as 24.3 kcal which is correctly below the experimental activation energy. Nitric oxide and nitrogen dioxide are paramagnetic molecules containing one unpaired electron each. By a suitable unpairing of electron spins the indicated dissociation becomes possible. A primary dissociation into three particles is unusual, however.

Equation 6 was considered at one time and rejected because the reverse reaction is known to occur very readily. Hence, it was believed that (6) could not be the slow step of an irreversible decomposition. The molecule NO_3 has not been conclusively identified but is believed to form when ozone reacts with nitrogen dioxide.[10]

$$NO_2 + O_3 \rightarrow NO_3 + O_2 \tag{7}$$

† For a discussion of such non-adiabatic reactions see reference 7.

A band spectrum not belonging to any other component in the mixture and attributed to NO_3 has been observed when oxides of nitrogen are treated with ozone.[11]

Ogg[9] has pointed out that all the facts concerning the nitrogen pentoxide decomposition can be explained on the basis of a quasi-unimolecular mechanism involving (6) as the first step. This reaction is rapid and reversible with the equilibrium lying well to the left. The slow step becomes a second-order reaction of NO_2 and NO_3

$$NO_2 + NO_3 \xrightarrow{k_3} NO_2 + O_2 + NO \tag{8}$$

This is followed by the rapid reaction

$$NO + NO_3 \rightarrow 2NO_2 \tag{9}$$

where the NO_3 as before comes from the dissociation of the pentoxide. The overall reaction then becomes

$$2N_2O_5 = 4NO_2 + O_2 \tag{10}$$

agreeing with equation 1.

The observed kinetics depends on the relative values of k_1, k_2, and k_3 and upon the concentration, where k_1 and k_2 are the rate constants for the dissociation and association,

$$N_2O_5 \underset{k_2}{\overset{k_1}{\rightleftharpoons}} NO_2 + NO_3 \tag{11}$$

Applying the steady-state assumption to NO_3 gives

$$d[NO_3]/dt = k_1[N_2O_5] - k_2[NO_2][NO_3] - 2k_3[NO_2][NO_3] = 0 \tag{12}$$

$$[NO_3] = \frac{k_1[N_2O_5]}{k_2[NO_2] + 2k_3[NO_2]} \tag{13}$$

The rate of decomposition of the pentoxide is

$$-d[N_2O_5]/dt = 2k_3[NO_2][NO_3] = 2k_3k_1[N_2O_5]/(k_2 + 2k_3) \tag{14}$$

so that the reaction will be first-order as observed with a rate constant

$$k_{obs} = 2k_1k_3/(k_2 + 2k_3)$$

The factor 2 comes in because two molecules of NO_3 and hence two N_2O_5 molecules react each time step 8 occurs. If k_2 is very much greater than k_3, as seems likely since k_2 refers to an association reaction of two free radicals and k_3 involves a breaking of bonds, then $k_{obs} = 2k_1k_3/k_2$.

The reaction is not a unimolecular one at all on this basis but involves a prior equilibrium followed by a bimolecular rate-determining step. Energetically the sequence is possible since the activation energy found experimentally will be equal to ΔE^0 for reaction 11 plus the activation energy for reaction 8.

$$24.6 \text{ kcal} = E_{\text{exp}} = \Delta E^0 + E_3 = E_1 - E_2 + E_3 \qquad (15)$$

And the only restriction put on the system by known thermal data is that

$$E_1 - E_2 + E_3 - E_4 = 24.3 \text{ kcal} \qquad (16)$$

which is possible if E_4, the activation energy for the reverse of reaction 8 has the low value of 0.3 kcal. Ogg assumes that E_2 is also approximately zero and that E_1 is about 21 kcal so that E_3 is about 4 kcal. This is consistent with k_3 being much less than k_2.

The mechanism correctly predicts the falling off of the rate constant observed at very low pressures.[2] The dissociation of N_2O_5 into NO_3 and NO_2 is a unimolecular process governed by activation through collision. At very low pressures, the number of collisions can become so low that the equilibrium concentration of activated molecules is not maintained. In the limit the rate of dissociation becomes equal to the rate of activation by collision. In the same manner, the association of NO_3 and NO_2 into N_2O_5 must have at low pressures a dependence on the total pressure. This follows from the principle of microscopic reversibility. Writing the mechanism in detail using the simple Lindemann-Hinshelwood theory:

$$N_2O_5 + M \underset{k_d}{\overset{k_a}{\rightleftharpoons}} N_2O_5{}^* + M \qquad (17)$$

$$N_2O_5{}^* \underset{k_A}{\overset{k_D}{\rightleftharpoons}} NO_2 + NO_3 \qquad (18)$$

where k_a is a collisional activation constant, k_d is a deactivation constant, k_D is a dissociation constant for the activated $N_2O_5{}^*$ and k_A is an association constant. At high pressures, k_1 of equation 11 is equal to $k_a k_D / k_d$ and k_2 is k_A. At low pressures k_1 becomes $k_a[M]$ and k_2 becomes $k_A k_d[M]/k_D$. Since k_1/k_2 is an equilibrium constant, the ratio must have a constant value at all pressures. This ratio is seen to be $k_a k_D / k_d k_A$.

Now considering the rate of decomposition of nitrogen pentoxide it was shown that $k_{\text{obs}} = 2k_1 k_3 / (k_2 + 2k_3)$. This may be tested experimentally by writing

$$\frac{1}{k_{\text{obs}}} = \frac{k_2}{2k_1 k_3} + \frac{1}{k_1} = \frac{1}{k_\infty} + \frac{1}{k_1} \qquad (19)$$

where k_∞ is the limiting value of the rate constant at high pressures. At low pressure we may write

$$\frac{1}{k_{\text{obs}}} = \frac{1}{k_\infty} + \frac{1}{k_a[M]} \qquad (20)$$

by considering the limiting form for k_1. The data of Hodges and Linhorst[4] fit just such a linear relationship at low pressures where M is nitrogen pentoxide itself. From the complex nature of k_∞ it can readily be seen why the falling off of the rate constant is not noticeable until very low pressures are reached, k_2/k_1 being so large that [M] must be very small before the second term of (20) becomes appreciable.

The data of Hodges and Linhorst at various temperatures permit a calculation of the collisional activation constant:

$$k_a = 10^{15.9}e^{-18,300/RT} \text{ liters/mole-sec} \tag{21}$$

The activation energy for k_1 is thus 18.3 kcal at low pressures. From the theories of unimolecular reactions (Kassel[5]) it is predicted that at high pressures the activation energy will be several kilocalories greater. Ogg estimates a value of 21–22 kcal. If the frequency factor for k_1 is normal, a value of k_1 at 25° may be predicted as $k_1 \simeq 10^{13}e^{-21,000/RT} \simeq 10^{-2} \text{ sec}^{-1}$ for high pressures.

There are several approaches to a direct study of k_1. If Ogg's hypothesis of a rapid equilibrium between N_2O_5, NO_2, and NO_3 is correct, it should be possible to demonstrate isotopic exchange in a system such as

$$N_2{}^{14}O_5 + N^{15}O_2 \rightleftharpoons N_2{}^{15}O_5 + N^{14}O_2 \tag{22}$$

Furthermore, the rate constant of such an exchange should be equal to k_1 and the reaction should be first-order in nitrogen pentoxide and zero-order in nitrogen dioxide. The rate-determining step would be

$$N_2O_5 \xrightarrow{k_1} NO_3 + NO_2 \tag{23}$$

This behavior has been demonstrated in the gas phase, using $N_2{}^{15}O_5$, and in carbon tetrachloride solution, using $N_2{}^{13}O_5$.[12] The rate constant in the gas phase at 27° is about 0.5 sec^{-1} at a total pressure of about 500 mm (CO_2 being added to bring up the pressure). Reducing the pressure to about 50 mm reduces the rate constant by a factor of 5. It thus appears that k_1 is a true unimolecular constant with a falling off at moderate pressures. This is in agreement with the theory of unimolecular reactions applied to a molecule as simple as nitrogen pentoxide.

The kinetics of the reaction between nitric oxide and nitrogen pentoxide offers another approach to k_1.

$$N_2O_5 + NO = 3NO_2 \tag{24}$$

From the mechanism of the nitrogen pentoxide decomposition, the mechanism of this reaction should be

$$N_2O_5 \underset{k_2}{\overset{k_1}{\rightleftharpoons}} NO_2 + NO_3 \tag{25}$$

$$NO_3 + NO \xrightarrow{k_4} 2NO_2 \tag{26}$$

with the dissociation the slowest step. The complete rate equation is

$$\frac{-d[NO]}{dt} = \frac{k_1[N_2O_5]}{1 + k_2[NO_2]/k_4[NO]} \tag{27}$$

This reaction has been observed to be first-order in nitrogen pentoxide and zero-order in nitric oxide.[2,13] There appears to be an inhibiting effect of nitrogen dioxide as indicated by equation 27. The observed first-order rate constant falls off with total pressure as would be expected for k_1. At the same total pressure the rate of the reaction with nitric oxide is the same as the rate of isotopic exchange (Ogg). The results of Daniels and Smith give a value of $k_1 = 10^{9.5}e^{-15,700/RT} \sec^{-1}$ at pressures of a few millimeters of Hg. This value of the activation energy is not in agreement with the low-pressure result of 18.3 kcal. The absolute value of the rate constant at 25° is about $10^{-2} \sec^{-1}$. This is not necessarily in disagreement with the data of Ogg, since the pressures used were much lower.

The reaction of nitrogen pentoxide with nitric oxide has been studied more carefully by Mills and Johnston,[14] at high pressures and at low pressures by Johnston and Perrine.[15] The results of Daniels and Smith were confirmed in general. The dependence of the observed first-order constant on the pressure was used to find the limiting values at zero pressure and infinite pressure. The results give $k_1{}^0 = 1.3 \times 10^{16}e^{-19,300/RT}$ liter/mole-sec, which corresponds to a rate constant of 1.0×10^2 at 27°, and $k_1^\infty = 0.29 \sec^{-1}$ at 27° with an activation energy of 21 ± 2 kcal. The activation energy of Daniels and Smith appears to be in error because of the presence of a heterogeneous reaction.

The interpretation of $k_1{}^0$ is that it is equal to k_a, the collisional activation constant for the decomposition of nitrogen pentoxide into NO_2 and NO_3 molecules. Its value at 27° is similar to that obtained from the low-pressure study of nitrogen pentoxide alone by Hodges and Linhorst[4] (1.50×10^3), and the activation energies are the same within experimental errors. This is as it should be according to the postulated mechanisms, since at low pressure activation becomes the rate-controlling step for both the decomposition and the reaction with nitric oxide. However, the values need not be identical, since different sets of molecules are involved in the two reactions.[16]

The high-pressure limit, $k_1{}^\infty$, is presumably equal to the ratio $k_a k_D/k_d$. Johnston[17] discusses this ratio further in a review of the reactions believed to involve the NO_3 molecule; that is, the decomposition of nitrogen pentoxide alone, in the presence of nitric oxide and in the presence of ozone, and the reaction of ozone and nitrogen dioxide to form nitrogen pentoxide. It is assumed that every collision of an activated molecule leads to deactivation. Hence k_d, the rate constant for deactivation, is equal approximately to 10^{11} liters/mole-sec on the basis of simple collision theory. This makes k_D, the rate constant for the decomposition of activated molecules, come out to be about 10^8 sec^{-1} and nearly independent of the temperature, as is k_d. An estimate is also made of the equilibrium constant for reaction 11

$$N_2O_5 \rightleftharpoons NO_2 + NO_3 \qquad (11)$$

as 10^{-10} mole/liter which gives an equilibrium concentration of NO_3 at ordinary pressures of N_2O_5 and NO_2 of the order of 10^{-6} mm. A low concentration is in agreement with the failure to observe the spectrum of NO_3 in the decomposition of nitrogen pentoxide at room temperature.

The similarity in the rates of decomposition of nitrogen pentoxide in the gas phase and in various solvents is not difficult to reconcile with the proposed mechanism, complex though it may be. An examination of the rate equation, equation 14, shows that all that is necessary is that the activity coefficients of nitrogen pentoxide and the activated complex change in the same ratio in going from solvent to solvent. Such a result is quite possible, since the same atoms are involved and the arrangement is probably not different enough to change the polarity appreciably. Plausible structures for nitrogen pentoxide and the activated complex are shown below, the dotted lines representing resonating bonds.

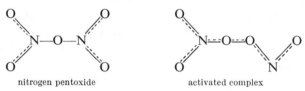

nitrogen pentoxide   activated complex

REFERENCES

1. F. Daniels and E. H. Johnston, *J. Am. Chem. Soc.*, *43*, 53 (1921).
2. R. A. Ogg, *J. Chem. Phys.*, *15*, 337 (1947); *18*, 572 (1950).
3. R. H. Lueck, *J. Am. Chem. Soc.*, *44*, 757 (1922); H. Eyring and F. Daniels, *ibid.*, *52*, 1473 (1930).
4. H. C. Ramsperger, M. E. Nordberg, and R. C. Tolman, *Proc. Natl. Acad. Sci. U.S.*, *15*, 453 (1929); E. F. Hodges and J. H. Linhorst, *ibid.*, *17*, 28 (1931); *J. Am. Chem. Soc.*, *56*, 836 (1934).
5. L. S. Kassel, *Kinetics of Homogeneous Gas Reactions*, Chemical Catalog Co., New York, 1932; N. B. Slater, *Theory of Unimolecular Reactions*, Cornell University Press, Ithaca, 1959.
6. M. Bodenstein, *Z. physik. Chem.*, *104*, 51 (1923).
7. S. Glasstone, K. J. Laidler, and H. Eyring, *Theory of Rate Processes*, McGraw-Hill Book Co., New York, 1941, Chapters III and VI.
8. W. F. Busse and F. Daniels, *J. Am. Chem. Soc.*, *49*, 1257 (1927).

9. R. A. Ogg, *J. Chem. Phys.*, *15*, 337 (1947).
10. H. S. Johnston and D. M. Yost, *J. Chem. Phys.*, *17*, 386 (1949).
11. H. J. Schumacher and G. Sprenger, *Z. physik. Chem.*, *136A*, 77 (1928); E. Warburg and G. Leithauser, *Ann. Physik*, *20*, 743 (1906); *23*, 209 (1907).
12. R. A. Ogg, *J. Chem. Phys.*, *18*, 573 (1950); *15*, 613 (1947).
13. J. H. Smith and F. Daniels, *J. Am. Chem. Soc.*, *69*, 1735 (1947).
14. R. L. Mills and H. S. Johnston, *J. Am. Chem. Soc.*, *73*, 938 (1951).
15. H. S. Johnston and R. L. Perrine, *J. Am. Chem. Soc.*, *73*, 4782 (1951).
16. M. Volpe and H. S. Johnston, *J. Am. Chem. Soc.*, *78*, 3903, 3910 (1956).
17. H. S. Johnston, *J. Am. Chem. Soc.*, *73*, 4542 (1951).

AUTHOR INDEX

SUBJECT INDEX

Absolute reaction rate theory, 77
Acetaldehyde, aldol condensation, 335, 345 ff.
  catalyzed decomposition, 203–204, 254–257
  hydrate, dehydration, 215, 221–223, 345
  pyrolysis, 254–258
Acetic acid, dimerization, 125, 126
  rate of ionization, 281
Acetone, enolization, 201, 215, 230, 341, 347
Acetophenone, enolization, 334
Acetylacetone, enolization, 230
Acetyl peroxide, decomposition, 129
Acidity function, 325–327, 329–334, 359
Aci-nitroethane, isomerization, 144, 212–213
Acrolein, condensation reactions, 32–34, 104, 128
Activated complex, 2, 3, 77
Activation energy, 22–24
  Arrhenius, 22, 99
  at absolute zero, 89
  estimation of, 106–108
Activity coefficient, of activated complex, 127, 132
  of ions, 150–151
  of neutral molecules, 132, 152–153
Adiabatic assumption, 78

Alcoholysis of esters and diketones, 218
Aldol condensation, 217, 335, 345 ff.
Alkoxymethyl esters, hydrolysis, 319
Alkyl bisulfate ions, hydrolysis, 287
Alkyl substituents, influence on rates, 115–116, 290–292, 304–305, 308, 311–312, 347
Allylic alcohols, esters of, hydrolysis, 319
Allylic resonance energy, 376–378
Amines, as catalysts, 223
  complexes with silver ion, 226
  exchange reactions, 278
Ammonium cyanate, reaction to form urea, 307–315
Anchimeric assistance, 298
Arrhenius equation, 22
Atom recombinations, 63, 97–98, 108–109, 239–241
Autocatalysis, 19–20
Autoxidation, 249–251
Azodicarbonate ion, reaction with hydrogen ion, 146, 216

Benzaldehyde, autoxidation, 249
Benzene, nitration of, 351–352, 362
Biacetyl as catalyst, 204–205
Bimolecular reactions, 61–68, 73, 92–95, 103–108
Biradical intermediate in pyrolysis of pinenes, 375–379

399